WESTERN PUBLIC LANDS

WESTERN PUBLIC LANDS
The Management of
Natural Resources in a
Time of Declining Federalism

edited by
John G. Francis
and
Richard Ganzel

ROWMAN & ALLANHELD
PUBLISHERS

ROWMAN & ALLANHELD

Published in the United States of America in 1984
by Rowman & Allanheld, Publishers
(A division of Littlefield, Adams & Company)
81 Adams Drive, Totowa, New Jersey 07512

Library of Congress Cataloging in Publication Data
Main entry under title:

Western public lands.

  Bibliography: p.
  1. West (U.S.)—Public lands. 2. Natural resources—
Government policy—West (U.S.) 3. Rangelands—Government
policy—West (U.S.) 4. Environmental policy—West (U.S.)
5. Conservation of natural resources—Government policy—
West (U.S.) I. Francis, John G., 1943-   . II. Ganzel,
Richard.
HD243.A17W47   1984   333.1'0973   83-19067
ISBN 0-86598-147-4

84 85 86/10 9 8 7 6 5 4 3 2 1

Printed in the United States of America

Contents

Tables and Figures

Figure

Introduction

JOHN G. FRANCIS
RICHARD GANZEL

Very early in his administration, President Carter notified Congress of his intention to review the need to fund nineteen western water projects. He justified his review by arguing that the massive water projects were costly to build, environmentally damaging, and only beneficial to relatively few people. At the time of his announcement, environmentalists warmly welcomed this challenge to the western water establishment. Potential termination of the projects, however, was denounced by nearly all western state officeholders as a threat to the economic well-being of the region. Carter's review was short-lived, and the vast majority of the projects continue to be funded today.

Near the end of his administration, President Carter announced an $88 billion energy program to develop synthetic fuels from oil shale, coal gasification, and tar sands. All of these projects posed potentially significant adverse environmental consequences. Moreover, in the same speech Carter proposed an Energy Mobilization Board to pace these proposed energy developments on a fast track. The board was to be permitted to circumvent existing environmental regulatory procedures and, in the process, to weaken the roles of states and localities in implementing siting regulations. Unlike their earlier support for Carter's announcement on western water projects, environmentalists greeted these proposals with great concern. Although federal money is still set aside for synfuels development, it remains largely unspent by the Reagan Administration.

These sharply contrasting proposals of the Carter presidency suggest the two observations that prompt this book. The first is the extent to which critical decisions or initiatives concerning the nation's resources and lands, particularly in the West, have been concentrated at the level of federal decisionmaking. That western leaders are forced to respond to these actions, either directly or through their congressional delegations, is a frequent stimulus of regional discontent. The second is the extent to which natural resource and public land policymaking is beset by contradictions and value conflicts.

The Reagan Administration came into office with the aim of promoting the development of energy resources through a good-neighbor policy developed in consultation with the affected states. This approach to development has also engendered controversy. The administration's accelerated coal leasing program and asset management program for the public lands, a program of limited disposition, have both been criticized for failing to involve state interests. Some are concerned that disposition represents an abandonment of the multiple-use doctrine of land management; others fear it will diminish environmental quality. Ironically, these difficulties have emerged despite the fact that the failure to pursue most of the synfuels projects proposed by the Carter Administration has avoided, in the main, the dramatic adverse environmental transformations associated with energy development.

The relationship between the states and the federal government is a complicated one in American politics. It is particularly complicated when it is compounded by natural resource and land policy issues. From the mid-1960s, and culminating in the mid-1970s, a consensus evolved that the federal government should play a preeminent role in natural resources regulation and public land use. During that period Congress asserted federal responsibility for developing and enforcing standards for air and water quality. It obliged developers to consider the environmental consequences of their proposed projects in all cases that involved major federal actions significantly affecting the environment. Congress committed the federal government to retain ownership of the public lands and to pursue a policy of active management and diverse uses for those lands. Congress also consolidated federal responsibility for energy development and planning.

Despite such an impressive federal commitment, no more than a few years later the federal government's commanding role in resource policy and public land use faced heavy challenges. Critics questioned the effectiveness and the costs of enforcing environmental standards through federal action. A number of groups criticized what they regarded as misplaced emphasis on environmentalist values guiding federal natural resource policy. Perhaps the most insistent of the critics were the politicians from western states, who pressed the federal government to transfer ownership of federal public lands to their respective states or at least give the states a larger role in deciding public land resource policy.

These challenges to the apparent consensus in Congress and to federal decisionmaking about natural resource issues should come as no surprise, however. Three central, intertwined issues continually complicate the politics of American natural resource decisionmaking. First, how much control over natural resource use decisions should be assigned to residents of the area where the natural resource in question is located? Second, what set of values is appropriate to guide natural resources and land use decisions? For example, what mix of developmental and environmental values should be applied to coal mining? Third, to what degree should resource and land use decisions be determined in the public sector rather than result from private sector choices?

These three issues cannot be separated. For example, opponents of nuclear waste disposal may seek to locate decisionmaking at the level of the local community if they believe the community will oppose the waste disposal. Schattschneider (1960) observed two decades ago that groups unable to realize their preferences at a lower level of government will broaden consideration of their issue concerns by seeking to advance them to a higher level of government. It is useful in the analysis of natural resource politics to extend Schattschneider's insight by recognizing the truth of its converse. That is, a group unsuccessful in gaining a favorable hearing for its issue position at a higher level of government may seek to shift the focus of decisionmaking to a lower level of government. Indeed, it may seek to shift from one branch of government to another or to remove the issue from governmental consideration altogether and place it in the private sphere.

Of these three issues interwoven into the fabric of contemporary natural resource politics, the areal division of power deserves to be considered first. The workings of American federalism have had major consequences for natural resource and land issues.

Areal Decisionmaking: Power to the Locals?

Long at the core of American politics is the issue of the division of powers by area: i.e., how much authority, and over what subjects, belongs to the states, local governments, and federal government? The difficulty with federalism—especially in the United States—is that governmental responsibilities have never been divided along clear–cut lines. Historically, state and federal governments have competed for control of natural resources, sometimes over ownership, and more frequently over questions of taxation and land use regulation. Commentators have competed as vigorously over the proper analysis to be accorded American federalism.

American federalism has a protean quality that may help explain the remarkable reliance on metaphor in the writings of commentators. One student of the federalist literature has recently compiled a list of 326 metaphors or models of federalism (Stewart 1982). This account of the problem of areal decisionmaking and natural resource issues will be mainly confined to the dessert metaphors that have been commonly employed in the literature, notably cake.

Federal–state relations earlier in this century were analogized to a layer cake of separate responsibilities assigned to the states and to the federal government. Grodzins (1974) argues that during the 1930s, as the federal government undertook new responsibilities in the economy and in society, the layer cake was transformed into a marble cake. No clear separation of responsibilities between state and federal government could be identified. Within a single policy area, an agent of the government could be charged with performing some duties that were imposed by both state and federal governments, some by the states alone, and others exclusively by the federal government. Thus emerged the marble cake metaphor which makes level of government no longer predictive of administrative or legislative responsibility. During the 1960s and 1970s, power flowed to Washington and states were left with many permitted duties but few independent responsibilities. Critics of this shift of power to Washington charged that it had turned the states into supplicants. Of late, Wildavsky (1982) has suggested that federal–state intergovernmental relations have come to resemble a fruitcake. The largess of federal programs is the candied fruit imbedded in the cake. By this metaphor, Wildavsky apparently means that as states compete for federal resources they do so with little concern for the costs that are distributed nationally.

These three metaphors both suggest the difficulty in distinguishing among the levels of American government and describe the accretion of power at the federal level over the past half century. As the brief historical survey which follows underlines, from the beginning of the Republic the federal government has played a major role in the pattern of natural resource use and development. That role has been in significant part defined by the interests of the states and to some extent substate governments as well.

The relationship between the states and the federal government, however, has not always been in the direction of accumulating federal resources. Richard Stewart (1982) has described the operation of American federalism as a "dialectical process with alternating surges of centralization and decentralization." Stewart believes that in the last few years the nation has been undergoing a surge of decentralization. Indeed, there is evidence that states and local governments have sought and gained some recent successes in achieving a larger role in natural resource issues.

But if the states are entering a period of strengthened responsibility, particularly in the areas of natural resource management and environmental regulation, what will be the implications for the future direction of policies? For example, will devolution of power to the states lead to increased or decreased environmental regulation of natural resource development? Will groups that have been influential on natural resource policy at the federal level now lose influence at the state level of government? A useful way to consider the possible consequences of an areal shift in power is to pose several likely scenarios for natural resource politics under a greatly strengthened state role. One scenario considered here, which may be called the Development Scenario, predicts an increase in natural resource development and a corresponding weakening of environmental regulation. Another scenario discussed below, the Diversity Scenario, anticipates that some states will strengthen regulation and engage in lessened resource development while other states will weaken regulation and intensify resource development. A third possible scenario, the Environmentalist Scenario, predicts that decentralization will result in increased regulation of undesirable externalities on the state or local level.

The Development Scenario

The Development Scenario argues that strengthening state governments at the expense of federal power will facilitate resource development. Proponents of increased resource development point out that even when the presidency is favorably inclined to increased development, its efforts are frequently checked in Congress or in the federal courts. Opponents of resource development, it is charged, have developed leverage at the federal level out of proportion to their influence in the resource–rich states. A further charge is that the concentration of power at the federal level in natural resource decisionmaking has allowed states in the Midwest and East to impose environmental and preservationist constraints on the states in the West, which are richer in resources. These constraints have been imposed ostensibly on environmental grounds but in actuality have functioned to protect the aging industrial base of the East. Thus devolution of power to the states is defended as a means to allow economic decisionmaking to become consonant with the preferences of the respective states.

Environmentalists, however, do not look favorably on the Development Scenario. Many conservationists and environmentalists have long argued that devolving power to the states may prove adverse to resource conservation and environmental protection. One reason advanced for the pessimistic predictions of the Development Scenario is the nature of state government decisionmaking. A long–standing concern of critics of state decisionmaking about natural resources is that state legislatures historically have been susceptible to the influence of developers and traditional resource users such as miners, ranchers, and farmers, and less receptive to the concerns of environmentalists and recreationists. Critics of state executives have argued that state agencies charged with enforcement of environmental regulations or state resource management have been understaffed, underfunded, and lacking in the political influence requisite to manage their operations in an environmentally sensitive way. Thus one explanation offered for the likelihood that a transfer of power to the states threatens environmental values is the inability or unwillingness of the states to pursue a vigorous regulatory policy toward natural resource and public land uses.

Another explanation for the anticipated decline in environmental regulation predicted by the Development Scenario is an application of the so-called free rider problem. If every state is responsible for environmental regulation, then the state imposing the most restrictive regulations is likely to lose that portion of its economic base that is sensitive to environmental and natural resource constraints. Industry will pull up stakes and move to states that pursue more relaxed regulatory strategies. Yet these states may continue to benefit from the policies of their environmentally more stringent neighbors. The costs of environmentalism are not evenly distributed, nor can they be exported. A decentralized political structure thus penalizes locally initiated environmentalist legislation and administration.

A third, related reason that decentralization may retard or erode regulatory objectives is that many environmental costs regulated by the federal government cannot successfully be contained geographically. Water pollution, air pollution, and acid rain easily extend over the boundaries of many states and indeed transcend national boundaries. They present a challenge to state, national, and international regulation. For a state to promulgate stiff ambient air quality standards while the contiguous state upwind encourages oil shale development can only demonstrate the unlikelihood of achieving locally promoted air standards. Problems of coordination also exist among states that share environmentalist goals. Although coalitions of like–minded states might form, there is little certainty that they would be geographically contiguous or politically effective. The overall conclusion drawn on the Development Scenario is that devolution of power from the center can only lead to deterioration of environmental standards.

The Diversity Scenario

The Diversity Scenario hypothesizes that an increasing shift of responsibility to the states will result in greater diversity of environmental policies. This scenario is suggested by advocates of increased restraint on federal growth. It predicts that some states will devote more resources to land use management and environmental regulation while others will choose to minimize their efforts. What is likely to account for the difference in response among the states that the

Diversity Scenario predicts? There is a trade–off between economic growth and environmental regulation: at some point, growth will decline in the face of the increased costs of regulation. The more prosperous the state, the greater the likelihood that the political community will discount the value of increased resource development and seek to avoid the costs to the quality of life that increased development would entail. One consequence is that wealthier states may experience a limited growth rate in order to maintain their amenities while poorer states would experience environmental deterioration in order to achieve a higher rate of growth. It is important also to recognize that states may have a diversity of development goals. Some western states may seek to promote tourism and/or recreation rather than, say, mineral development. Recreational development is particularly likely to increase if user fees are adopted.

Such diversity in response to a contracting federal role would yield sharply different environmental and income situations among the states. The notion of a minimum standard of living or of a minimum standard of environmental quality would no longer be recognized as a national standard—then again, neither would economic development. As the states distanced themselves from one another on environmental quality, presumably there would be sharply different consequences for the populations living within the several states. Such differences might promote increased migration by people seeking economic possibilities or environmental conditions congruent to their needs and wants. To many, this federalist diversity represents a desirable form of market choice. To others, it would block misguided federal development. To still others, however, diversity would only contribute to greater inequality and hardship for those without the resources to move. Moreover, those states that pursue economic development at the expense of environmental regulation cannot ensure that the undesirable externalities of air pollution and water pollution will be confined to their geographic boundaries. (Externalities are defined in this context as activities that impinge on the environment but are outside the pricing system.) These states, therefore, would be likely to contribute to the degradation of the air and water in adjacent states that prefer strengthened environmental protection at the expense of economic growth.

The Environmentalist Scenario projects that decentralization would lead to stiffened environmental regulation and preservation of open spaces. Two images of society have been advanced for this prediction.

The first image is the hope that local communities, once given some measure of autonomy, will support restrictions on development out of affection for the landscape around them. This conception has antecedents in a venerable tradition of political theory which argues that small states are a superior form of political organization in that they can elicit the will of the entire community. Frederick Jackson Turner predicted that the longer people stayed in an area the greater their attachment to their environment and therefore their willingness to maintain the quality of their local environment (Steiner 1979). A contemporary illustration would be a seaside community resisting offshore oil drilling on the grounds that such drilling would transform the seascape that has nurtured the community. Local control can cut two ways, however. A number of environmentalists such as Rene Dubos (1972) and Barry Commoner (1972) have argued that local communities must ultimately be given much more extensive selfgovernment if the objective of environmental balance is to be achieved. But if local values, poverty, or other factors lead to disregard for environmental values, much harm can occur before Turner's prediction comes true.

A second image is that small–scale political organizations are less likely to promote large–scale projects that transform the landscape adversely. This view regards smallness of scale as a positive virtue because it narrows the range of adverse possibilities from governmental action. Large nation–states may possess the resources, the capacity, and the intent to promote large–scale economic projects, or military projects such as the MX missile, and to advance policies that bring about major transformations in the land. These policies may lead to substantial depletion of natural resources and, as a consequence of unpriced externalities, to serious deterioration in the quality of the landscape and the environment. An illustration would be the proposed synfuels program of the Carter Administration, which apparently required the federal government to underwrite the major portion of its capital. It could not be launched or even reach the stage of organizational possibility without a system of greatly strengthened federal responsibility at the expense of the states in the areas of resource development and environmental regulation. There is also an argument that centralization of authority, particularly in energy development, impedes technological innovation. Decentralization may produce more in the way of "soft path" solutions to the provision of energy such as reliance on conservation.

These three scenarios—Development, Diversity, and Environmentalism—suggest the importance attached to the areal allocation of power in American politics specifically with respect to natural resource and land use issues. In two out of the three scenarios, the predicted result would be the weakening of environmental regulation over the nation as a whole. It is only in the third scenario, the vision of some environmentalist writers, that smallness of government facilitates considerations of simplicity and reduction of scale in economic and natural resource development. The debate over centralization of political decision making is thus linked to the values conflict over natural resource and land use policy.

The Values Conflict and Natural Resource Policy

The second major theme that is interwoven in contemporary natural resource politics concerns the values that should guide natural resource and land use policy making. There are, of course, firmly held values and preferences that come into play and often into conflict over natural resource policy. Such debates reach to basic value differences. Positions taken rest on conflicting interpretations of the fundamental relationship between humanity and nature.

One important conception of this relationship that has long been influential in this nation's history places humanity at the focal point of nature. This location of man at the center stage in nature is often expressed in the imagery of overcoming nature, of transforming nature's resources into socially useful products. Images such as "taming the West," controlling wild rivers, and building pipelines in inhospitable climes are legion in the history of American economic development. The recent roots of this conception are apparent in the writings of John Locke, the 17th century English philosopher who argued approvingly that the right to property arises when man mixes his labor with the natural world, given by God to mankind in common (Locke 1967). This view of man as nature's moving force transforming natural resources into the building blocks of political and economic organization may be labelled the "developmentalist" orientation.

The developmentalist perspective, however, is not alone in locating man at the center of nature. One important theme for the conservationists was the concept of sustained yield, cultivating nature's bounty for generations to come. The forest management schemes begun with the creation of the Forest Service in Theodore Roosevelt's Administration are an example of this concept.

Although it is perhaps less obvious, preservationists such as the novelist Edward Abbey (1975) and the legal scholar Joseph Sax (1980) also assign humanity pride of place in nature. These preservationists see in undeveloped nature the source of deeply important spiritual, aesthetic, and recreational values for society. They seek to preserve mankind's natural surroundings as psychological restoratives to the complaints engendered by contemporary society. Although developmentalists and preservationists may share the framework of regarding nature as for humanity, they may not agree on much else. In particular, they are worlds apart on how nature can and should be used to sustain mankind. The classic conflict between developmentalists and preservationists is personified by the debate between John Muir, the founder of the Sierra Club, and Gifford Pinchot, the creator of the United States Forest Service. Pinchot supported the flooding of Hetch Hetchy Canyon near Yosemite National Park, in order to supply San Francisco with water. This was in his view a desirable use of a renewable resource to build urban civilization. Muir dissented, arguing that Hetch Hetchy was an area of great aesthetic beauty and should be preserved. Muir lost that battle, but this sort of conflict regularly recurs in natural resource politics between these two interpretations of nature's value for men.

Center stage, however, is not the only place that mankind has been assigned in nature. A fundamentally different framework regards humanity as simply one part of nature, inextricably dependent on the fortunes of the larger biological community. Many of the contemporary environmentalists who share this orientation have been influenced by such writers as Aldo Leopold (1949), Rene Dubos (1972), and Barry Commoner (1972). These writers stress the interrelatedness of man and nature. Man properly understood is not at the focal point of nature but must be understood as a part of the biosphere that sustains life. The biosphere is a narrow, fragile strip that extends several hundred feet in the air and a few dozen feet into the earth. These environmentalist writers fear that when society undertakes major projects to channel natural resources for economic development, it will produce ecological disturbances so adverse as to threaten human society itself. Many environmentalists subscribe to Leopold's land ethic for guidance in devising natural resource policy. The land ethic stresses that people must live together as members of a larger community that includes not just other people but soils, waters, plants, animals, and the land itself. A proposed resource use must be evaluated not just in terms of its potential to help the human community but in terms of its consequences for the much larger community of all living things.

On the plane of practical politics, the shifting coalitions seeking to influence public land and natural resource policy frequently combine groups with very different orientations to the natural order of things. An important part of the animation of these groups, however, is to be found in their members' respective orientations to the meaning and value of nature. In executive and legislative debates over policy, environmentalists are often taken to include leaders of such organizations as the Sierra Club, the Audubon Society, the Wilderness Society, and the National Wildlife Federation. Politically such organizations are frequently found together in support of more stringent restrictions on natural resource development projects in areas judged to be sensitive to environmental disturbance or of great recreational or aesthetic value. Often described as opponents of the environmentalists are organizations that wish either to maintain existing patterns of commodity use, such as timber cutting or livestock grazing on the public lands, or to intensify resource development such as coal or hardrock mining. But the environmentalists and the traditional user groups are not opponents in every conflict. The National Cattlemen's Association and the American Mining Congress may share the goal of weakening the Endangered Species Act, yet the Sierra Club and the Cattlemen may work together to further certain provisions of the Surface Mining Control and Reclamation Act. That resource user groups favor loosening environmental regulatory constraints on their respective activities does not mean they favor weakening all regulatory restraints on all uses of natural resources and the public lands. If it did, legislative outcomes would be far more predictable.

Natural Resource Use and the Public-Private Debate

A third persistent theme in land and natural resource politics is the debate over whether resource use decisions should be made by society, that is, by governmental allocation. A contrasting possibility would be to consign such decisions to the market—to decisions by private individuals and groups. A useful perspective is to consider this issue of societal decisionmaking on a continuum of governmental intervention ranging from complete governmental determination of natural resource and land use allocation to complete abstinence from governmental intervention in free market allocative forces. All along the continuum, the existence of government is presupposed. A free–market allocative system requires the existence of a government that legally enforces a system of defined and assigned property rights, which in itself may indirectly help to shape allocative decisions. The issue at hand is not whether government should exist at all, but whether governmental or free–market choice should determine resource and land use allocation. In recent years, the debate over the form and extent of governmental intervention in natural resource use has been particularly vigorous in the two issue areas of environmental regulation and natural resource development.

Debates about the proper role for government in these two issue areas must begin with an understanding of the nature of the goods at stake. In particular, a discussion of these two issue areas requires some understanding of the economist's concept of a public good and the possibility of treating many natural resources as public goods or at least of understanding them as having a strong component of "publicness" to them. A public good is thought to have the following two important characteristics. First, an individual's consumption of the good precludes its consumption by others. The air we breathe is an example of a public good, while an avocado that, once consumed, cannot be eaten by others is an example of a private good. Second, a public good is a good whose provision is not restricted to selected individuals, such as those who pay for it. Improvements in the air quality of an area, for example, will benefit all who live there, regardless of whether they pay any of the costs. By contrast, allocation of gas masks only to those who pay for them would be provision of a series of private goods. Because access to them is unrestricted, public goods pose especially serious problems for fairness in the distribution of the burdens needed to produce them.

A natural resource, when it is understood as a public good, is a resource whose consumption is not or cannot at present be restricted (Freeman, et al. 1973). Current examples include the air, fishing in the high seas, and the much–cited application of public good analysis in the literature of livestock grazing on a commonly held pasture. The argument is that when a natural resource is treated as a public good it is likely to be abused or overutilized. There are several pertinent examples. Air pollution can result from the release into the atmosphere of the unappealing byproducts of industrial processes, at little or no cost to the producer of the pollution. Fishing areas on the high seas may become overfished if access is unrestricted: the rational behavior of any one fishing boat is to maximize its own catch since to do less is simply to leave more fish for others to catch.

Grazing land is the most widely discussed example of the projected deterioration of natural resources when they are treated as public goods. Hardin (1968) and others have argued that pasturage held in common will be overgrazed—the so–called "tragedy of the commons." The tragedy of the commons argument is spelled out as follows. When livestock are added to a commons, the costs of increased forage consumption are borne by all users, but the benefits of an enlarged herd accrue to its individual owner alone. The advantages to the grazer of increasing the size of his or her herd are reinforced by the recognition that all other grazers are likely to make similar increases, so that decreased forage will be available to the grazer without an augmented herd. An individual grazer operating on a common pasture thus acts rationally to add additional livestock to his or her herd. This enjoyment of the benefit of the commons, however, will be tragically temporary if the result of all the individual decisions to augment their respective herds must be the ultimate deterioration of the common pasture to the point that all users suffer.

If one accepts this argument that treatment of a natural resource as a public good will lead to abuse or overutilization, what can be done? Sets of solutions to this tragedy can be located along the continuum of degree of governmental intervention. At one end are solutions that rely on a combination of private property and market–choice strategies. At the other extreme are solutions that require an active program of government intervention in the common interest.

Proponents of the first set of solutions argue that the only way to resolve the problem of the misuse of resources is to assign clearly delineated property rights that can be judicially enforced as are other forms of property. They would transform the common pasture into parcels of privately held grazing land. Advocates of privatization contend that firmly linking the user to ownership of the resource would not allow the grazer to displace the costs of augmenting his or her herd onto the larger community. The resource owner will come to understand that depletion of the resource at issue will result in a reduction of his or her yield or income. Moreover, a depleted resource reduces the market value of property holdings. Thus property rights will force users to come to terms with the benefits and costs of their resource use, and presumably lead them to husband the resource for a strategy of long–term yield.

Proponents of social decisionmaking for natural resource use are skeptical of whether privatization will foster such a look at the long term. Private interests, they contend, will exploit natural resources for short-term returns. Exploitation will be encouraged still further by the extent to which private holders are allowed to displace the costs of their activities onto others, as by freely discharging pollutants into the atmosphere. Opponents of privatization thus argue that the only way to avoid the abuse or overuse of natural resources is to strengthen the governmental agencies responsible for natural resource and land use management. They contend that a responsible strategy of resource use can only be realized when the government is strengthened so that it has the power to identify socially acceptable uses and to manage, monitor, and enforce them.

The recent sharp debates over the specific issue areas of natural resources development and environmental regulation have taken place against this backdrop of the public goods problem and proposed solutions to it. Historically, federal involvement in natural resource development is as old as the Republic. Large–scale governmental responsibility for the condition of the environment, on the other hand, is of fairly recent origin.

Extensive land holdings greatly facilitated the federal government's role in natural resource development. That role, however, varied greatly. In some natural resource areas, the federal government merely surrendered its holdings to private activity. In hardrock mining, for example, the federal government ratified into the 1872 Mining Act the informal rules that had evolved among California gold miners during the California Gold Rush in the 1840s. In effect the miners had treated the federal commons as unowned land and had established their own procedures to stake claims. These procedures are in large measure still in effect today.

One of the best known examples of active government support for development was in agriculture. The government made extensive land grants for farming and for the creation of universities and colleges for agricultural research and the training of farmers. Another important example of federal underwriting of infrastructural development in the nineteenth century was the land grant subsidy for railroad construction. Later in the nineteenth and early twentieth centuries, the federal government began to assume responsibility for the active management of large portions of the nation's forest lands. In 1902, with the passage of the Newlands Act, the federal government assumed a central role in financing water projects in the western states. These projects provide irrigated water, most notably for agriculture in southern California, and they are increasingly used for flood control and hydroelectric power.

The commitment to realize public purposes by governmental intervention has been apparent for years in the Forest Service and more recently in the Bureau of Land Management (BLM). Both agencies have been charged by Congress to promote actively the public's interest in the national forests and on the public rangelands. Historically, the mission of the natural resource agencies was to salvage the forests and rangelands that had been depleted or overgrazed by private users. Later the task of the natural resource agencies was to determine the appropriate set of multiple uses for forests and the public lands through elaborate planning machinery animated by congressional statute. The challenge for such agencies is to attempt to achieve public purposes in sorting out the multitude of often conflicting uses proposed for a particular resource—or tract of land. It is compounded by debates over the nature of the staff and policies needed to sustain these uses.

Governments can employ many ways to promote economic development, some more actively interventionist, and hence further along the continuum, than others. A government may create tax incentives, loosen regulation, or grant subsidies directly. Even more comprehensively, it can own and manage specific resource projects controlling every aspect from water rushing down a river to hydroelectric power for a city many miles away. The Tennessee Valley project of the 1930s combined energy development, flood control, and agricultural enterprise that over the years have made the federally owned TVA one of the nation's largest corporations. Supporters of TVA argued that government intervention was necessary to the economic development of the region because the project's returns appeared speculative and very much in the future. On the other hand, government investment in hydropower in the Columbia River basin did not evolve in the same comprehensive fashion as did TVA.

It is obvious that government promotion of economic development has not been universally well received. Critics have argued that the government ought not to intervene in the free market allocation of resources. Other critics such as Niskanen (1971) have contended that the United States Treasury functions as a common pool, into which congressmen and bureaucrats dip to satisfy their various constituencies without restraint, because their respective constituencies can capture the benefits while distributing the costs to all taxpayers. If the specific congressman or bureaucrat does not dip into the Treasury, others will. Thus, the critics conclude, it is rational for all to act to the detriment of fiscal responsibility. Natural resource projects will be funded that are too localized in their economic effects and are, therefore, little more than political patronage.

For the most part, direct federal involvement in resource management has been oriented toward development. Supporters of direct federal investment in natural resource projects have claimed that federal involvement has increased agricultural productivity, produced the energy for economic growth, and further integrated the nation's economy. In the last two decades, however, criticism from environmentally concerned groups has charged large–scale federal projects with contributing to environmental degradation, and it soon became apparent that governmental authorities were, on the one hand, still charged to promote economic development but, on the other hand, expected to maintain or achieve standards of environmental quality.

The challenge to national policymakers has been to devise a regulatory approach that will protect the environment without weakening the role of state and local governments. Congress has assumed a measure of responsibility for the condition of the environment in the areas of air and water quality particularly. The National Environmental Policy Act, the Clean Air Act, the Clean Water Act, and many other environmentally sensitive pieces of legislation have proliferated federal regulatory agencies, which in turn have proliferated rules and standards. For the most part, the regulatory model employed by the federal government in the search for environmental quality has been drawn from prior experience in the regulation of economic activity. This regulatory model, which evolved during the New Deal, has contributed to the federal government's central role in the economy. Under the model, Congress uses a statute to authorize a federal agency to articulate standards, which are then enforced by the appropriate federal authorities. For the public lands, this approach has meant grafting the regulatory model onto existing agencies.

Congress has been progressively dissatisfied, however, with the New Deal regulatory model. By the beginning of the 1970s, Congress had begun to abandon the New Deal model of assigning discretionary authority to an independent regulatory commission (Ackerman and Hassler 1981). This trend is apparent in the Clean Air Act and in natural resource regulatory legislation such as the Surface Mining Control and Reclamation Act of 1977. In the effort to control agencies, Congress in these statutes articulated detailed standards and set out methods to achieve them. Frustration over the repeated failure to attain air quality, however, has led critics to conclude that the Clean Air Act with its emphasis on standards rather than economic trade–offs has generated an elaborate administrative and legal effort but has produced relatively little clean air. Similar charges have been leveled at Forest Service and BLM planning as well. Thus opponents of the existing regulatory standards model argue that this form of governmental intervention has not solved the problem of the abuse of public goods such as the air or the range.

Lave and Omenn (1981) among others have advocated yet a different model, the use of economic incentives, to improve air quality. Such incentive models with their strong reliance on the extension of the pricing system have been suggested for a whole range of natural resource and land issues. These models reflect the market solution to the tragedy of the commons. As applied to environmental regulation, the incentive model works on the assumption that pollution or any other cost imposed on others by economic activity should be assigned a charge. If a firm or individual wishes to release pollutants into the air, for example, it may do so by paying a charge "which no moral or legal prejudice attaches to the fee itself or the action on which or for which it is paid" (Schelling 1983). Its supporters characterize the incentive model as a market–allocation system, less dependent on governmental intervention. They argue that it extends the market–pricing system to the problem of undesirable externalities. Individuals and firms can decide whether they wish to pollute or not based upon their willingness to pay the assessed fees.

Implementing the incentive model involves three tasks. One is to set prices for particular levels of pollution or harmful effects. A second is to determine who has contributed how much to pollution over time. The third is to find adequate ways of allocating compensation to those who may be harmed. The situation is relatively clear for simple cases such as an upstream polluter of a river, but much less so for a power plant emitting sulfur dioxide that eventually falls as acid rain. The incentive model may lessen damages to private productivity but it is doubtful whether it can reduce the role of government in many cases.

The appeal of the incentive model is enhanced by the public controversy that surrounds the question of environmental regulation. There is widespread popular sentiment for maintaining some level of environmental quality. This sentiment is found in Congress and reflected in congressional legislation. Generalized feelings, however, often encounter intense resistance to specific regulatory efforts, particularly those involving trade–offs that include the reduction of an identifiable industry's output.

The difficulty of environmental regulation is intensified by charges that Congress and the Executive branch are particularly susceptible to cross–pressures from the demands of private organizations and regional interests. Thus, according to Ackerman and Hassler (1981), attempts to clean up the air resulted in scrubbers being required for not only high sulfur eastern coal but low sulfur western coal in order not to confer a price advantage on western coal producers. Sanders (1981) argues persuasively that natural gas regulation, as it became more comprehensive, became more susceptible to conflicting regional and industry interests. Thus the demand for environmental regulation has not infrequently resulted in increased political frustration as efforts to establish and to enforce standards reveal the political vulnerability of the federal government to private, state, and regional interests.

The federal government's dual role as active promoter of natural resource development and enforcer of standards of environmental quality commits it to efforts that sometimes appear to be mutually at odds. The case of the federally held lands aptly illustrates this dilemma. The federal government owns one–third of the nation's lands directly. It claims to ownership and the right to do as it wishes with the lands are far greater than those of private land owners—a judgment confirmed and reaffirmed by the federal courts. Yet the position of the federal landowner is analogous to that of a landlord of an apartment building in a city with a strong tenants' rights statute. The landlord not only cannot evict tenants it finds undesirable, it cannot refuse to rent to prospective tenants of whom it does not approve.

To complicate the situation in the apartment building, some tenants dislike one another thoroughly, and engage regularly in disputes with one another over maintenance and decor. The federal government has granted a variety of users from livestock grazers to ski resort operators de facto permanency on the federal lands. Efforts by the government to restrict the conditions of their use have repeatedly proved ineffectual. Moreover, some users, such as recreationists and preservationists, have resented what they regard as the subsidies given by the federal government to other users such as grazers, whose operations in the view of the preservationists interfere with other uses of the lands. Stockmen and miners retort that wilderness is administered for an elite of wealthy users who pay no fee at all. On one hand, commodity users have regularly called for the elimination of active federal management or indeed the end of any form of federal ownership of the public domain. On the other hand, a number of recreationists and many environmentalists have demanded tighter, more active federal management in order to reduce the level of private development concessions on the lands. The reverse attitude toward an active federal role is found on the question of the promotion of resource development. Many environmentalist groups have sought to diminish the federal government's role in advancing the capital necessary for large–scale water and energy projects, while a number of private firms and communities have welcomed an active federal role in such activities.

These issues concerning values, areal decisionmaking and governmental intervention in natural resource politics are explored by the contributors to this collection. These issues have arisen from the distinctive relationship between federal possession of vast unsettled lands and the competition over the development and use of the resources found on these lands. The section that follows provides the historical context for the issues examined by the contributors.

Public Lands and Natural Resources in Historical Perspective

Throughout the history of the American Republic, issues such as water rights, mining, timber, agriculture, and recreation have all been part of the larger picture of federal land policy. At the beginning of the Republic, the federal government was given the western lands claimed by the original thirteen states. The understanding was that as the lands stretching from the Blue Ridge to the Mississippi River came to be populated they would be divided into territories and later admitted to the union as states. In accord with this understanding, it was the policy of the federal government to dispose of its holdings.

These holdings were enlarged greatly by purchase (from France in 1803), by conquest (from Mexico in 1845), and by annexation. As early as 1850, however, 300 million acres of the public domain had been disposed of by the federal government. Land disposition was promoted for a variety of objectives—settlement, new sources of revenues, and as rewards for deserving groups such as veterans—and achieved by a number of means. The advancement of economic development through land grants for railroad construction was an important factor in the middle of the 19th century. Support for public education was deemed an important goal from the late eighteenth century, with regular federal donations of school sections to each state as it joined the union.

A major theme in land policy was the creation of an independent yeomanry of small landholders. Congress in the 1862 Homestead Act mandated a maximum of 160 acres for dispositions to homesteading settlers. Larger grants were opposed for many years, partially on the ground that they risked creation of great landed estates and thus would be undemocratic. As a consequence of federal homesteading policies, however, land was quickly settled near water and transportation routes. The acreage restriction discouraged settlement of arid and mountainous land, where small ranches were not economically viable. Successive expansions of the acreage provision were still not sufficient to relieve the federal government of much of its landholdings west of the 105th meridian (roughly, west of the Rocky Mountains), a region of limited rainfall.

Federal land disposition policy was always controversial but became even more complicated as time went on. As new states entered the union, varying portions of the land within their borders remained under federal ownership. At admission, new states were obliged to include in their constitutions a disclaimer clause surrendering any claims to the federal land. In return, the new states were granted varying amounts of acreage to support state services, most notably education. Paul Gates (1968), the public lands historian, argues that tension developed between the older states without federal land and the new western states with substantial federal holdings. The eastern states' leaderships regarded the federal lands as won by their states' blood and treasure and therefore wished the federal government to profit from the lands. In sharp contrast, the federal lands states' leaders believed that the lands should be turned over to their respective states to promote economic growth and settlement. As evident in the essays in this volume, this tension between state control and federal benefit continues today.

Even in the mid nineteenth century, however, disposition was not the only approach to federal land policy. The early land disposition laws retained for the federal government certain lands on which were located precious minerals or other resources deemed to be in the national interest. In some cases, as discussed earlier, federal land policy simply accepted what had already taken place. The best illustration is the California Gold Rush where miners staked out claims on the public domain. The practices were ratified into federal law by 1872.

Other bases for reserving land from private ownership began to appear as early as the 1830s with the reservation of lands of great natural beauty so that, undisturbed, they would provide the public with pleasure. Arkansas Hot Springs, for example, was established from four sections in 1832. In 1867, the federal government gave Yosemite Valley to California as a state park. Later the area was returned to the federal government and redesignated as a national park. The most spectacular land reservation occurred in 1872, a time of great interest in resource development: some two million acres along the upper Yellowstone River were set aside as a national recreation area (Huth 1957).

In the late nineteenth century, the land disposition programs came under increasing criticism (Dana and Fairfax 1980). With respect to timber, critics charged that privately owned forests were poorly run and the private entrepreneurs looked only to immediate profits and therefore were depleting one of the nation's greatest natural resources. Beginning in the Harrison Administration, continuing in the Cleveland Administration, and on a large scale in Theodore Roosevelt's presidency, some 234 million acres of timber land were turned into national forests. In 1905, Roosevelt gave Gifford Pinchot the responsibility for creating a professional forest service to administer the lands. During the same period, another 80 million acres of land were reserved as potential mineral lands, particularly for coal and oil. Again the concern was that inefficient private developers would rapidly deplete the nation's patrimony.

The turn of the century was the age of the conservationists, personified in Theodore Roosevelt and Gifford Pinchot. The conservationists argued that private developers of nation's resources were motivated principally by immediate profits and lacked the skills necessary to develop critical resources rationally. Instead, the conservationists advocated direct federal responsibility for natural resource development. The conservationists did not wish to forgo development. Rather, they believed that scientific management of natural resource development would provide for the needs of generations to come in a prudential and efficient manner (Hays 1959).

The conservationist movement coincided with the Progressive era in American politics. This association with Progressivism helps to explain the conservationists' reliance on the federal government rather than the states to take the initiative in resource management policy. The Progressives criticized state and local governments for being too often dominated by party political machines that misappropriated tax revenues and were corrupt in the provision of governmental services. The Progressives also judged state and local governments to be far more susceptible to manipulation by dominant but questionable local interests. This Progressive critique lent support to the conservationist commitment to establishing expert civil service agencies, at the federal level, to administer the nation's resources. The creation of the Forest Service under Pinchot exemplified this early federal commitment to public resource management removed both from the private sector and from the states.

At the state level, the Progressives supported institutional changes designed to increase direct democracy, such as the referendum and the initiative. Paradoxically, this Progressive attack on government dominated by party political machines may have contributed in the long run to the weakening of state governments. The undercutting of political parties as a basis of political organization made it more difficult to mobilize support for a program in the executive and legislative branches of state governments. Progressive ideas on party reform and governmental organization were particularly influential in the western states. A reasonable speculation is that this weakening of state political institutions may have made it more difficult for individual states to develop their own active natural resource policies.

Although states may have possessed neither the interest nor the capability to undertake the management of the federal lands, they were, nonetheless, politically influential in capturing a large share of the returns on resources developed within their respective boundaries. By 1920, Congress had worked out a compromise between the public lands states and the federal Treasury. The contours of that compromise are still evident in today's division: 50 percent of revenues from an energy lease (oil, gas, or coal) goes to the state of location, 10 percent goes to the federal Treasury, and the remaining 40 percent goes to the Reclamation Fund but, in effect, goes to the Treasury as well. In addition, western states and to some extent Indian tribes are able to impose severance taxes on energy leases.

By the second decade of the twentieth century, it had become apparent that the age of homesteading and settlement of the public domain was coming to an end. One–third of the nation remained in federal ownership. Half of the land in the eleven contiguous western states, even excluding Alaska, was federal property. The pattern of nonfederal land ownership in the western states was a complicated checkerboard of private land, usually near water; scattered state sections given as grants for schools; and railroad–grant land. All these lands appeared as islands in a sea of federal land. It is important to emphasize, however, that the federal lands were not uniformly administered under the same statutes or by the same agencies. Indian reservation lands enjoyed a quasi–autonomous standing, national parks and monuments were administered for preservation and recreation, and national forests were managed for sustained yield.

The vast remainder of the lands in the federal domain that for many years had been barely touched by federal management were the unappropriated lands known as the public lands. The public lands were the responsiblity of the General Land Office. The GLO lingered on, albeit with diminished responsibilities, until 1946. Politically the most important surface users of the public lands were and continue to be the livestock growers of cattle and sheep. During the nineteenth century the stockmen were almost the only users of the vast, arid tracts of western lands. They used their own private ranches, which usually had water, as the bases for their respective operations. The adjacent public grazing lands were used at several times in the year for forage. No real effort was made by federal land officials to regulate public land use through much of the nineteenth century. A major shift occurred during Theodore Roosevelt's administration in 1905, with the creation of the vastly enlarged national forest system. Many of the national forests, before 1905 known as forest reserves, included grazing lands. The Forest Service imposed grazing fees for its lands in 1905, which led to sharp conflicts between the ranchers, supported by members of western congressional delegations, and the administration. This conflict presaged the battles in the 1940s and the 1970s between the Bureau of Land Management and the stockmen using the public lands.

During the 1920s, the conservationist achievements of the early years of this century were neither rejected nor advanced. The Harding Administration was marked by one of the great political scandals in American history: Teapot Dome. The Teapot Dome scandal involved administration officials illegally leasing to private interests oil fields in Wyoming that had been reserved for naval purposes in times of military need. The scandal may have been one reason why the subsurface mineral deposits were not included in the only major public land initiative of the Hoover Administration—the proposal to turn the surface estate of the public lands over to the states of their location. This offer was rejected by the states, which were beset by drought and depression.

The administration of Franklin Roosevelt, however, introduced a new era of federal activism in both land and natural resource development issues. The overriding values of the Roosevelt Administration remained conservationist. In the related areas of water management and electric power, the federal government greatly expanded its efforts. The New Deal initiated the Tennessee Valley Project that was its principal experiment in regional economic development through water control and the provision of electricity. A less comprehensive federal hydroelectric project was established in the Columbia River basin in the Northwest during the same time period.

During the New Deal, the public land ranchers, long at the center of political disputes involving the federal lands, both favored and feared active management of the public domain. They favored the creation of a grazing service to improve the deteriorating condition of the public range, but feared that the fees for such a service would weaken them economically. In 1934, the Taylor Grazing Act created the Division of Grazing, later renamed the Grazing Service in 1939. The Service and the General Lands Office divided responsibility for the administration of the public lands. Initially, the Service operated by delegating authority to the ranchers themselves. In 1939, the Service took a more activist line in arguing for the imposition of grazing fees to cover federal management costs. The grazing fee controversy lasted throughout the 1940s. It brought down the Grazing Service and led to the creation of the Bureau of Land Management in 1946 (Peffer 1951). The Bureau took over the responsibilities of both the Grazing Service and the General Land Office.

Prior to the recent Sagebrush Rebellion, the last great political battle over the disposition of the public lands grew out of this attempt to achieve active federal management of the public lands in the late 1930s. Just after World War II, in 1946, representatives of stock growers and some mining groups met with western congressional leaders to consider legislation to strengthen the position of public land ranchers. One expressed goal was transfer of the grazing lands of the Forest Service and the BLM to the states of their location. Conservationist groups opposed the plan. Such groups were weaker in the 1940s than they were in the 1970s. Nonetheless, aided by federal land managers and the skilled essays of Bernard de Voto, conservationists foreclosed congressional action on the western proposals (Stegner 1974).

In the decades after World War II, the debate over the federal government's role in natural resource management on the public lands became increasingly complicated. On the one hand, the long tradition of promoting natural resource development, greatly expanded during the New Deal, continued to be fostered by the federal government's involvement in the peaceful applications of atomic energy. On the other hand, there was growing interest in the recreational possibilities of the public lands. The concept of "multiple use" had been gradually evolving during much of this century as a guiding principle of federal land management and it expanded as recreation became a major activity in the western states during the 1950s. This expansion of uses was not the extent of the complication, however. The environmentalist movement began to gather influence with its radically different conception of the stewardship of natural resources. During the 1960s, recreationists, preservationists, and environmentalists became increasingly influential on federal legislation concerning natural resources and the public lands.Indeed, their impact was felt at every level of government. Although local and state authorities expressed increasing interest in determining land use decisions, the federal role predominated. Federal laws in the area of natural resources, pollution, and air quality control began to affect land use planning to a major extent.

The frontispiece of federal legislation inspired by environmentalist values was the Wilderness Act of 1964. The act celebrates the values of land untouched by man. Other environmentalist legislation included the National Environmental Policy Act of 1969, which required preparation of environmental impact statements for major federal actions significantly affecting the environment, the Endangered Species Act of 1973, the Clean Water Act, the Clean Air Act and successive amendments to it in 1970 and 1977, the Surface Mining Control and Reclamation Act of 1977, and the Federal Land Policy and Management Act of 1976. These statutes are but a subset of legislation that both strengthened the federal role in natural resource and public land decisionmaking and deepened the federal government's express commitment to environmentalist values.

Although much of the major natural resource and land management legislation antedated the Carter presidency, that administration confronted the task of implementation at a time of increasing challenges to federal preeminence in natural resource decisionmaking. Challenges came from private groups, notably the oil and gas industry, from grazing and timber interests, and from western state governments. The Carter Administration, as observed above, did not speak with a consistent voice on natural resource issues. Its policies reflected diverse sets of interests, just as the policies of past administrations had done. But for some critics of the administration, particularly those urging increased oil and gas exploration, the Carter Administration symbolized the coming of age of the environmentalist movement. A number of long–standing leaders of the environmentalist organizations received appointments in the administration and these officials were perceived as unsympathetic to developmentalist interests.

By the mid–1970s, state–level challenges to federal policy initiatives and federal management had mounted. Many of the western states were very different entities than they had been in the 1930s. States such as California,

Arizona, and Colorado had experienced massive population growth. Other states such as Wyoming and New Mexico possessed new–found wealth in oil, gas, and coal. In general, state officials and state leaderships had grown in political sophistication and competence. Western congressional delegations had long been adept in securing federal funds and often policies favorable to western resource users; by the 1970s many states themselves wished to play a larger role in decisionmaking about natural resource use and in securing new sources of revenue directly from fossil fuels and strategic minerals. This new impetus was particularly evident in the formation of the Western Governors' Policy Office.

In the late 1970s, traditional resource users, notably the stockmen, once again began to organize a protest over federal grazing policies. From one perspective the "Sagebrush Rebellion," as the movement came to be known, was simply a recurrence of the range wars that had flared up periodically for the past century between the grazers on the public lands and the federal authorities. But from another perspective, the demand of the grazers and their allies that the public lands be transferred—or "returned"— to the states drew attention to the many changes that had occurred in the West since the last public land rebellion in the 1940s.

Many of the public lands, specifically their subsurface fossil fuel resources, were worth much more than before. The states themselves had become increasingly active in resource issues and management, particularly in seeking to retain the benefits of resource development while dispersing or displacing some of the costs. The federal government did not enjoy the standing it had had a generation earlier when the claims of scientific management fostered by the conservationists were more widely accepted. Finally, this is an age of many competing values and concerns over natural resource policy and land use that broaden the meaning of the Sagebrush Rebellion.

Privatization of the public lands—more precisely, sale of some portions of federally owned land—became a major controversy for the Reagan Administration in 1982. The privatization controversy rivaled the Sagebrush Rebellion in intensity and suggested the extent to which federal land policy had become politicized nationally. In January 1982, Senator Charles Percy (R–Ill.) introduced a resolution calling for an inventory of federal real property and the liquidation of unneeded assets (SB 231). The monies raised were to be assigned to reducing the national debt. On February 25, 1982, President Reagan established a Property Review Board to review federal real property acquisition and disposal policies. Federal agencies were to identify unneeded property. Property so identified was to be sold and, consonant with Senator Percy's resolution, the proceeds were to be applied against the national debt. One estimate projected receipts from the sale averaging $2 billion a year. But estimates varied. So did the amount of real property under consideration for sale. Some observers suggested that redundant land meant isolated parcels of BLM land in urban areas. Others such as Steve Hanke, an economist with the Council of Economic Advisors until the spring of 1982, seemed to indicate that all of the grazing land could be up for sale (Hanke 1982).

Political support for large–scale privatization of the public lands in the western states, however, was not forthcoming. Western governors such as Arizona's Bruce Babbitt opposed large scale privatization as a threat to the nation's heritage. Governor Herschler of Wyoming testified that privatization would require the direct role and voice of the western states. The National Association of Counties pointed out that the administration proposed to discontinue the low–cost transfers of federal land to state and local governments for public purposes, a proposal they firmly rejected. Huey Johnson, then California Secretary for Resources, argued that if the federal government were to receive fair values for the use of its natural resources, it would not need to sell off land for revenue (Hearings Before the Senate Committee on Energy and Natural Resources 1982).

By the fall of 1982, the Reagan Administration downplayed the scale of the privatization contemplated. Officials in the Department of the Interior went out of their way to distinguish between privatization— which they described as large–scale land sales—and "asset management" envisioned by the administration. The asset management program appeared aimed at limited land disposal (Stroup 1982).

The Sagebrush Rebellion and the privatization debate that closely followed have sparked extensive discussion of the nature and purpose of federal land-ownership and natural resource policy. The debate by no means has been confined to the question of federal ownership of lands and resources, but has raised serious institutional questions over how natural resource policies are to be determined and the appropriate managerial strategies for their realization. It is a debate that has intertwined issues of areal decisionmaking, contending values, and the role of the public sphere in policy. This collection has grown out of the fundamental debate on land and natural resources. In their respective discussions, the contributing authors explore the themes of this debate: land management, water, energy, environmental regulation, and recreation.

References

Abbey, Edward. 1975. *The Monkey Wrench Gang*. Philadelphia: Lippincott.

Ackerman, Bruce A., and Hassler, William T. 1981. *Clean Coal, Dirty Air*. New Haven and London: Yale University Press.

Commoner, Barry. 1972. *The Closing Circle: Nature, Man and Technology*. New York: Alfred A. Knopf.

Culhane, Paul J. 1981. *Public Lands Politics: Interest Group Influence on the Forest Service and the Bureau of Land Management*. Washington, D.C.: Resources for the Future.

Dana, S.T., and Fairfax, S.K. 1980. *Forest and Range Policy: Its Development in the United States*, 2nd ed. New York: McGraw–Hill.

Dubos, Rene. 1972. *A God Within*. New York: Charles Scribner and Sons.

Fradkin, Philip L. 1981. *A River No More*. New York: Alfred A. Knopf.

Freeman, A.M., Haveman, R.H., and Kneese, A.V. 1973. *The Economics of Environmental Policy*. New York: John Wiley.

Gates, Paul W. 1968. *History of Public Land Law Development*. Washington, D.C.: Government Printing Office.

Grodzins, Mortin, et al. 1974. *A Nation of States: Essays on the American Federal System*, 2nd ed. Edited by Robert A. Goldwin. Chicago: Rand McNally.

Hanke, Steve. 1982. The privatization debate: An insider's view. *The Cato Journal* 2: 653–62.

Hardin, Garrett. 1968. The tragedy of the commons. *Science* 162: 1243–48.

Hays, Samuel. 1959. *Conservation and the Gospel of Efficiency: The Progressive Conservation Movement, 1890–1920*. Cambridge: Harvard University Press.

Huth, Hans. 1957. *Nature and the American: Three Centuries of Changing Attitudes*. Berkeley: University of California Press.

Lave, Lester, and Omenn, Gilbert S. 1981. *Clearing the Air: Reforming the Clean Air Act.* Washington, D.C.: The Brookings Institution.

Leopold, Aldo. 1949. *A Sand County Almanac and Sketches Here and There.* New York: Oxford University Press.

Locke, John. 1967. *Two Tracts on Government.* London: Cambridge University Press.

McConnell, Grant. 1954. The conservation movement—past and present. *Western Political Quarterly* 7: 463–78.

Niskanen, William. 1971. *Bureaucracy and Representative Government.* Chicago: Aldine, Atherton.

Sanders, M. Elizabeth. 1981. *The Regulation of Natural Gas: Policy and Politics.* Philadelphia: Temple University Press.

Sax, Joseph. 1980. *Mountains Without Handrails: Reflections on the National Parks.* Ann Arbor: University of Michigan Press.

Schattschneider, E.E. 1960.*The Semi--Sovereign People: A Realist's View of Democracy in America.* New York: Holt, Rinehart and Winston.

Schelling, Thomas C., ed. 1983. *Incentives for Environmental Protection.* Cambridge, Mass.: M.I.T. Press.

Stegner, Wallace. 1974. *The Uneasy Chair: A Biography of Bernard de Voto.* Garden City, N.Y.: Doubleday.

Steiner, Michael C. 1979. The significance of Turner's sectional thesis. *Western Historical Quarterly* 10: 437–66.

Stewart, Richard B. 1982. The legal structure of interstate resource conflicts. In Kent A. Price, ed., *Regional Conflicts and National Policy.* Washington, D.C.: Resources for the Future, pp. 87–109.

Stewart, William H. 1982. Metaphors, models and the development of federal theory. *Publius* 12: 5–24.

Stroup, Richard. 1982. Comments on a paper by Joseph Sax. Portland, Ore.: National Workshop on Rethinking the Public Lands sponsored by Resources for the Future.

U.S. Congress. 1982. Hearings Before the Senate Committee on Energy and Natural Resources. 97th Congress., 2d Session.

Wildavsky, Aaron. 1982. Birthday cake federalism. In Robert B. Hawkins, Jr., ed., *American Federalism: A New Partnership for the Republic.* San Francisco: Institute for Contemporary Studies, pp. 181–91.

PART ONE
THE PUBLIC LANDS

Overview:
Policies for Public Lands

An enormous body of public land legislation was enacted during the 1960s and 1970s. This new regime for the public lands has fundamentally changed the processes through which decisions now are reached, altered the substance of those decisions in sometimes equally fundamental ways, and modified the relationship between the federal government and the states in ways not yet fully understood. One might therefore anticipate a period of consolidation as the consequences of new policies become apparent through their implementation.

To be sure, much attention has been given to implementation by both the Carter and Reagan Administrations. Nevertheless, the Carter Administration toiled diligently   although with limited success and at the high cost of contributing strongly to ignition of the "Sagebrush Rebellion" and other forms of assertiveness by western states—to strengthen federal and executive control over public land policy. Most observers would conclude that, on balance, Carter's thrust was toward increased ecological sensitivity, conservation, and preservation, though the impact of resource development programs must be factored into any final assessment. The Reagan Administration in its turn has installed James Watt at the head of a team mandated to emphasize resource development on the public lands. Meanwhile, members of the Reagan team have talked expansively of selling off substantial federal landholdings. It also seems clear that support has been given to ideologically oriented advocates outside the administration who bear the burdens of justification for sales while the administration stresses its own moderation. Relatively modest legislative changes have been enacted since 1976, but only because neither Carter nor Reagan has been able to translate discontent into legislative reform. It seems clear that opinion has polarized during this interlude.

The Reagan Administration's commitment to accelerated natural resource development, approached primarily through revision of regulations and reorganization of administrative responsibilities and budgetary priorities, and its interest in sales of public lands, have left it in the midst of an intergovernmental thicket remarkably similar to that which plagued President Carter. Western state governmental officials claim that their preferences are being ignored and that they have not been adequately consulted regarding land sales, leasing, and other initiatives. The common difficulties of these administrations in formulating public land policies suggest that the interplay of intergovernmental and value conflicts has intensified in recent years.

The contributors to Part One have different views of the mix of values that should guide public land decisions, and even more sharply divergent views on the location and the structure of decisionmaking institutions that should determine land use patterns. Contributors have been selected because each approaches the issues in ways that clarify the implications of existing policies, reveal unresolved difficulties, and/or formulate alternatives to present policies. The aim of this section is a contribution to debate over the future of public lands that avoids reliance on emotion–laden metaphor and the constraints on choice that overarching ideologies have long exercised on the analytical public lands literature.

In Chapter 1 John G. Francis examines the roots of pro–Sagebrush Rebellion sentiment. Nevada's prototypical approach crystallized discontent in the late 1970s by asserting ownership to the federal lands administered by the Bureau of Land Management. Other states followed suit. Francis focuses on the extent to which constituency characteristics in state legislative districts explain support for Sagebrush legislation. A finding of minimal relationship would suggest a "West against the Rest" interpretation of Sagebrush sentiment. Instead, Francis links support for Sagebrush to political parties and to rural/suburban district characteristics. Francis's conclusion is that a shift to sustained commitment to development at the federal level might lead proponents of environmental values to support greater responsibility at the state level, while proponents of development rest more content with federal ownership.

Another source of discontent with federal land management can be gleaned from Robert Nelson's analysis in Chapter 2 of the Bureau of Land Management's halting efforts during the 1970s to settle upon criteria for assessing alternative resource management strategies. To attain legitimacy, decisionmaking processes must be widely perceived as competent, efficient, and fair by those with interests or values at stake. Nelson traces the Bureau's sojourn in search of a scientific basis for asserting competence, the successful challenge to its approach mounted through the courts by the Natural Resources Defense Council, and the continuing difficulty of the BLM in moving very far along its "learning curve" toward an accepted approach. He identifies unresolved conflicts between range ecology and economism as alternative scientific beacons as well as institutional conflicts that run deep. The Forest Service shares some aspects of the BLM's quandary. Recently, the contest has returned to the courts for both agencies. Perhaps more than anything else, Nelson's piece challenges the long–cherished Progressive assumption that government could bring scientific management to public land management.

Sagebrush rebels emphasized language in the 1976 Federal Land Policy and Management Act (FLPMA) calling for retention of most BLM–administered lands in perpetuity and argued that this abrogates a long tradition of disposals to states and to citizens, even as it symbolizes a centralizing thrust of policy.

In Chapter 3 Sally Fairfax not only challenges this key assumption of Sagebrush rebels but also identifies an alternative for those unhappy with centrally directed BLM decisionmaking. Fairfax argues that the essential feature of the FLPMA is its mandated procedural emphasis upon federal consistency with state and local preferences. Assertive states and localities consequently have many opportunities to be decisive voices in the federally structured decisionmaking process. That option is open regardless of the political goals espoused, a lesson that Interior Secretaries Andrus and Watt have come to appreciate through chastening experience.

But should local and state values prevail regardless of their content or consequences in public land agency policy implementation? Preoccupation with BLM lands may turn attention away from essential public values, affected parties including governments and Indian tribes as well as private interest groups, and resource–specific issues. As one turns to these specific policy implications, is greater consensus possible?

In Chapter 4 Charles F. Wilkinson approaches these questions by asking what kind of public land law is good law. His answer is that good law recognizes and protects the respective interests of states, Indian tribes, and the general national public as represented by the federal government. The rub, of course, is that each of those interests is subject not simply to intrinsic needs but also to both development and expansion at the expense of the others. Many will find provocative Wilkinson's support for the existing set of interests protected by federal law. All will find his piece essential legal background, succinctly and clearly summarized, for an understanding of conflicts over such issues as state taxing authority, federal assertions of water rights, public land resource management goals, and the restive claims and efforts to achieve control over resources being asserted by Indian tribes and coalitions.

Christopher K. Leman has been immersed in studies of the administrative practices of federal land management agencies, which have led him to appreciate the enormous difference between vague, shifting congressional mandates to public land agencies and abstract economic models of efficient delivery that assume clearly ranked uses. He argues in Chapter 5 that this crucial distinction is ignored by the school of libertarian/public choice economists, many of whom have inspired if not guided Reagan Administration thinking on the desirability of land disposals. Leman's central thesis is that, by failing to contend with multiple and frequently competing agency mandates, these economic analyses are reduced to mere propaganda tracts that satisfy only the ideologically committed. Leman does not deny the vagaries of federal land management, yet is optimistic that improved management can help to realize the promise of the public lands without privatization.

Richard Ganzel offers a sharply contrasting perspective in Chapter 6, although his analysis also is critical of ideas underlying the Reagan Administration. Rather than taking vague multiple–use agency mandates as given, Ganzel argues that they were actively sought by both the Forest Service and the BLM as means to increase agency power over recreational and wilderness claimants. He contrasts multiple use with the much more directive mandates that guide specialized land management agencies and also the screening process for wilderness designation. Ganzel then distinguishes different groups that may have interests in particular public lands and constructs a land allocation approach designed to maximize those particular interests. Its implementation, illustrated through selected examples, would involve public land zoning as well as transfers to state governments of recreational lands that serve limited clientele and lands to be sold to accommodate urban expansion.

In Chapter 7, the final contribution to Part One, John Francis also advocates selective disposals of the public lands. However, in seeking to link regulatory means to public land administration he advocates selling use rights rather than the land itself and he proposes to delegate administration of many public lands to the states under federally established guidelines. These recommendations are derived from an analysis that begins by isolating four central values that public land management has attempted to achieve, noting that the mission given to the BLM is not only impossible to achieve but erosive of a sense of public legitimacy, and that many lands managed by the agency cannot contribute much to these public goals. To the extent that states are willing to assume implementation responsibilities, he argues, the federal government should accommodate those desires. Moreover, land should be made available for urban expansion in any case.

These seven approaches suggest a rich range of possible futures for the public lands. Without specifying whether their view extends to all federal lands, Wilkinson and Leman persuasively argue that a federal role is essential to protect national values and to fulfill trust obligations toward Indian tribes. By contrast, Fairfax, Francis, Ganzel, and Nelson argue that discharge of federal responsibilities can accommodate substantial modifications of the respective roles of federal agencies and state agencies, and can also accommodate changes in land tenure and in the substance of land policy.

1

Environmental Values, Intergovernmental Politics, and the Sagebrush Rebellion

JOHN G. FRANCIS

Between 1979 and 1981, in what became known as the Sagebrush Rebellion, fifteen states in the western half of the nation considered legislation that challenged the federal government's extensive ownership of land in the West. Approximately one–third of the nation's land is in federal hands. About half of the federal lands are in Alaska; of the remainder, 90 percent are located in eleven of the western states (see Table 1.1). The western states' legislative challenges to federal land ownership ranged from Wyoming's enactment that claimed as state property all federal lands within Wyoming administered by the United States Forest Service and the Bureau of Land Management (BLM), to Oregon's defeat of a bill that would have established a commission to consider transfer of some federal lands to state ownership.

The Sagebrush Rebellion has been regarded as another skirmish in the recurring struggle between western states and the federal government over who should control the vast, resource–rich public lands. This conflict over control is as old as the American frontier (Gates 1969). The Rebellion can also be viewed, however, as a conflict over the values that are to govern public land management and use—such as the conflicts between livestock grazing and wildlife management, extractive resource development and landscape preservation, and hard and soft recreational uses.

Schattschneider (1960) and, more recently, Kochen and Deutsch (1980) have argued that a group's support for locating decisionmaking for a particular policy area at a specific level of government is importantly related to the group's perception of its political success at that level of government. Disputes over values, therefore, may be played out as disputes over the structure of governmental decisionmaking. This study examines whether support for the Sagebrush Rebellion may best be understood as a conflict in land–use values whose respective supporters perceive that they have unique influence at different levels of government.

The analysis focuses on the positions taken by members of western state legislatures about ownership of the public lands. It is argued that if patterns of support for the Rebellion are congruent with partisan and urban–rural divisions in the legislatures, the Rebellion may be interpreted as a conflict over land–use values. If, on the other hand, neither party nor district location, but the proportion of state land under federal ownership predicts legislative support for the Rebellion, then it is reasonable to interpret the Rebellion as a federalist conflict over landownership rather than a conflict in the values that are to govern the public lands.

The findings presented here support understanding the Rebellion as a conflict over the values that should govern land management. This conflict is expressed as a conflict between state and federal management responsibilities. If the prospects for a shift in federal land use policies made current by the Reagan Administration are sustained, however, the historical conservationist and environmentalist support for locating public land decisionmaking with the federal government may not persist.

I. The Rebellion

The Sagebrush Rebellion is the most recent incarnation of the western states' periodic campaigns to gain control of federally held land within their borders. It began in 1979, when Nevada passed AB415, claiming all land administered by the BLM as state property. Interest in the Nevada bill expanded rapidly in the 1979–1980 legislative year. Endorsements of state ownership of the public lands came from a number of western regional associations, including the Western Conference of the Council of State Governments, the Western Conference of Attorneys General, and the Association of Western State Land Commissioners. By the end of 1980, Nevada, Utah, Wyoming, New Mexico, and Arizona had enacted bills claiming federal public lands as state property. Washington had passed similar legislation contingent on ratification by referendum. Hawaii, Alaska, and Idaho had passed supportive resolutions.

As described in the introductory chapter to this volume, conflicts between western resource users, notably stockmen, and federal land managers have a long history. Western political leaders have regularly requested major new grants of federal land to their respective states. Nonetheless, the question remains of why the Rebellion emerged suddenly in the late 1970s, especially because Congress apparently had resolved the issue of public landownership in favor of federal retention in the 1976 Federal Land Policy and Management Act. Moreover, why did the Rebellion appear to gain as much political support and attention in the press as it did? Three factors have been mentioned frequently by both observers and participants in the Rebellion: the growing insecurity of traditional user groups, particularly the grazers; the growing influence of the environmentalists and the land managers; and western political perceptions of the Carter Administration.

First, livestock operators on the public lands have long feared that changes in federal managerial policies would affect adversely their economic viability. Such fears were exacerbated by the settlement of *Natural Resources Defense Council* v. *Morton* (1974), in which the Department of the Interior agreed to prepare 212 Environmental Impact Statements assessing the impact of grazing on the public lands. By the late 1970s, draft impact statements were being circulated. Recommendations for some areas involved reductions in livestock grazing that could last for several decades. These proposals produced consternation among stockmen and led to congressional hearings critical of the impact statements.

Second, the growing body of federal natural resource legislation has expanded the meaning of the multiple–use doctrine for the public lands. For example, the Federal Land Policy and Management Act (FLPMA) formally incorporated environmentalist and recreationalist values in guiding management policy. The implementation of such statutes, particularly in the later 1970s, increased the representation of noncommodity user groups in advisory roles to the Bureau of Land Management but reduced representation from traditional user groups such as ranchers and miners (Cawley 1981). Public land administrators used the growing influence of environmentalist organizations to gain independence from traditional resource user groups. Administrators consciously manipulated the environmental threat to obtain user compliance with the restrictions agencies themselves wished to impose on land use (Culhane 1980).

Third, the Carter Administration was perceived by many western politicians as politically inept or hostile to western state resource development, or both. Western politicians were shocked by the threatened withdrawal of federal support for many water projects in the West. They were concerned about the appointment of well known environmentalists to natural resource responsibilities in the administration. Their political doubts about the Carter Administration were heightened by the realization that the president had not carried a single western state and enjoyed little support among the increasingly Republican western congressional delegations. Communication between the Carter Administration and the West can be described as limited at best.

The Sagebrush Rebellion did not burst forth suddenly from the Nevada state legislature in the spring of 1979. For example, Nevada had long sought land transfers from the federal government. In the 1960s, Nevada had requested the federal Public Land Law Review Commission to consider federal land grants to the state. In the early 1970s, the state set up a legislative commission to consider means of deriving additional benefits for the state from the public lands. The bill claiming BLM lands as state property that was signed into law on June 2, 1979, was six years in the making, according to its author.

Of the eleven western states, Nevada is the premier illustration of sustained support for the Rebellion regardless of a legislator's party or district location. Between the spring of 1979 and the summer of 1981, the Nevada legislature passed seven pieces of Sagebrush legislation that ranged from the state's claim of ownership to restrictions imposed on potential state sales of the "newly acquired" public lands. Opposition to these bills never went beyond three state senators and one representative, even though only eight of the forty Nevada assembly districts lie outside of Reno and Las Vegas. Nevada's Sagebrush Rebellion has been built on a coalition of urban and rural legislators; no other state has exhibited such coherent, enduring support for the Rebellion.

The heart of the coalition was an alliance of northern Nevada rancher legislators and Las Vegas Democrats. For example, Democratic Assemblywoman Karen Hayes, the urban leader of the Rebellion in Las Vegas, saw federal ownership as the obstacle to growth in the Las Vegas Valley. Defenders of Sagebrush in Nevada argued that land transfers to the state would permit urban growth. In a 1979 report to the Nevada legislature, the Division of State Lands proposed that the state could meet existing levels of expenditure for public land maintenance by periodically selling public land in the Las Vegas Valley.

The urban–rural coalition in Nevada has been stimulated by dynamic changes in federal land use in the form of new military reservations and new parks. Proposed uses such as the siting of the MX missiles have not helped the cause of federal landownership in Nevada. Defenders of Sagebrush have argued that transfer of federal lands to the state would give the state greater autonomy over land use decisions. Finally, in addition to the vast size of the federal landholdings and the vigor of the state's interest in acquiring the lands, the ideological conservatism of urban Democrats in both Reno and Las Vegas was an important factor in state legislative support for the Rebellion.

The political irony of the Rebellion's success in Nevada is that the state is neither rich in fossil fuels nor a major livestock producer. The impetus for the Rebellion rests on the unexamined assumption that the state would not continue to grow unless it expanded the nonfederal sector of land in order to allow intensified land use under state direction.

Outside of Nevada, Sagebrush bills introduced during the 1979–80 legislative session were met with much more success than were bills introduced in the 1980–81 session, when opposition in a number of states increased. During the spring of 1979, when the Sagebrush bill was under consideration in Nevada, California Assemblyman Hayes, a Republican, introduced a bill in the California Assembly to require the State Land Commission to examine whether lands managed by the BLM should by constitutional right be vested in the state. The State Land Commission was regarded by many as sympathetic to increased natural resource development. The bill was unanimously passed by both Houses but vetoed by Governor Brown on the advice of his Secretary for Natural Resources, Huey Johnson, a leading critic of the Rebellion. A compromise bill, worked out in February of 1980, commissioned a study of the public lands with less specific objectives. To allow both proponents and critics to have a voice in the final report, the study was divided among three agencies, which reflected different orientations toward the public lands issue.

The Washington state legislature gave early consideration to two Sagebrush bills: a Nevada–style land tranfer bill and a proposed amendment to the state constitution repealing the state's disclaimer of the unappropriated public lands. Both bills were enacted virtually unanimously. As in Nevada, the Washington legislature sought only lands managed by the BLM, not lands administered by other federal agencies. Less than 1 percent of the state is managed by the BLM, however, and the bills would have left 28 percent of the state in federal hands.

During the winter and spring of 1980, resolutions in support of the Rebellion were passed in Alaska and Hawaii. Idaho and Colorado passed resolutions calling for studies of the feasibility and consequences of the public land transfer to the respective states. By May 1980, Arizona, New Mexico, Utah, and Wyoming had joined the Rebellion, each by enacting a Nevada–style bill. Wyoming's bill included Forest Service land in addition to BLM lands, on the ground that these lands are the source of the state's water supply. In the summer of 1980, the

Rebellion was at its peak with six states claiming the public domain as their respective state properties.

Thereafter, opposition to the Rebellion became increasingly more organized. Environmentalist groups, including the Sierra Club, the Wilderness Society, and the Isaac Walton League created an umbrella organization called Save Our Public Lands. This network of opponents was supported by the League of Women Voters in a number of western states. Moreover, American Indian leaders testified against the Rebellion before legislative committees in states where the issue was still under consideration. Public awareness increased although the issue continued to provoke a large "don't know" response in many newspaper opinion polls (*Deseret News*, May 3, 1980). The survey results were mixed, although some regional findings suggest greater opposition than support for the Rebellion (Behavior Research Center 1979).

In 1980, the Rebellion was on the ballot in Washington and in two Nevada counties, Pershing and Mineral. One Pershing County official predicted an eight to one majority in favor of the Rebellion (*Nevada State Journal*, Nov. 2, 1980); in fact, opposition was around a third of the voters in a county virtually owned by the federal government (*Nevada State Journal*, Nov. 5, 1980). In Washington, the repeal of the disclaimer clause was submitted to the voters because it was a constitutional amendment. It met with widespread electoral opposition; 59.9 percent of the voters overall, and a majority in every county, rejected repeal. Although the first statewide test of public support for the Rebellion proved a setback, proponents of the Rebellion explained that because most of the federal land in the state was administered by the Forest Service, which enjoyed a good reputation, the referendum was not really an appropriate test of sentiment on the public lands issue.

The Rebellion also became an important campaign issue in the 1980 contest between Church and Symms for the United States Senate seat in Idaho. In October, incumbent Senator Church was advised by his pollsters of growing concern among voters that land transfers to the states would result in loss of recreational access (Barton 1982). Church emphasized his opposition to the Rebellion and it appears that Symms was placed on the defensive (*Idaho Statesman*, Nov. 1, 1980). Church's position in the polls improved, although not sufficiently to secure his reelection. It is likely that the perception of public unease over the Rebellion contributed to the resistance of Idaho legislators to a Nevada–style Sagebrush bill.

During the 1980–81 legislative term, Sagebrush bills were considered in California, Oregon, Idaho, Montana, and Colorado. Only in Colorado was a Nevada–style bill passed by both houses, but the Assembly was unable to override the governor's veto. By the end of 1981, the Rebellion in its present form appeared to have run its course. The election of Ronald Reagan brought an end to the disliked Carter Administration and, for many Rebellion supporters, introduced a president seen as sympathetic to the traditional resource users. The movement lost impetus and began to fissure between advocates of privatization—i.e., sale or transfer of public land title to individuals or groups—and those who continued to press for state authority over at least some portion of the public lands.

During the course of the Rebellion, four principal arguments were offered for land transfers to the states. First, proponents urged that transfers of the federal lands would facilitate timely and efficient development of their natural resources and promote economic growth more generally. A second, related contention was that state ownership would allow the respective states to capture a greater portion of the benefits produced from the lands, and thereby increase state revenue. Third, the Rebellion's supporters contended that transfer of the public lands would strengthen state autonomy in the important areas of land use and environmental regulation. Finally, Sagebrush rebels argued that public land use decisions should be made by citizens living in proximity to the lands at stake.

Opponents of the Rebellion expressed concern that land transfers to the states would result in major losses of public recreational access, given the likelihood that states would strengthen the positions of commodity users on the lands. Opponents likewise contended that the western states' past performance in state land management indicated insufficient concern for environmental protection and the goal of multiple use. Opponents additionally argued that transfer of the lands would diminish federal funds for land management and federal payments in lieu of taxes to western counties, with correspondingly increased burdens on state treasuries. Finally, opponents of the Rebellion viewed the federal lands as owned by the American people rather than by those who happened to live nearby. They feared that land transfers to the states might ultimately devolve control of a national heritage to private hands.

In every state where the Rebellion became a legislative issue, major supporters included stockgrowers' associations, wood products organizations, mining groups, and (notably in California) off–road vehicle associations. Representatives of these groups testified extensively before state legislatures. Their testimony stressed the burdens imposed on their respective user groups by restrictive federal regulations and the delays caused by federal administrative procedures.

In contrast, environmentalist organizations and soft recreationists such as hiking and backpacking organizations opposed land transfers by arguing that federal administration voiced their interests and concerns. Indeed, in every western state where environmentalists compiled annual ratings of state legislative voting records, votes on Sagebrush Rebellion bills were included as important indicators of environmentalist support.

It is reasonable to predict that environmentalists would oppose transfer of the public lands to the states. Historically, conservationists have supported federal responsibility for natural resource policymaking (McConnell 1954). Recent evidence indicates that environmentalists continue to support determination of public land and natural resource issues at the national level (Andrews 1980). Arguments for federal responsibility rest on efficiency and administrative ease, as well as resolution of the public goods problem in environmental regulation. Moreover, there is evidence that western state legislatures in the past were more responsive to livestock and agricultural interests than to urban recreational needs (Mann 1969). So today, there are indications that developmental interests have more legislative influence than environmentalists in the states of the four corners region (Ingram, Laney, and McCain 1980).

II. Hypotheses

These hypotheses are designed to elicit the value–conflict and states' rights–conflict dimensions of the issue of control of the public lands. They do so by examining patterns of support for the Rebellion among western legislators. If, on the one hand, patterns of support are explained by the extent of federal landholdings in an area, it is proposed, the Rebellion is best viewed as a states' rights conflict. If, on the other hand, patterns of support for the Rebellion reflect partisan or urban–rural divisions, the Rebellion would appear to manifest dispute over the values that are to govern management of the public lands.

Hypothesis One: The greater the land area within a state held by the federal government, the greater the state legislative support for state claims to the federal lands. As a corollary, the greater the land area within a state managed by the BLM, the greater the state legislative support for state claims to the federal lands.

Legislative victories for Sagebrush bills depend on coalitions of urban and rural legislators. Where federal landholdings are extensive, the likelihood of forging such coalitions should increase. In general, the more intrusive and inclusive the federal land presence, the more likely urban areas and rural communities would combine to gain control over land use decisionmaking or at least to realize additional benefits from use of the public lands. The BLM historically has been the most controversial federal land management agency, if for no other reason than that the stockmen, its principal surface client group, are at the focal point of public land politics. At least in comparison to the Bureau, the Forest Service has suffered less western criticism, although Wyoming included forest lands in its claim to the public lands. Distrust of federal land management has been the rallying point of those who regard Sagebrush as a conflict between West and East; if this hypothesis is not rejected it will provide support for interpreting the Rebellion as a states' rights movement.

Hypothesis Two: The more densely populated a legislator's district, the greater the likelihood of the legislator's opposition to state claims to the federal lands.
Hypothesis Three: If a legislator is a Republican, the greater the likelihood that the legislator will support state claims to the federal lands.

On the basis of the preceding discussion, it is likely that legislative opposition to the Rebellion would be strongest in districts populated by voters sympathetic to environmentalist positions on the public land issue, or at least disposed to think of the public lands primarily in terms of low–cost recreational use. Fry (1979) argues that urban dwellers are more interested in high environmental quality and recreational access than rural residents. There is also some evidence that in postindustrial areas,—i.e., areas characterized by a diminishing industrial work force and an increasing educated, professionally employed class,—will stress values of personal freedom and quality of life concerns such as aesthetic surroundings and recreational access (Inglehart 1977; Barnes, Kaase, et al. 1979; Watts and Wandsforde–Smith 1981).

Furthermore, some limited evidence suggests a relationship between education, occupation, and environmentalism (Van Liere and Dunlap 1980). Far stronger relationships emerge between political and social liberalism and environmentalist concerns (Buttel and Flinn 1978). Among officeholders, Democrats are more likely to support environmental issue positions than Republicans. At least on public lands issues, the differences between the parties appear to be long-standing; from the nineteenth through the mid–twentieth century, with the exceptions of the Roosevelt and Taft Administrations, Republican Party platforms reflect greater support for disposition than their Democratic counterparts (Englebert 1961). The adoption of environmental legislation in the western states in the 1970s tended to correspond to the strength of moderate to liberal Democratic representation in the legislature.

Partisan division on environmental issues is reinforced when ideological differences coincide with party membership (Dunlap and Gale 1974; Lester 1980). There is, for example, evidence that Democrats are restrained in their support for environmental policies because such issues arouse suspicions among working class supporters (Buttel and Flinn 1978).

III. Findings

The data for this study are drawn from the legislative voting records of the eleven contiguous western states. The hypotheses were tested against compilations drawn from state legislative votes on Sagebrush Rebellion bills. A summary of votes on selected Sagebrush legislation is presented in Table 1.1. The bills selected are the strongest Sagebrush bills in each state subjected to a vote in either house of the state legislature.

The Sagebrush Rebellion illustrates symbolic politics in action. No state was prepared to seize the public lands even after it had enacted legislation claiming them as state property. Thus, it was certainly plausible for legislators to support the Rebellion, without believing that the lands would be transferred forthwith and without viewing votes for the Rebellion as very meaningful. Yet it became apparent, as the public lands question was debated over a two–year period, that votes on Sagebrush were carefully watched by groups deeply interested in the public lands. The symbolic dimension of the Sagebrush issue, therefore, highlights rather than undercuts the importance of votes on Sagebrush bills as a test of legislators' orientations to the public lands question.

An examination of actions on Sagebrush bills taken by state legislatures reveals that the proportion of a state's land in federal ownership is not a consistent indicator of legislative support for the Rebellion. (See Table 1.2 for a summary of federal landholdings in the western states.) Three of the four states in which state and private land holdings occupy only about a third or less of the land area—Nevada, Utah, and Arizona—did enact bills claiming the public lands as state property. In the fourth, Idaho, the Sagebrush bill was blocked successfully in the upper house. In the other western states, however, no clear relationship emerges between size of the federal domain and Sagebrush support. Of the four states with between 42.4 and 53.7 percent of their land area in federal control or in federal trust, Wyoming and New Mexico exhibited substantial support for the Rebellion, but California and Oregon did not. Finally, the three states in which the federal domain occupies between one–quarter and one–third of the land likewise diverged in patterns of support for Sagebrush bills. Washington's bill laid claim to only .7 percent of its land area. Montana, with the least land under federal control, experienced the greatest legislative opposition to the Rebellion in the Mountain States. Colorado, in contrast, had substantial legislative support for claiming the 12 percent of the state's land area under BLM control as state property, although the Colorado House was unable to override the governor's veto.

Table 1.1 Sagebrush Legislation in the Eleven Western States, 1979-1981

	Bill number	Lower House	Upper House	Outcome
Nevada[a]	AB415 (1979)	38-1	17-3	Enacted
Washington[a]	AB3593 (1980)	90-7	42-1	Void by referendum
Wyoming[b]	HB6 (1980)	48-13	21-8	Enacted
Utah[a]	SB5 (1980)	57-11	20-7	Enacted
New Mexico[a]	HB79 (1980)	48-18	25-13	Enacted
Arizona[a]	SB1012 (1980)	47-11	20-10	Governor's veto overridden
Idaho[a]	HB425 (1981)	44-23	Blocked in committee	Failed
Colorado[a]	SB170 (1981)	39 17	25-19	Vetoed
Montana[a]	SB123 (1981)	Blocked in committee	28-22	Failed
California[c]	AB694 (1981)	43-34	22-8	Vetoed
Oregon[d]	HB2987 (1981)	27-31	Not considered	Failed

[a] Bill based on Nevada AB415 claiming public lands as state property.
[b] Wyoming bill similar to Nevada bill but including U.S. Forest Service Lands.
[c] Bill requiring state to continue payments in lieu of taxes to local governments if public lands transferred to State of California.
[d] Bill establishing commission to consider reduction of federal landholdings within Oregon.

More to the point than size of the federal domain in understanding the relationship between federal landholdings and the success of the Rebellion is the distribution of federal lands among federal agencies. The Rebellion included Forest Service lands in only one state, Wyoming, and even there the inclusion apparently was grounded on the perceived need for the state to secure its water supply rather than on managerial factors per se. Where the Forest Service was the major federal land manager, or even where the BLM acted as forest manager, as in Oregon, the Rebellion ultimately collapsed. Many state and local governments have enjoyed the generous financial terms under which the

Table 1.2   Federal and Indian Lands in the Eleven Western States   (by percent of state)

	BLM	Forest Service	Total federal land	Indian trust lands	Federal and Indian lands
Arizona	17.3	15.5	44.0	27.4	71.4
California	15.3	19.9	46.6	0.5	47.1
Colorado	12.5	21.5	35.5	1.2	36.7
Idaho	22.4	38.1	63.8	1.5	65.3
Montana	8.6	17.8	29.7	3.4	33.1
Nevada	68.4	7.2	86.1	0.1	86.2
New Mexico	16.4	11.6	33.3	9.1	42.4
Oregon	25.4	25.1	52.4	1.2	53.6
Utah	51.7	14.8	63.6	4.2	67.8
Washington	0.7	20.8	29.2	5.7	34.9
Wyoming	27.8	14.8	48.6	3.0	51.6

Source: Department of Interior statistics, 1980.

Service administers its responsibilities and appear reluctant to undertake forest management. The voters in Washington and the state legislatures in Oregon and Idaho—all states with major Forest Service holdings—exhibited extensive opposition to Sagebrush. Likewise, California and Montana are both states where the Forest Service is a major presence and where Sagebrush was defeated. It must be noted, however, that in Colorado the Forest Service has twice the holdings of the BLM, but there was substantial legislative support for the Rebellion.

The obvious focal point of the Rebellion was the public lands managed by the BLM. In those states in which the BLM was the major federal land agency and its holdings were used for grazing or mining, legislative support for the Rebellion was especially high. The states that meet these conditions are Arizona, Nevada, New Mexico, Utah, and Wyoming. The long tradition of rancher political activity in conjunction with the increased value of the lands' natural resources, worked to the advantage of the pro–Sagebrush forces.

An examination of opposition to the Rebellion within states yields evidence to support the hypotheses that Democrats and legislators from the more densely populated areas would be more likely to oppose the Rebellion. Republicans were consistently more supportive of the Rebellion in every state save Nevada. Table 1.3 shows the vote by party in the lower house in each state on the selected Sagebrush legislation. (In Montana, where the lower house did not consider a Sagebrush bill, data from the upper house are used.) In no state did Democratic opposition fall below one–quarter of the Democrats voting on Sagebrush bills; in six states, Rebellion legislation was opposed by a majority of Democrats. In contrast, Republican opposition exceeded 20 percent in only two states: Idaho (22.6 percent) and Oregon (48 percent). This does not mean that consistent majorities of Democratic state legislators opposed the Rebellion, but that where opposition was found, it occurred more often among Democrats. For example, although a majority of Democrats in New Mexico supported the Sagebrush bill, a markedly greater level of support was exhibited by New Mexico Republicans.

Table 1.3 Percentage of Legislators by Party in Opposition to Selected Sagebrush
Rebellion Bills

	Democrats	Republicans	Index of partisanship[a]	Outcome
Montana (Senate)	81.8	14.3	67.5	Blocked in House
Idaho (House)	78.6	22.6	56.0	Blocked in Senate
Colorado (House)	70.0	8.0	62.0	Vetoed by governor
California (House)	69.6	6.0	63.6	Vetoed by governor (very weak bill)
Oregon (House)	54.3	48.0	6.5	Defeated
Wyoming (House)	52.6	7.0	45.6	Enacted
Arizona (House)	36.8	10.3	26.5	Enacted
New Mexico (House)	35.8	12.0	23.8	Enacted
Utah (House)	26.1	11.1	15.0	Enacted
Nevada (House)	4.0	0	5.0	Enacted

[a] Index is created by subtracting percentage of Republicans in opposition to the Sagebrush bill from the percentage of Democrats in opposition to the bill.

Furthermore, the greater the unity among Democrats opposing a state's Sagebrush bill, the less likely it was to pass. In Montana, Idaho, Colorado, and California, the Rebellion was a clear partisan issue. The index of partisanship reveals that at least 50 percentage points separated the two parties (see Table 1.3). Among states defeating the bill, the issue was partisan in all save Oregon, where the 45 percent of Republican legislators approached the 54.3 percent of Democratic legislators who voted against the bill. In the five states in which Sagebrush bills were enacted, only in Wyoming did just over half of the Democrats in the lower house oppose the measure; Democrats, however, held only about one–third of the seats and consequently were in a weak party position. It is, therefore, fair to conclude that in all western states save Oregon, support for the Rebellion was stronger among Republican than Democratic legislators.

A cartographic depiction of support for the Sagebrush Rebellion in the eleven Western states would reveal vast geographic commitment to claiming the public lands as state property. Table 1.4 shows the pattern of legislative opposition to Sagebrush bills within states by party and by district location. California is omitted from the urban—rural analysis because of its size and the difficulty in many cases of drawing urban–rural delineations. Nowhere is the geographic division over Sagebrush more apparent than in Montana, where opposition was principally confined to the cities while the countryside overwhelmingly favored the Rebellion.

One major finding in this study is the highly partisan nature of the legislative response to the Sagebrush Rebellion. Eight states were selected on the criteria of comparable population size and available data. Table 1.5 calculates an average percentage in these states of support for Rebellion legislation. The eight–state average reveals overwhelming support among Republican legislators for land

Table 1.4   Legislative Opposition by Party and District Location to Selected Sagebrush
Rebellion Bills (by percent)

		Democrats		Republicans	
Arizona	Urban	50	(6)	13.5	(3)
	Maricopa County	20	(1)	0	(0)
	Pima County	71.5	(2)	40	(2)
	Rural	14.3	(1)	14.3	(1)
Utah	Urban	31.25	(5)	13.5	(5)
	Salt Lake City	71.42	(5)	0.9	(2)
	Rural	16.66	(1)	0	(0)
Wyoming	Urban	83.3	(10)	0.7	(10)
	Rural	5.9	(1)	0	(0)
New Mexico	Urban	47.6	(10)	5	(1)
	Bernalillo County	57.1	(8)	12.5	(1)
	Rural	23.8	(5)	12.5	(2)
Colorado	Urban	81.25	(13)	10.7	(3)
	Denver/Boulder	92.86	(13)	5.9	(1)
	Rural	25.0	(1)	0	(0)
Idaho	Urban	100	(5)	45	(9)
	Rural	33	(6)	9	(3)
Montana	Urban	100	(5)	42.8	(3)
	Rural	33	(2)	4.8	(1)
Oregon	Urban	69	(16)	70	(7)
	Portland	65	(13)	100	(13)
	Rural	30	(3)	33.3	(5)

transfers to the states. In contrast, the multistate average for Democratic legislators indicates deep divisions within the party; nearly half opposed the land transfer, while half were in support of the Rebellion. To an important extent the party differences at the state level mirror historical party differences at the national level. Republicans have been much more likely to support land transfers to the states or to private individuals and groups. Democrats at the national level have been much less inclined to do so.

Patterns of support for the Rebellion were clearly affected not only by the legislator's party but by district location as well. Indeed, legislative district location appears to have intensified as well as constrained inter–party differences on the land transfer issue. In Table 1.6, legislators are divided by whether their districts are located in areas of high population density (cities, suburbs, or large towns) or in low population areas (rural areas). The contrast in levels of opposition to the Rebellion between Democrats from high–density districts and low–density districts is dramatic. On average over three–quarters of Democrats from high–density districts opposed the Rebellion, yet 70 percent of Democrats

Table 1.5   Average Percentage of Legislators by Party in Opposition and in Support of Sagebrush Legislation in Eight Western States

	Democratic legislators	Republican legislators
Support Sagebrush bill	49	86
Oppose Sagebrush bill	51	14
Total	100	100

from low density districts supported the transfer of the public lands to the states. The pattern of Republican support for the Rebellion is also sensitive to district location, although less dramatically so. In low population density districts, on average 10 percent of the Republican legislators opposed the Rebellion, while in high density population areas Republican opposition averaged 25 percent. Thus the major inter–party conflict over the Rebellion was to be found in the more urban areas. In the rural districts of these eight states both Republican and Democratic legislators were more inclined to see the merits of the Rebellion. As rural legislators they may have indeed been more sensitive to the concerns of traditional resource users such as the grazers.

Table 1.6   Average Percentage of Legislators by Party and District Location in Opposition to Sagebrush Legislation in Eight Western States

	Low density population legislative districts		
	Democrats	Republicans	% Diff.
Supported Sagebrush bill	77.3	91.8	14.5
Opposed Sagebrush bill	22.7	9.2	11.5
Total	100	100	
	High density population legislative districts		
	Democrats	Republicans	% Diff.
Supported Sagebrush bill	29.7	74.9	47.2
Opposed Sagebrush bill	70.3	25.1	45.2
Total	100	100	

Further analysis of the respective positions of legislators from inner city and agricultural areas reveals the value conflicts underlying their different positions on the public lands issue. Inner city Democrats, with the exceptions of those from Reno, Las Vegas, and Phoenix, were consistently the most vigorous in their opposition to the Rebellion (see Table 1.4). Cities in which strong opposition to the Rebellion was found included Denver/Boulder, Albuquerque, Tucson, and Salt Lake City. Moving out of the inner city to the suburbs, and to the centers of recent, rapid population growth, underlines the importance of partisan composition to Sagebrush support. Growing metropolitan counties such as Dona Ana in southern New Mexico, El Paso (Colorado Springs) in Colorado, or Maricopa (Phoenix) in Arizona, solidly favored the Rebellion; these counties are heavily Republican. The major exceptions to the pattern of legislative support for the Rebellion in Republican areas of population growth were areas that are the most postindustrial: university cities. In cities where universities are located, such as Fort Collins, Boulder, Provo, and Laramie, opposition to the Rebellion among Republican legislators was greater than in other areas.

Opposition to the land transfers proposed by Sagebrush was greatest among legislators from central city districts—districts characterized by greater concentrations of minorities, lower income levels, and lower educational levels than the surrounding suburbs. Such districts do not conform to the profile of the postindustrial middle class citizen leaving behind material concerns in support of quality of life issues. Indeed, this level of inner city opposition to state control of the public lands suggests rich research possibilities. There is a long tradition among the inhabitants of the western states to use the public lands for hunting, fishing, and low–cost recreation. It may be that the less well–off see existing federal landownership policies as the foundation for an important component of the western way of life, with its emphasis on outdoor recreation at relatively low cost. It may be conjectured that less well– off urban residents opposed the Rebellion from the fear that changes in public landownership would restrict access. More affluent suburban residents may have either discounted the costs of restrictions following land transfers to state ownership, or perceived reasonable recreational alternatives.

Legislators from agricultural backgrounds were the reverse image of inner city legislators in their opposition to Sagebrush. Agricultural legislators gave the Rebellion nearly universal support. An impressive number of western legislators list ranching or farming as their occupations. In the three states that lack major metropolitan centers—Idaho, Montana, and Wyoming—the percentages of legislators in the lower house who list agricultural occupations are 38, 33, and 33 respectively. Such large percentages far exceed the actual population strength of farmers or ranchers. Nonetheless the number of legislators in the western states who are ranchers suggests the importance of the rancher at least as a symbol in western politics. The rancher's way of life symbolizes traditional frontier values of independence from society, enjoyment of the out–of– doors, living off the land, and taming nature by hard work. Ranchers support the Rebellion overwhelmingly and epitomize the values espoused by proponents of the Rebellion for the public domain.

Despite the overwhelming rancher–legislator support for the Rebellion, one source of rural opposition must not be forgotten. The Indians were an important exception to the pattern of rural support for the Rebellion. In both northwestern New Mexico and northeastern Arizona, Republican and Democratic legislators alike opposed transfer of the public lands to state control. This area contains the largest concentration of American Indians in the nation. Indian population growth and increased political activity, particularly on the part of the Navajos, has produced an electoral force of some magnitude in the legislative districts of the southern half of the Four Corners region. It is not surprising that legislative opposition to the Rebellion was apparent in predominantly Indian constituencies, in the face of the historic conflicts between many reservations and the western states over taxation, law enforcement, and water rights.

In summary, there is a limited support for the first hypothesis that when a vast preponderance of a state is under the management of the Bureau of Land Management, the state legislature endorsed the Rebellion. Far greater support exists for interpreting the Sagebrush Rebellion as a conflict in public land use values. Party differences in support for the Rebellion were readily evident in all but one of the western states. Democrats were far less supportive than Republicans of proposed land transfers. Such party differences were intensified when conjoined with rural–urban divisions. The more urban a legislator's district, the more likely he or she opposed the Rebellion.

The Sagebrush Rebellion is best understood as a conflict in land use values rather than as a conflict over states' rights. This is not to suggest that the size of federal landownership is unimportant as a factor in western politics in general or in the resurgence of interest in states' rights. The magnitude of federal landownership in Nevada has produced a recurring debate over the meaning of statehood. In other western states, proponents of the Rebellion enjoyed success in coupling the Rebellion to the issue of self–rule. But in the majority of Western states, the issue is not federal landownership per se, but how and for what purposes the lands are managed.

In spite of the colonialist rhetoric of alien policies imposed on the western states from the banks of the Potomac, it is much more likely that the best interpretation of this symbolic rebellion is the strength of pro–development sympathies in a number of the Western states. Kochen and Deutsch remind us that a group's support for centralization or decentralization of political decision-making may be a consequence of the group's influence at a particular level of government:

> If a disfavored group or class is weak on the average across the aggregate domain of a service system but strong in some smaller region, it should gain from decentralization; and a favored group, predominant on average but weak in some localities, should gain from policies of centralization and amalgamation. As a group over time shifts one of these conditions to the other, its attitudes towards decentralization are likely to reverse.

If the environmentalists have been influential on national policy but less influential on policy in some western states, then it is likely that traditional resource users will seek to shift power over land policy to the states. Similarly, American Indians oppose the Rebellion in part because they have less influence locally and greater influence nationally. If the present policies of the Reagan Administration continue for a long period, then a reasonable prediction is that some western states will seek a greater say in land policy while other states would become less interested in the land control question.

IV. Aftermath of the Rebellion

The Rebellion in the state legislatures had largely come to an end by the summer of 1981. The Rebellion enjoyed less success in Congress and in the federal courts. But it should not be concluded that the efforts to augment state holdings at the expense of federal landownership was without consequence or that the goal has disappeared from political debate within the western states. Although the grand scale of the Rebellion is no longer evident, efforts to secure land have taken new and sometimes quite creative forms. What follows is a brief survey of the Reagan Administration's efforts to transfer selected lands to the states and of the ongoing politics of the land transfer issue in three states: Arizona, Nevada, and Utah.

The Reagan Administration came to power committed to defusing the Rebellion through better land management and closer cooperation with the western states. In February 1981, Secretary of the Interior James Watt adopted a Good Neighbor policy that offered western states and local governments the opportunity to buy small parcels of land for public purposes such as schools and parks. As provided by statute, the parcels were offered at prices substantially below market rates. Interior received requests to purchase in excess of 900,000 acres. The program was blocked by the privatization debate that took place in the administration in the winter and spring of 1982. Interior did accelerate granting states their indemnity selections, that is, lands still owed them since admission to the union but for various reasons never actually transferred.

Three western states have had ongoing legislative or electoral efforts for or against acquisition of the public lands since the decline of the Sagebrush Rebellion in the fall of 1981. In Arizona, Governor Babbitt had vetoed the Rebellion bill. The legislature overrode his veto. Opponents of the Rebellion took heart and organized a petition drive to place an initiative on the November 1982 ballot to repeal the Sagebrush statute. The drive appears to have been unevenly organized: it required a court order to place the initiative, proposition 203, on the ballot. By the fall of 1982, Sagebrush was no longer much of an issue in the state. The repeal proposition lost 57 to 42 percent. In the same election Governor Babbitt won re–election handily by a two–to–one margin. The Arizona result did demonstrate, however, that the Rebellion could withstand the test of popular opinion in at least one state.

In its 1983 session, Nevada's state legislature continued to demonstrate the ingenuity exhibited by past Nevada legislatures in devising fresh arguments to secure additional federal land for the state. State legislative leaders argued that the aftermath of the Rebellion, in conjunction with the Reagan Administration, had produced a climate in Washington favorable to requests for the transfer of some federal lands to the state. In three bills and two resolutions, the state legislature memorialized Congress to grant the state additional lands. The state legislative actions reconfirm Nevada's sustained commitment to securing fresh land transfers.

Utah's efforts to secure federal land are quite different from those of other states. If realized, they would provide a powerful precedent for intergovernmental relations in public lands politics. Governor Matheson has labeled his proposal "Project Bold." The proposal is to consolidate state landholdings by a comprehensive, congressionally endorsed, land exchange between the United

States and Utah. The state's proposals, endorsed by the state legislature in 1983, would enable the state to select lands owed Utah since admission to the union and to exchange state inholdings in federal parks, military land, Indian reservations, and other federal holdings for federal parcels in proximity to existing state land. The objective of "Project Bold" is to allow the state to realize greater rents from resources on state lands. The major problem of state land consolidation is that the lands exchanged are required by statute to be of equal value. Equalization is a difficult proposition given surface acreage considerations and subsurface mineral wealth. The exchange problem is compounded by the concerns of federal land lessees that existing land use policies will not be changed in a way that adversely affects their interests. No prognostication exists for the chances of federal approval. Implementation of "Project Bold," however, would radically alter the distinctive checkerboard pattern of landownership in the West.

References

Andrews, Richard. 1980. Class politics or democratic reform: Environmentalism and American political institutions. *Natural Resources Journal* 20: 221–42.

Babbitt, Bruce. 1980. An alternative to the Sagebrush Rebellion. Dillon, Montana: Address to Man and His Resources Symposium.

Barnes, Samuel, and Kaase, Max. 1979. *Political Action: Mass Participation in Five Western Democracies.* Beverly Hills: Sage Publications.

Barton, Andrew J. 1982. The 1980 Election in Idaho: A Case Study in New Right Intervention. Unpublished undergraduate thesis, Oxford University.

Behavior Research Center. 1979. Poll on the Rebellion in the eight states found opposition to the Rebellion. Phoenix: Behavior Research Center.

Benedict, Robert C., and Francis, John G. 1981. Political activists in Utah: A study of the delegates to the 1980 Democratic and Republican state conventions.

Buttel, H. Frederick, and Flinn, William L. 1978. The politics of environmental concern: The impacts of party identification and political ideology on environmental attitudes. *Environment and Behavior* 10: 17–36.

Cawley, R. McGreggor. 1981. The Sagebrush Rebellion. Unpublished dissertation, Colorado State University.

Clark, Cal, and Walters, B. Oliver. 1981. Rising Republicanism in the West: A regional tide or harmonic state waves? *The Social Science Journal* 18: 1–6.

Culhane, Paul J. 1981. *Public Land Politics: Interest Group Influence on the Forest Service and the Bureau of Land Management.* Baltimore: Johns Hopkins University Press.

Dana, Samuel Trask, and Fairfax, Sally K. 1980. *Forest and Range Policy: Its Development in the United States.* New York: McGraw Hill.

Dunlap, Riley E., and Gale, Richard P. 1974. Party membership and environmental politics: A legislative roll call analysis. *Social Science Quarterly* 55: 670–90.

Echohawk, Larry. 1981. The Sagebrush Rebellion—a matter of concern for Indian tribes. Spokane: NCAI Midyear Meeting.

Englebert, Ernest. 1971. Political parties and natural resources policies: An historical evaluation, 1890–1950. *Natural Resources Journal* 1: 224–56.

Everett and Associates. 1980. Survey document: Withdrawal of public lands from access to minerals and fuels. Washington, D.C.: prepared for the Public Lands Study Group.

Fry, Rene. 1979. Interregional welfare comparisons and environmental policy. In H. Siebent, I. Walter and K. Zimmerman, eds., *Regional Environmental Policy: The Economic Issues.* New York: NYU Press.

Gates, Paul W. 1979. *History of Public Land Law Development.* New York: Arno Press.

Hays, Samuel P. 1959. *Conservation and the Gospel of Efficiency: The Progressive Conservation Movement, 1980-1920.* Cambridge: Harvard University Press.

Herger, Wally. 1981. News from the Office of Assemblyman Wally Herger, March 9, 1981. Sacramento, Calif.: State Capitol.

Hibbard, Benjamin Horace. 1924. *A History of the Public Land Policies*. Madison: The University of Wisconsin Press.

Inglehart, Ronald. 1977. *The Silent Revolution: Changing Values and Political Style Among Western Publics*. Princeton: Princeton University Press.

Ingram, Helen M.; Laney, Nancy K.; and McCain, John R. 1980. *A Policy Approach to Political Representation: Lessons from the Four Corners States*. Baltimore: Johns Hopkins University Press.

Kirschten, Dick. 1979. The federal landlord in the middle. *National Journal* 10: 1928–31.

Kneese, Allen U., and Brown, F. Lee. 1981. *The Southwest Under Stress: National Resource Development Issues in a Regional Setting*. Baltimore: Johns Hopkins University Press.

Kochen, Manfred, and Deutsch, Karl W. 1980. *Decentralization: Sketches Toward a Rational Theory*. Cambridge, Mass.: Oelggschlager, Gunn & Hain.

League for the Advancement of States Rights (LASER). 1980. Agenda for the eighties: A new federal land policy. Salt Lake City, Utah: Proceedings of the National Conference on States Rights, the Sagebrush Rebellion, and Federal Land Policy.

Lester, James P. 1980. Partisanship and environmental policy: the mediating influence of state organizational structures. *Environment and Behavior* 12: 101–31.

Mann, Dean. 1969. The political implications of the migration to the arid land states of the U.S.A. *Natural Resources Journal*: 212–27.

Nevada Department of Conservation of Natural Resources, Division of State Lands. 1979. Preliminary estimated state costs for administration of BLM lands (SB240, AB413). Carson City: Nevada State Government.

Peffer, Louise E. 1951. *The Closing of the Public Domain: Disposal and Reservation Policies, 1900–50*. Stanford: Stanford University Press.

Schattschneider E.E. 1960. *The Semisovereign People*. New York: Holt, Rinehart and Winston.

Trueblood, Ted. 1980. They're fixing to steal your land. *Field and Stream* 44: 166–67.

Van Liere, Kent D., and Dunlap, Riley E. 1980. The social bases of environmental concern: A review of hypotheses, explanations and empirical evidence. *Public Opinion Quarterly* 44: 181–97.

Voight, William. 1967. *Public Grazing Lands: Use and Misuse by Industry and Government*. New Brunswick, N.J.: Rutgers University Press.

Wattenberg, Martin P., and Miller, Arthur H. 1981. Decay in regional party coalitions: 1952–1980. In *Party Coalitions in the 1980's*, Seymour Martin Lipset, ed. San Francisco: Institute for Contemporary Studies, pp. 341–67.

Watts, Nicolas, and Wandesforde–Smith, Geoffrey. 1981. Postmaterial values and environmental policy change. In *Environmental Policy Formation*, Dean E. Mann, ed. Lexington, Mass.: D.C. Heath & Co., pp. 29–42.

Western Governors' Policy Office (WESTPO). 1980. Issues and considerations in the management of the public lands. Park City, Utah: background paper prepared for the third annual meeting of WESTPO.

2

Economic Analysis in Public Rangeland Management

ROBERT H. NELSON

There is wide criticism that the public lands are managed inefficiently.[1] Too much is invested in some places and too little in others; decisions to allocate land among contending uses do not accurately reflect the relative values of the uses; rates of federal resource production in some cases are too rapid and in other cases too slow. One proposed remedy would require that public land agencies consistently base their decisions on economic analysis. Agency economists would estimate the social benefits and social costs of alternative actions; the action finally selected would necessarily be the one calculated to maximize net social benefits minus costs.[2]

This approach would be a return to the original Progressive era concepts that motivated and justified retention of public lands in public ownership. The conservationism of Gifford Pinchot, the founder of the Forest Service, emphasized scientific management to achieve efficient use of natural resources. Indeed, Samuel Hays characterized conservationism as the "gospel of efficiency."[3] Conservationists, however, never developed an operational definition of efficiency; as a result, the commitment to efficiency was more rhetorical than real. But now some economists and other critics of public land management are saying that the original conservationist ideal of efficient management should be implemented; they suggest that the way to do this is to employ the concepts and tools of modern economics.

Skeptics, however, doubt that the public lands can ever be managed efficiently,

This paper, originally entitled "Basis Issues of Public Range Management in the 1980s," is adapted from sections of Robert H. Nelson, *The New Range Wars: Environmentalists Versus Cattlemen for Public Rangelands* (Draft), Office of Policy Analysis, U.S. Department of the Interior, 1980. The Interior Department does not necessarily agree with the analysis or conclusions of this paper.

as long as they remain in public ownership. The original conservationist concept presumed a separation of politics and management under which management would be left to the professional experts—professional economists under the recent proposal. In practice, however, professional management has regularly been undermined by politics to the detriment of efficiency. Instead of efficiency, political decisionmaking puts a much higher emphasis on distribution of benefits to various constituencies that use the public lands. If economic efficiency is really the aim, these observers conclude, sale of public land into private ownership and management is a necessary step.[4] Nonmarket objectives such as environmental protection could be achieved by traditional regulatory mechanisms, or by easements included in the terms of public land transfers to the private sector.

Still other public land students question the paramount importance of efficient management.[5] Other social goals—equity for one—may reasonably require the sacrifice of efficiency. Moreover, incrementalists point out that radical reforms seldom achieve their stated intent—instead, they often end up serving purposes never conceived by the original architects. The incrementalist strategy emphasizes caution in making major changes, and concentration instead on improving the existing system.[6] Rather than a wholesale turn to economic analysis in public land management, it might be better to adopt economic methods case–by– case.

This paper seeks to shed light on these issues by examining a small episode in the history of public land management—some agency efforts during the 1970s to make greater use of economic analysis in public rangeland management. Most of the public rangelands are administered by the Bureau of Land Management (BLM) in the Interior Department. The most important legislative guidance for their management is provided by the Taylor Grazing Act of 1934 and the Federal Land Policy and Management Act of 1976. The BLM rangelands include 170 million acres—about 9 percent of the area of the lower 48 states— that are grazed each year by around 2 million cattle and another 2 million sheep and goats. Together, these livestock consume around 10.5 million "animal unit months" (AUMs) of forage.

Livestock grazing is conducted on the public rangelands under a permit system; private ranchers have grazing privileges—effectively for an indefinite period—to graze certain areas of public rangeland at certain times.[7] In recent years the lands have also become much more valuable for recreational use. For example, around 23 million acres are currently being examined for wilderness designation. Indeed, it was partly the difficulty of resolving a growing conflict in the 1970s between traditional grazing uses and new recreational uses that generated a demand for greater economic input into rangeland decisions.[8]

The prospects for making greater future use of economic analysis in public land management depend significantly on political, institutional, and even broad social and cultural factors. Economics is one perspective which must compete with others to influence agency thinking. The role of economics must be justified by a broader ideology or theory of government. For example, in a world of interest group politics, what is the political constituency to ensure that an identifiably efficient result is in fact followed? Will the estimation of monetary values be socially acceptable even for nonmarket uses such as wilderness that have heavy symbolic or even religious overtones? If economic efficiency is not the sole basis for agency actions, how can efficiency be compromised with other considerations? Such questions confront the economic analyst of public range-land issues.[9]

Scientific Management:
The Inadequacy of a Biological Objective

As noted above, Gifford Pinchot and other architects of the public lands system envisioned that political influence would be excluded from most management decisions.[10] Politics was to set only the broad policy direction; implementation would then fall to the professional experts. Following this concept, a profession of range science was created to advance and maintain rangeland expertise, eventually resulting in the founding of the Society for Range Management in 1948. On the public rangelands, the progressive prescription for scientific management has consisted of the application of the principles of range science, as taught by the profession of range management.

The bulk of the subject matter of range science consists of biological and other studies. What are the various classifications of rangeland plants? How do plants respond to different climates, soil types and other varying site characteristics? What is the effect of livestock grazing on the plant composition of the range, etc.? All of this information is useful in range management, although some more so than the rest. However, it does not answer the question of what the basic management objective on the public (or private) rangelands should be. This question actually is passed over quickly in the field of range management. It has not been the subject of great debate or analysis among range professionals.

There is, however, an answer provided. The objective of range management is to maintain a high and permanently sustainable level of productive capacity of the rangelands. Before the recent emphasis on a variety of range uses, this was considered to consist of maintaining a high sustained yield of forage available for domestic livestock grazing. Thus, a long time leading text stated that range management was: "The science and art of planning and directing range use so as to obtain the maximum livestock production consistent with conservation of the range resources."[11] More recently, the need to allocate forage for use by wildlife and for watershed purposes has been more widely recognized. A more contemporary version states that: "Prior to the 1960's range research was designed primarily to maximize forage production for domestic livestock. Current trends in range research are geared to optimize the functioning of the entire range ecosystem."[12]

The basic philosophy of a BLMer is that "as resource managers, our first consideration is to preserve and improve the production potential of the land while using its resources."[13] The major BLM statements presenting its range management program in one way or another include achievement of a high sustained yield of forage production as the fundamental objective of BLM management. The programmatic grazing EIS released in 1974 stated that "the specific objectives of the livestock grazing management program are to maintain and improve vegetative resources through management actions and supportive measures, to aid biological processes which will result in improved vegetative conditions and greater stability of soil, a sustained yield of livestock forage, more productive wildlife habitat and enhancement of aesthetics."[14] The goal of a high sustained yield of the resource is not just held by BLM, but by Congress and most other leaders in public land management. The Federal Land Policy and Management Act of 1976 states that "management be on the basis of multiple use and sustained yield."[15] The Public Rangelands Improvement Act of 1978 states a policy to "manage, maintain and improve the condition of the public rangelands so that they become as productive as feasible for all rangeland values."[16]

Yet, despite their frequency and wide acceptance, statements of a goal of high sustained yield cannot be accepted at face value. By pouring greater resources and investments into the range, we could increase its productive capacity enormously. For example, much of the range could produce at far higher levels if only water could be supplied. Reclamation efforts in fact have just this objective, with spectacular results in some areas such as the Imperial Valley in California, a former desert. Less extremely, the range could be greatly improved by physically removing existing low productivity shrubs and plants that have resulted from past overgrazing and replacing these plants with more desirable species. There are a whole variety of such "vegetative treatments" available. Even without treatments of this kind, extensive fencing and water facilities could be installed to give very exact management control over livestock movements in small areas.

However, in practice, few if any range managers propose doing everything feasible to raise the productivity of the rangelands. They recognize that there must be a limit to social expenditures on rangelands. On the other hand, they seldom discuss in any explicit way how this limit is to be determined—or even acknowledge formally its existence. The fact that range professionals continue to state an objective of maximum range productivity suggests instead a desire to avoid questions of real objectives. Range managers may want to preserve a maximum discretion for professional management. Alternatively, they must consider that questions of ultimate objectives for rangelands are beyond the scope of their professional expertise and must be decided in the political arena.

However, political decisionmakers have done no better than range professionals in formulating operational objectives for rangeland decisions. As noted above, guiding legislation tends simply to repeat the objectives stated by the range professionals themselves, to maximize sustained yields of forage and other rangeland outputs. In practice, lacking any clear standards, public rangeland management has often been determined in response to various special interest pressures. Such pressures are exerted by stockmen, hunters, general recreationists, wilderness proponents, off–road vehicle riders, and other interests. In short, even if seldom acknowledged explicitly, range management in practice generally turns out to be management by interest group competition and accommodation.[17]

One might ask why range scientists do not acknowledge this result more directly, instead of disguising the interest group role in the vague generality of achieving a high sustained level of forage production. Perhaps the answer is that range scientists feel the need for higher aspirations for their profession. Maintaining professional pride, establishing a professional identity, in general holding together the profession, all may depend on a higher purpose than advancement of one or another special interest. It would also appear that range scientists prefer to stay out of the contentious disputes over the distribution of forage and other benefits of the public rangelands. Directly entering these disputes might require taking one side against another, no doubt creating enemies on the losing side. In effect, range scientists have maximized the long run sustained yield of their own professional advisory services by avoiding political questions of the distribution of rangeland benefits.

To be sure, interest group politics is hardly an outcome unique to management of public rangelands. Indeed, competition among interest groups appears to be an accurate description of the general workings of the American political system, at least during ordinary times. Moreover, some social theorists have gone beyond the question of factual accuracy to suggest further that interest group competition generally leads to a desirable social result. In the 1950s John Kenneth Galbraith wrote approvingly of a system of countervailing powers that he saw emerging.[18] Political scientist Theodore Lowi argues that the dominant intellectual creed of the post–World War II period has seen competition among contending interest groups—in his terms "interest group liberalism"—elevated to the basic principle for the intellectual and governing elites for organizing and justifying the political system:

> It may be called liberalism because it expects to use government in a positive and expansive role, it is motivated by the highest sentiments, and it possesses strong faith that what is good for government is good for the society. It is "interest group liberalism" because it sees as both necessary and good that the policy agenda and the public interest be defined in terms of the organized interests in society. In brief sketch, the working model of the interest group liberal is a vulgarized version of the pluralist model of modern political science. It assumes: (1) Organized interests are homogeneous and easy to define, sometimes monolithic. Any " 'duly elected' " spokesman for any interest is taken as speaking in close approximation for each and every member. (2) Organized interests pretty much fill up and adequately represent most of the sectors of our lives, so that one organized group can be found effectively answering and checking some other organized group as it seeks to prosecute its claims against society. And (3) the role of government is one of ensuring access particularly to the most effectively organized, and of ratifying the agreements and adjustments worked out among the competing leaders and their claims. This last assumption is supposed to be a statement of how our democracy works and how it ought to work. Taken together, these assumptions constitute the Adam Smith "hidden hand" model applied to groups.[19]

As a social organizing principle, interest group competition suffers from the defect that it rewards pure self–seeking—hardly the most attractive human quality—most highly. In the case of direct economic competition in the marketplace, there is at least a credible theory of how such competition may lead to a socially desirable result. However, in the case of interest group competition in the political arena, there is no such theory. The expectation that the result will be socially preferable seems as much based on wishful thinking as any carefully reasoned analysis. Lowi is especially critical: "Interest group liberal ideology merely reflects the realities of power and rationalizes them into public policies. Criticism is irrelevant."[20]

As Lowi suggests, there is a great need for some objective standard for government decisions—some way actually to say what constitutes the public interest. While range science professionals have not been able to provide such a standard, professional economists have more recently suggested that economics can meet this need.[21]

Scientific Management: The Economic Concept

In a very general way, it is not difficult to discern what the actual objective of range management should be—maximization of the net value to society obtained from the range. Once the concept of net social value is introduced, consideration of costs is required. The fundamental deficiency in the concept of achieving a high physical sustained yield is that it fails to consider costs.

Net social value will increase as long as further investments or other expenditures on the rangelands yield greater social benefits than the costs incurred. Hence, maximization of net social value leads to the rule that one continues to spend more on the rangelands until the social costs begin to exceed the social benefits. In this rule benefits and costs include not only marketable inputs and outputs but also all nonmarket benefits and costs. The latter include environmental impacts, aesthetic values, existence values, option values and a number of other benefits never encountered in direct market transactions. Put in this general way, maximization of net social value is no more operationally useful than a goal of high sustainable forage production. However, in concept at least, it is possible to value the social benefits and the social costs of range improvements. Although the profession of range scientists has shown little interest in making such an analysis, economists and other social scientists more familiar and comfortable with evaluation problems have taken the lead in some attempts at such analyses.

While members of BLM generally see themselves as professional range managers, the economics profession is well represented in other parts of government decisionmaking, most importantly by the Office of Management and Budget (OMB). Other budget and analysis offices within the Interior Department, and in recent years within the BLM itself, also take an economic rather than a biologic view of range management. Since each profession professes to have the objective truth, the disputes between economists and range or forestry scientists sometimes look a little like a religious war. There has been a long tradition of basic disagreement between OMB and the public land agencies.

OMB recognizes that achieving a high sustained level of rangeland production is not a satisfactory goal by itself. It cannot provide an objective, nonpolitical standard for public rangeland management. OMB sees the attachment of the public land agencies to such biologic goals as an attempt either to stake unlimited claims on public resources or, more likely, to preserve maximum administrative discretion and leave the field open for political maneuvering. OMB thus sees itself as the genuine defender of scientific professionalism; the agency's biological concept of scientific management is in OMB's eyes an imposter within which masquerades the real opponent, the surrender of the public lands to domination by various special interest groups. In the case of BLM the early history of domination by ranchers after the Taylor Grazing Act of 1934 helped to confirm OMB in this perception.

OMB thus has put strong pressures on the public land agencies to perform more economic analyses. It has sought to resist interest group politics in favor of a greater role for professional expertise—by its preference, economic expertise. Economic analyses should explain the allocations of land among uses, show whether the benefits of investments exceed the costs, and give the basis for other important agency decisions. In the face of limited success, OMB has often responded by taking an attitude that it will try to restrict agency budgets to the most promising items and, as one OMB official put it, "cut our losses on the rest."[22] However, OMB pressures on occasion have helped to push the public land agencies into efforts at economic analysis.

Early Uses of Economics in Public Land Management

The first major use of economics in public rangeland management was for the purpose of determining a proper grazing fee. Largely due to OMB insistence, a major study of western livestock grazing was conducted in the mid–1960s.[23] This study examined fees on privately leased grazing lands and the adjustments in the private fee necessary to derive a comparable public land grazing fee. The fee system developed using this economic approach was adopted in 1969 over rancher opposition.[24]

The public land agencies did not oppose this application of economics because it did not infringe on any of their key decisionmaking responsibilities. Range science itself did not offer any theory of the proper grazing fee. The agencies correctly saw higher fees as a step towards assertion of greater agency managerial control. A higher fee also would make OMB happy, might relieve some of the contentiousness between OMB and the agencies, and thus might lead to higher budgets. If economists could be helpful in raising the fee, as proved to be the case, their contribution would be warmly received.

The next big application of economics to range problems was a major study led by the Forest Service and involving seven other agencies in the Departments of Agriculture and Interior.[25] The study examined the different sources of range forage in the United States, public and private, and alternative means of achieving various levels of total forage production. It made extensive use of systems analysis and linear programming methods to handle the large masses of data and complex calculations required. The study closely adhered to traditional prescriptions of scientific management. It was comprehensive, looking at the efficient allocation of forage production from the perspective of the nation as a whole. It set a clearly defined objective, a given level of total forage production, and then examined various alternatives for achieving this objective. Finally, it had a clearly defined basis for selection of the preferred alternative, minimization of the total cost to the nation of achieving the desired level of total forage production.

However, when published in 1972, the results were disconcerting. It turned out, according to the calculations, that livestock grazing in the western United States was economically unjustified in many areas. Other regions had higher potential for investment in forage production at lower costs. Within the West, grazing should be concentrated in certain areas with high investment potential and should be discontinued in many other areas.

For example, the study calculated the lowest cost, most efficient way to meet the projected forage demand of 320 million AUMs in the year 2000. Because range investments and intensive management would be concentrated on the most productive grazing lands, the total area grazed in the United States would fall by 49 percent, from 835 million acres in 1970 to 429 million acres in 2000. Although acreage grazed would decline in every region, it would fall by the greatest amount in the West, from 86 percent of western rangelands grazed in 1970 to only 23 percent in 2000. Associated with these shifts would be changes in the composition of forage output as described in the study:

> The production of animal unit months on the Western Range ecogroup would remain essentially unchanged at 56 million. . . . Production on the Western Forest ecogroup would decrease from 11 to 8 million animal unit months while on the Great Plains it would drop from 93 million to 80 million. These changes would take place in the face of an overall increase in animal unit months of 50 percent. The remaining ecogroup, Eastern Forest would produce the increase. Production in this ecogroup would increase from 53 million to 175 million animal unit months, a gain of 226 percent. This again represents a major shift in the location of grazing.[26]

These were radical conclusions apparently based on a scientific approach— minimize costs to achieve a given objective. However, they were also clearly unacceptable politically and institutionally. Hence, the Forest Service set about constraining the extent of changes that could occur in seeking to minimize overall forage production costs. One main alternative included limits on the intensity of investments and on the acreage that could be newly converted to forage production. The intent was, as the study acknowledged, to pick constraints that "were deliberately restrictive and to a large extent prescribed the distribution of grazing," essentially preventing any radical departures from the status quo.[27] With such constraints added, instead of a 49 percent decline in acreage grazed, the decline was only 5 percent. The savings in forage production costs were also much less. The unconstrained solution produced a 59 percent reduction in average forage costs, from $4.03 per AUM to $1.66 per AUM. However, when constraints were imposed to maintain existing patterns of forage production, there was only a 19 percent decline in costs.

It might be thought that the cost–constraints would have greater adverse environmental effects. However, because grazing was eliminated from so much land under this solution, it also had fewer adverse environmental impacts.

The unconstrained cost–minimizing solution thus was preferable in terms both of costs and environmental impacts; under the assumptions of the study, it represented the scientifically rational distribution of grazing. However, a proposal to eliminate grazing on more than half the rangelands in the West would be totally unacceptable to western ranchers. In deciding to be more practical and to constrain the least cost calculation to stay within reasonable bounds, Forest Service economists made a decision that departed from the scientific realm and became essentially political. The bounds they specified showed their estimate of the degree of change that might be politically tolerable over the next decade or two. Forest Service economists lacked any scientific or other solid basis for this judgment; in fact, there probably was no such basis. In effect, what had been a scientific management question had been transformed into a political one.

Although the constrained solution largely gave back the constraints imposed, it was still useful from the Forest Service viewpoint to have such a solution formally calculated. The existing grazing pattern could be described as the solution to a large and complex cost minimizing calculation. In effect the approximate grazing status quo could be legitimized in a scientific dress.

In 1974 an interagency work group in the Department of Agriculture published a widely distributed study, "Opportunities to Increase Red Meat Production from Ranges of the USA."[28] Much of the technical data and other backup for this publication was drawn from the 1972 Forest Service study just described. The surprising result of the earlier study had been the maldistribution of grazing from an efficiency standpoint, in particular excessive grazing of western rangelands, mostly federally owned. But by the 1974 study the discussion had been turned around to the point that it now justified grazing on federal lands:

> . . . elimination of grazing from Federal range would require a shift of a considerable proportion of total range livestock numbers and their production to other lands and, in essence, would remove a sizeable proportion of this resource from the Nation's productive resource base. More animals on non–Federal lands would require more intensive use of private range, increases in pastures, harvested forages and feed grains, more acres in cultivation, and greater dependence on feedlot feeding for meat production. Although proportionately small in relation to non–Federal range livestock production, loss of Federal grazing would upset the supply–demand situation for beef cattle and materially affect the sheep industry.[29]

While the Forest Service at first watered down and then ultimately ignored the results of its own large–scale model for national forage production, this behavior was not necessarily inappropriate. The usefulness and reliability of large computer models of economic systems is often greatly exaggerated by the public. The accuracy of such models is particularly suspect when used to draw conclusions about small parts of a large system. Thus, a computer model result showing that grazing is uneconomic in a particular area should certainly not be taken as decisive in itself. In short, the Forest Service can be criticized for using large models to impress a gullible public, but the agency nevertheless was probably wise not actually to base its own decisions on these models.

The alternative is to examine the economics of livestock grazing in a more site–specific way, perhaps for each individual ranch. Is a public subsidy necessary to sustain grazing in an area? If not, and if private ranchers bear the costs and want to continue livestock grazing, it would make little sense to tell ranchers that their decision is economically unjustifiable. While the Forest Service made its first 1970s foray into rangeland economics on a national scale, the major BLM economics initiative adopted a narrower focus. The BLM in the mid–1970s developed procedures for economic analysis of rangeland investments.[30]

Benefit-Cost Analysis of Range Investments

Public land managers see economists as clearly a different breed. In opening a conference of BLM economists, the Director of the BLM Range Division remarked that "I'm a little uncomfortable when I'm around economists because I don't really understand their language."[31] In his heart, the true BLMer feels that "it is morally, ethically, and professionally right to institute management practices that stop erosion, grow better forage and vegetation, and improve rangeland condition and trend. We should not have to economically justify these management practices."[32]

Economists thus tend to be placed in a slot separate from the rest of the agency. They are called on when the wider society demands an economic justification for an agency program, but otherwise they are not expected to play much of a role in formulating agency policies. If they expect to fit in, they thus come under strong pressures to tailor their economic analysis to provide the support desired by superiors.

Responding to growing public concerns, the BLM in January 1975 submitted a report to the U.S. Senate concerning range conditions on BLM lands.[33] In terms of use of economics, the report was significant in that it contained one of the first BLM efforts to perform a benefit–cost analysis for its range program. The BLM estimated long–term total benefits of about $400 million and costs of about $225 million for a full scale program of range rehabilitation. Not surprisingly, BLM strongly advocated the congressional authorization and funding of such a program.

The economic analysis, however, had been done after the fact and had basic deficiencies. For instance, the analysis included increased retention of soil and soil stability as one major benefit and increased future productive capacity of the land as measured in future forage production as a second major benefit. Since the foremost purpose of increasing soil stability is to increase future productive capacity of the land, the same benefit had in effect been measured and counted twice, simply in different forms. In this case as well, economics was used to provide a justification for the outside world.

By 1975 the Office of Management and Budget could see growing pressures to make more investments on the public rangelands. It suspected, however, that many if not most of these investments would not be economically justifiable. At a minimum, it preferred that money spent should be allocated first to the investments with the highest payoffs; even if many projects were not justified, the least bad should be undertaken first. Hence, OMB requested that BLM analyze the benefits and costs of range investments. Policy and budget offices at the Secretarial level in the Interior Department had concerns similar to those of OMB and agreed that BLM should undertake some economic analysis of its range investments. As a result, an interagency work group was formed to examine the question of the extent to which the benefits of range investments were greater than their costs. The specific plans for range investments were generally provided in plans for individual allotments grazed by particular ranchers. These "micro" plans are called "allotment management plans" or "AMPS."

From BLM's point of view, the primary purpose in making benefit-cost estimates was to show OMB and other concerned parties that investments on the public rangelands could be paying propositions. In this objective the undertaking was successful, somewhat to the surprise of several participants. Most of the investment plans examined in fact had benefits which were estimated to exceed their costs. A later BLM canvas of the benefit–cost studies done up to 1979 indicated that 80 percent of AMPs showed benefits greater than costs.

In a number of cases the initial benefit–cost results had been less favorable but the AMP had subsequently been modified to improve its benefit–cost ratio. Although this might appear to be an abuse of the process, it actually is a desirable procedure. It showed that BLM was searching for the best investments to include in AMPs. Ideally, many investments would be examined until the one with the highest payoff was finally selected.

Because BLM had previously discouraged rancher investments, but had only limited funds available itself, there probably was a considerable backlog of high return investments waiting to be undertaken. Many areas of public rangeland had been seriously overgrazed at one point, and had the potential for significant forage increases if properly managed. For ranchers such forage increases translated into income boosting growth in their herd size and resulting beef and lamb production. In some cases, range investments improved forage quality as well as quantity, causing increased calving rates and more rapid calf and cow weight gains on the range, both of which would be valuable to ranchers.

The estimates of benefits and costs were hardly precise, however. Estimates of benefits depended on rather uncertain estimates of the actual rangeland forage and livestock response to new grazing systems. A further major uncertainty was the proper value to use in evaluating the increased forage, hunter days, and general recreation increases resulting from the investments.

There may have been a tendency in some cases to be too optimistic. This would not be surprising; even with the best of intentions, agencies often see their proposals through rose–colored glasses. The system in effect expects that agencies will be advocates and put proposals in their best light. For example, in early uses of the benefit-cost analysis, improvements in calf weights and calf/cow percentages were one of the most significant gains projected from the range investments. Yet, within two years there was sufficient doubt about the actual existence of such gains that BLM economists recommended that, as a general rule, they should not be included as benefits.[34]

The Implementation of Investment Analyses

Although BLM had initially been skeptical about the usefulness of benefit–cost analyses, other than for justification purposes, it gradually decided that in fact economics might actually be useful in setting its investment priorities. As long as there were not enough funds to make all investments, it made sense to allocate the available funds into the investments with the highest returns. Moreover, the exercise of requiring field personnel to specify economic impacts of investments on ranchers and to estimate the values of nonmarket outputs produced a healthy critical attitude towards many investment plans. It led to better invest-ments, from a range science as well as an economic efficiency standpoint. In 1976 BLM decided to require benefit–cost analyses of all range investments. In transmitting the benefit–cost procedures for field use, BLM Associate Director George Turcott explained the purposes of the newly required analysis:

> Economic analysis at this stage will provide considerable assistance in: 1) determining whether proposed AMP's are economically feasible; 2) providing a basis for consideration of other alternative levels of management and investment; and 3) providing one additional method of comparing AMP's. . . .
> Following the ES process, the results of the economic analysis should be reviewed to determine whether, for environmental reasons, the proposed action has been modified or changed to the extent that such changes will affect

the AMP economic analysis. After review, the results of the economic analysis should be arrayed with environmental and other information to assist the decision maker in comparing and analyzing the proposed action and alternatives and selecting a final decision, determining priorities, scheduling implementation, and programming/budgeting.

It is imperative that AMP's for the grazing ES program be of the highest quality and represent the Bureau's best effort in placing before the public our proposals for the range program. To meet the test for quality, AMP's must be not only technically and environmentally acceptable, but also realistic, feasible, and economically justifiable.[35]

Although many BLM range managers still regarded economists with suspicion, the benefit–cost analyses were on occasion used by managers to help them in forming their decisions, a rare role for economists in the history of public land agencies. For instance, in the Rio Puerco environmental impact statement (EIS) for grazing, the proposed action included consolidation of certain allotments in order to create larger pastures and thereby to hold down on costs for fencing and water facilities. Ranchers much preferred individual allotments and pressured BLM to abandon the consolidation plan. In explaining why consolidation was necessary BLM cited the economic benefits: "There are several justifiable reasons for combining allotments," including "economics—each AMP had to arrive at a benefit-cost ratio approaching one to one."[36]

By 1979 a number of the first generation of grazing EISs had been released. With a large number of proposed investments entering the pipeline OMB once again expressed its doubts about the economic merits of investments being proposed for funding:

One policy position is that the highest priority for use of range investment funds should be to offset direct adverse local economic impacts, such as changes to incomes and employment, that result from livestock grazing reductions in terms of actual current use. Our argument is that basically range investment funds are a type of income redistribution, and that since most ranchers are economically wealthy (in terms of ranch value even if not in terms of annual income) our priority should be one of offsetting adverse economic impacts rather than subsidizing and increasing rancher wealth through investments in the range to increase grazing.

A broader policy issue is whether range investments should be made at the expense of the ranchers. The benefits of such investments are increased income and increased asset value which accrue to the ranchers. Although the Federal Government receives grazing fees from ranchers (at less than fair market value rates), 50% of those receipts go to fund range improvement investments, and most of the rest goes to the State for benefit of the counties where the lands are located. (Some counties spend the funds on range improvements). Since ranch operators are for the most part substantially more wealthy than the average Federal taxpayer, publicly funding range investments transfers income and wealth from those with less to those who already have more. Range investments also benefit wildlife and the argument can be made that the national public ought to pay for such benefits. A contrary argument is that there would be no need to make such investments to benefit wildlife, if the lands were not grazed by ranchers' cattle.[37]

BLM responded to OMB's questioning by reemphasizing the role of benefit–cost studies. A set of criteria for making range investments was developed in which the benefit–cost ratio played a prominent role.

As the investment example shows, the acceptance of economics increased in the public land agencies during the 1970s. In 1979 the Secretary of Agriculture issued a statement on range policy in which he stated the policy to "administer the range resources of the National Forest System for the benefit of the American public through costeffective management and development of the range."[38] A 1979 rangelands symposium gave a prominent place to economics in the summary report. It stated the need to "apply economic analyses so that the better, more cost–effective conservation and other practices can be applied. Widespread application of practices on a cost–effective basis could lead to more responsive and favorable public consideration of budget requests for range and associated conservation programs."[39]

Nevertheless, economic concepts still occupy a distinctly secondary place to the biologic concept. In 1979 the BLM Range Division published a major policy paper assessing the basic future of rangeland policies.[40] The paper scarcely mentioned the need for benefit–cost or other economic analyses. Besides the factors already mentioned, the resistance to economics has some other bases. To the extent that the results of economic analysis are expected to be the actual determinant of decisions, the economist tends to displace the professional land manager as the effective final decisionmaker. Land managers who suspect economists of seeking this role naturally are not very supportive of economic studies. Public land agencies also tend to see their role to advocate as much as they can get and OMB's role to resist them. The biologic approach to range management is much more congenial to this view, since the goal sought is maximum productivity of the range. Economic analysis in effect asks the agency to curb its appetite voluntarily.

Although not examined so far, another major consideration is simply the difficulties of economic analysis, especially in circumstances where there are many nonmarket benefits and costs.

Difficulties in Evaluating Range Investments

In considering an investment, a business would estimate the increase in its net revenues resulting from the investment and then compare these with costs incurred. Differences in the time stream of revenues and costs are taken care of by discounting future revenues and costs to a lump sum "present value." A benefit–cost analysis by a public agency follows the same procedure, but "social benefits" substitute for private business returns and "social costs" for private business costs.

In a typical business investment analysis, revenue projections would be based on an estimate of demand and future sales. Costs also can often be estimated straightforwardly. But in the public sector social benefits and social costs frequently are much harder to estimate. For range investments, costs usually do not pose a great problem, but benefits are often difficult to establish.

Much of the benefit of a range investment consists of increased forage provided to livestock. One way to evaluate this forage would be to use the public grazing fee, which by law is based on an estimate of fair market value. But most ranching students consider that the grazing fee is well below the true value of public rangeland forage, which rules out this option.

The availability of BLM forage determines the feasible total size of the operation for many ranchers. Hence, it might be possible to value the increased availability of public forage by valuing the increase in size of the ranch made feasible by this greater forage. The common practice in the West is to appraise ranches by the number of livestock the ranch will support; a ranch might typically be worth around $1,000 per "animal unit" in the herd. Hence, if an increase in BLM forage due to a public investment would make possible a 20–animal unit increase in ranch herd size, the added forage could be valued at $20,000.

But this apparently reasonable approach is confounded by the peculiar behavior of ranch values. Almost all economic studies of the returns and costs to ranching have concluded that ranch values far exceed any reasonable estimate of the capitalized value of the net income that can be earned from ranching. In fact, one researcher concluded that the economic returns to ranching were so low that one must view "cattle ranching as a consumer item comprised of many components, including the utility obtained from consumption of such intangibles as 'love of land' and 'love of rural values.' "[41] Hence, another candidate for evaluating forage increases must be ruled out.

The simplest method would be to evaluate new forage at the going lease rate for private forage in the area, seeking to find private forage similar in quality and seasonal availability to public range forage. In recent years, such private forage typically has been leased for between $5 and $10 an AUM.

Annual data on private lease rates, however, often are for forage that does not correspond with public range forage. Moreover, the role of public forage varies greatly according to individual ranch circumstances. In the Southwest, where the public range may be the sole source of forage and livestock grazing occurs all year long, there will be a close correlation between public forage availability and livestock numbers and rancher returns. In other parts of the West, however, public land forage typically provides only a portion of the total forage needs; hay and private lands supply the rest. In these circumstances, the forage on the public lands can vary from critically needed to incidental to the ranch operation. Forage supplied at particular seasons may be especially valuable; public land grazing in the early spring is often much sought by ranchers because it relieves them of the necessity to buy expensive hay.[42]

These considerations suggest that the only accurate way to evaluate forage increases is to examine the economics of each individual ranch. Revenues and costs would be projected, with and without the availability of greater public land forage. But the public land agencies lack information on such details of individual ranch budgets and operations. Ranchers might be unwilling to provide it and resent any efforts to acquire it as an intrusion on their privacy. Moreover, it would be costly to survey and analyze individually each of the thousands of ranches using public rangelands.

One might then classify ranches by various ranch types. For each basic type, a model of the ranch operation, including revenues and costs, could be constructed. By adding extra forage at a certain time of year, the effects on ranch operation and ranch income can be calculated by means of the model. This approach is particularly suited to use of linear–programming techniques.

However, it still suffers from the defect that individual ranch differences are not fully accounted for; certain ranches may not fit any of the generalized ranch types very well.

A number of these approaches have desirable features. But different approaches will produce different answers, no one of which appears definitive. The BLM state economist for Wyoming reported in one instance that "the BLM estimate of the return per AUM was $2.85. The University of Wyoming on the other hand, estimated the returns to be $10.67 per AUM for cattle and $14.91 per AUM for sheep. . . . Without a clear and concise explanation of the divergence between the two values, . . . economic impacts derived from BLM values are unacceptable to the academic community."[43] In practice, the prices of comparable privately leased forage (the "commercial value") have most often been used as the estimate of the value of public land forage. Linear–programming models of representative ranches have also been employed where available.

Evaluation of recreation benefits on public lands is still more difficult. There are two main techniques.[44] First, a survey can be made of recreational users to ask how much they would have been willing to pay and still participate in the recreation activity. The major drawback of this approach is that responding to a survey and actually paying are much different things; what someone says he is willing to pay may be a lot different from what he would actually pay. The wording or phrasing of the question may itself produce large differences in response. If the respondent fully understands the purpose of the question, he may choose to answer strategically; hunters may overstate the value of hunting hoping to justify greater wildlife expenditures. Other hunters are offended by the question and refuse to give any answer.

Another recreation method relies on an actual demonstrated willingness, the expenditure of time and money by recreationists who have traveled to a particular site. Based on examination of trips to different sites, an estimate can be formed of the general willingness to pay for recreational experiences. The greatest drawbacks of this approach are the uncertain value to put on time spent traveling—for some it may be all pleasure—and the fact that the same trip often combines a number of destinations, making it difficult to say what the proper share for any given trip segment should be.

A third less common method of evaluating recreation use of public lands would be to find comparable private recreation experiences which are marketed. There are many places where private landowners are able to control access to hunting or fishing and charge for such access. In some cases landowners control key entry points into areas of public lands and impose substantial charges to pass their gateway point. Where such prices can be found, they provide a good estimate of the true willingness to pay for the product. However, not many such studies of private recreation markets have been done.

There has been a great deal of work by economists to apply the survey and the travel cost methods of estimating recreational use values. However, the quality of data available combined with the element of judgment required to employ these techniques have caused the actual estimates derived to vary widely. Some economists are still using improper methods—including counting of hunter expenditures in the local economy —in estimating hunter day values. For a deer hunter day, it is possible to find estimates of value varying from $10 to $100 per hunter day. Large variations could also be found for other types of hunting and fishing. Although some of these variations are valid reflections of local variations in hunter day values, a greater part is no doubt due to differences in data or techniques of the researcher.

Compared with hunting and fishing, there is much less information about values for activities such as rock hounding, bird watching, or ordinary hiking. There is almost a total absence of any accepted methods to estimate a value where this value is placed on the existence of the species. Yet, for the nation as a whole the total value of antelope may be as much in the knowledge of people throughout the country that antelope are commonly found in the West as in the hunting of antelope by individual hunters. For some species, such as wild horses and burros, the "existence value" of their presence on the public range is the only value—at least under current law that prohibits commercial use. The values of a wilderness area, archeological site, historic site, and a number of other public rangeland outputs have the same feature that a significant portion of this value is due purely to existence: the value is derived by people who may never visit the public lands.

Similar issues are encountered in the third main area of benefit of range investments—watershed impacts. Some watershed impacts can be evaluated fairly easily: reduced flooding may allow predictable reductions in expenditures required to repair flood damage to a road or other structure located downstream. Reduced sedimentation may lengthen the useful life of downstream reservoirs. Changes in water yields may be evaluated at existing prices where water rights have been bought and sold in the area. But where the impact is a contribution to river salinity that is only a small part of a much larger waterway system, it would be difficult to estimate the value of this salinity reduction. The same is true of sediment disbursed into a wide system.

While estimates of hunter and often recreational day values do not inspire much confidence, they provide positively hard data in comparison with existence values, options values, and some of the other less direct nonmarket values of the public rangelands. For such nonmarket outputs, economic evaluation is still largely unexplored terrain.

Taking all these analytical difficulties together, it must be recognized that estimates of the benefits of range investments will be a rough approximation. A further major uncertain element affecting the results of the economic analysis is the choice of the proper discount rate. In recent years government agencies have used discount rates varying from 4 to 10 percent (in real terms). My own rough guess, having prepared some benefit–cost estimates, would be that, most likely, the true benefits typically lie somewhere between one–half and three times the estimated benefits. However, I would hardly be shocked if some future study were to reveal that this range is too narrow.

The importance of being able to limit the actual range of benefits even to a rough extent should not be underestimated. Unless there has been some attention already paid to economic considerations, it is not uncommon that a project will turn out with benefits less than half the costs. In an early trial run of the benefit–cost methods developed for range investments, fully one–third of the initially proposed investments had benefits less than half their costs. Many projects also will have benefits more than three times the costs, making it quite likely that these are actually paying propositions.

Even if estimates are very rough, the higher the estimate, the higher is likely to be the true ratio of benefits to costs. For the purpose solely of establishing priorities among investments, it is not especially important that the benefit–cost ratio be accurate in an absolute sense; it is much more important that relative magnitudes accurately represent the true relative investment priorities.

Use of market prices and estimation of "willingness to pay" in benefit–cost analysis also raises some basic philosophical questions not mentioned thus far. Most important, it acts to perpetuate the status quo of the income distribution. The same output has a higher value if it is consumed by a person with a higher income, because he would be willing to pay more for it. Hence, other things equal, if public benefits are estimated by private willingness to pay, public outputs would be oriented to higher income groups. This is particularly unsatisfactory when there is no charge imposed for these outputs—a common circumstance for recreation and other public land outputs.

The accuracy of benefit–cost estimates depends, of course, on the quality of the data and the caliber of the analyst using it, as well as the quality of the methods of analysis available. One might ask, why not simply improve the accuracy of benefit–cost estimates by putting much greater resources into them. But the exercise of conducting benefit–cost studies can itself be subject to economic analysis; in many cases it may simply not be economically justifiable to improve the accuracy of benefitcost studies.

The Economics of Planning

The subject of the economic benefits of more accurate and sophisticated information gathering and analysis has been much neglected. Consider a case where a decision must be made and the worst possible result would be a loss of $10,000, while the best possible result a gain of $20,000. In this circumstance it can never be economically rational to spend more than $30,000 in information–gathering and analysis to try to determine how to achieve the best possible result. In fact, the expenditure that should rationally be made on information gathering and analysis normally would be far less than $30,000, which would consume all the gains from the action itself, leaving society no better off than if it had taken no action.

Ideally, one should keep spending more on information and analysis until the resulting expected improvement in the decision is worth less than the cost of the additional information and analysis. A little information and analysis often goes a long way; the rate of return on further information and analysis may fall rapidly. If society's stake in the decision is not large, in many cases it will be economically rational to get by on some "quick and dirty" studies.

The scientific management approach encounters a basic logical quandary in considering the question of the proper amount of information and analysis. In order to determine how much information and analysis is required, scientific management would say to gather it as long as the returns exceed the costs. But in order to determine this point, further information and analysis is required. The question then arises how far to go in this second type of information and analysis. One might once again say, continue until the returns exceed the cost. However, this again requires a third type of information and analysis. Since the pattern could be repeated indefinitely, at some point, it obviously becomes necessary to say "go no further." But how can it ever be known where this point lies without further information and analysis?

A theoretical possibility exists that some rational procedure could be devised to cut off information–gathering and analysis but does not itself require any further information or analysis. This is a largely unexplored subject at present; such a formula certainly is not available as a practical matter to the working analyst. Hence, the decision to cut off further information and analysis can only be made on an intuitive basis—it is essentially an exercise of professional judgment. Even at a basic conceptual level it may be theoretically impossible to achieve complete scientific rationality.

Until the 1970s there had not been a great deal of money spent on studies of the public rangelands of the United States, reflecting the considered low value of their outputs. With much greater public interest in the rangelands in recent years, it makes sense that there should be increased resources devoted to studying their problems. When the courts mandated a comprehensive set of environmental impact statements for the BLM rangeland program in 1975, there was in fact a sudden sharp increase in the resources available for range-land studies.

There clearly is a limit, however, to the amount that should be spent. There was little discussion at the time concerning the economic rationality of so much information–gathering and analysis of the public rangelands. Just as important, there also was little discussion about the alternative ways in which this money could be spent, assuming it was to be spent.

The 1975 court order in *NRDC* v. *Morton* determined the basic guidelines for the great bulk of the funds to be made available for future rangelands studies. First, the money would be spent in a site specific fashion for 212 (later reduced to 144) separate analyses of particular public range areas where livestock grazing was occurring. The courts specifically rejected a "programmatic" or national focus for future rangeland studies. Second, the focus of the studies would be on management decisions, how much livestock grazing, how much forage for other uses, and what investments to make. Federal Judge Thomas A. Flannery directed that the EISs should "discuss in detail the environmental effects of the proposed livestock grazing, and alternatives thereto, in specific areas of the public lands that are or will be licensed for such use."[45]

The orientation toward management decisions largely dictated that the graz-ing EISs would have to be land use plans. Initially, BLM prepared two separate documents, a formal land use plan and then a grazing EIS. The land use plan was intended to be completed before the grazing EIS was begun. However, as matters turned out, the preparation of the grazing EIS usually raised so many questions not covered in earlier plans that a whole new round of planning had to be undertaken. The grazing EISs thus became, de facto, the land use plans.[46] In 1979 BLM resolved this problem by combining the grazing EIS and the land use plan into one document.

The land use planning system developed in the BLM in the 1970s required large amounts of data and analysis. It reflected the widely held philosophy that planning must be comprehensive; it must analyze all the issues and the decisions made must resolve these issues. Although an unrealistic expectation, the basic idea was to provide a full blueprint for the future.[47] Such a system is expensive and the court order effectively decided that large amounts of money would have to be spent on land use planning.

Preparation of grazing EISs in fact turned out to be very expensive. A BLM study of nine of the first EISs completed found that $5.7 million had been spent in direct preparation costs—an average of $630,000 per EIS.[48] However, the direct preparation costs were only a fraction of the total EIS costs; for example, they did not include most of the inventories and land use planning required to lay the groundwork for writing the EIS. The same BLM study estimated that the direct preparation costs of the EIS were only about 10 percent of the total costs associated with completion of each grazing EIS. On this basis, the nine EISs could have cost as much as $50 million, or more than $5 million per statement. Seven million acres were covered by these grazing EISs, so the total cost per acre was perhaps as large as $7 or $8 per acre.

There exists a limited market in grazing rights to public land in which one rancher may sell his rights to another rancher. In this market the rights to public land grazing might typically sell for anywhere from $30 to as much as $100 per AUM. Since the average AUM on BLM land requires about 15 acres, even using the highest purchase price of $100 an AUM, permanent grazing rights probably would be worth no more than $7 per acre. Hence, for the first nine grazing EISs, it appears that, if costs are completely accounted for, the total costs involved in grazing EISs approached the total value of the forage to ranchers for livestock grazing. To put the matter another way, if the government had instead used the EIS money to buy out grazing rights to public land, it might well have been able to buy out a significant part of the grazing rights in the EIS areas for no more than the costs to prepare the EIS.

The costs of the first generation of EISs were higher than later ones and BLM's estimates of indirect or supporting costs for EISs may well have been too high. But there is a strong case that the early expenditures on grazing EISs far exceeded what could be justified on any economically rational basis. It seems likely that in total all the BLM grazing EISs will end up approaching $100 million in direct preparation costs and considerably more if all the inventory, planning, and other associated and indirect costs are included. The resulting sum is likely to be at least as great as the total expenditures for range improvements authorized in the Public Rangelands Improvement Act over the twenty years from 1980 to 2000.

As the levels of funding for preparation of early grazing EISs and associated activities rose, the money available for on–the–ground improvements fell. Partly, this was because many new improvements could not legally be undertaken under the court order until a grazing EIS had been completed in an area. In 1977 budget discussions in the Interior Department, the Assistant Secretary for Policy, Budget and Administration pointed out that from 1975 to 1977, funding for on–the–ground capital improvements had declined from $8.3 million to $5.8 million per year, while "paper work" expenditures for inventory, planning, and EIS writing had risen sharply from $3 million to $13 million. The question was raised in these budget discussions as to whether adequate rationales existed for the greatly increased expenditure for information gathering and analysis. A past tendency to undertake these activities for their own sake was noted. "In practice due to the difficulty in determining whether or how much decision making is improved and how to value such improvement, there often exists a tendency to make activities such as inventories, planning and environmental analysis ends in themselves." The theoretical necessity of considering benefits and costs of information and analysis was also raised: "Greater expenditures for information gathering can be justified the more important the decision is and the more is at risk in making it." Skepticism was shown in particular concerning the large amounts of expensive inventory data BLM was proposing to collect: "It is important to ask certain questions relating to whether such a level of inventory effort is needed."[49]

These doubts, however, carried little weight in the face of the prevailing enthusiasm in 1977 to do everything necessary to rehabilitate the public range. By 1980 BLM expenditures on rangeland inventories alone equalled $26.1 million, more than the total revenues collected from grazing fees.

A number of observers questioned the expenditures on information– gathering and analysis for the early grazing EISs. Three years after the court order, NRDC, disappointed at the large expenditures being made at its instigation, concluded that "the Bureau has expended very substantial sums of money and manpower on these EISs but has produced little or nothing to show for it." NRDC stated further that "the BLM has so far failed to complete a single, adequate EIS concerning livestock grazing."[50] In its view the basic problem lay in BLM's incompetence: "We believe that the EIS effort to date has been character- ized by mismanagement, misrepresentation, unnecessary delay and the failure to comply with the most rudimentary standards for the contents of grazing EISs under NEPA."[51] Moreover, backing away a little from earlier comments, NRDC now felt that inventory and other data were not the problem: "The evidence reveals that the inadequacy of the draft EISs released to date by the BLM were caused by its failure to address the issues which are central to proper grazing management, rather than insufficient data."[52]

BLM had prepared a second Challis grazing EIS in large part in order to include much better soils and vegetation data. But NRDC still doubted the extent to which improvements in the analysis for the Challis EIS had resulted. With respect to new soils data, for example, NRDC observed that, despite its much greater availability, "nothing is done with this data."[53] BLM's internal evaluations also indicated that conceptual, personnel and other deficiencies sometimes led to a failure to employ information and analysis.

> Evaluation reports indicate that such deficiencies contribute to superficial treatment of significant issues surrounding decisions relating to the use of the Public Lands, and to the compilation of uninterpreted data whose implica- tions for resource decisions are obscure. . . .
>
> [G]uidance as to what constitutes analysis, as well as what constitutes *analytical* procedures for social information is uneven, often rudimentary, and in some cases, nonexistent. This uneven guidance creates a situation which one BLM Regional Economist noted as often leading to "inappropriate utilization of existing data, and at times, lack of awareness of the existence of such data."[54]

BLM also examined why information and analysis might not be used by managers. One problem was that the way in which managers formed decisions was very imperfectly understood, making it difficult to know what information and analysis would be of help to them:

> Interviewers revealed that managers found it difficult to articulate just how they used information of any sort, but it appears that several factors act to shape the output of the conversion process. Most prominent is that manager's

own values appear to "filter" decisions through a "screen" of local values and preferences. Managers tended to reflect local community attitudes, values, and preferences in their decision making process. Larger concerns did have weight, particularly in decisions related to energy development, but local values and orientations predominated. Such "filtering" is expectable, since the backgrounds and values of managers in BLM are substantially congruent with those of the communities of the rural West.[55]

The success of data–gathering and analysis is likely to be greatest when it is built on a solid theoretical foundation. There should be a good idea what analyses will be used in making decisions and what data are required to prepare these analyses. Such a foundation appears to be missing in the area of rangeland management. A 1976 report prepared for the Council on Environmental Quality on range issues indicated that there had not really been very much research in the past: "Research on rangelands and their management has generally been scant and disorganized, with narrowly based objectives. . . . The rangeland area to research is so vast, the need so great, and the money available so small that an organized approach has been impossible."[56]

The CEQ report indicated that major gaps in basic knowledge about rangeland biology existed. "Fundamental knowledge of plant response to grazing is incomplete and fragmented. We have not progressed beyond the hypotheses of Stoddart and Smith's (1943) first edition of their range management text. In fact, we have not even tested many of the hypotheses they proposed because 'intuitively they feel good.' " Even the concepts of rangeland trend and condition, on which much is based in management, have a suspect theoretical foundation: "Techniques for measuring range condition are imprecise at best, even today."[57]

The 1979 Rangelands Symposium noted above offered some similar reflections. There were suggestions in the summary statement of the need for basic rethinking of the concept of rangeland condition: "Not all the range data base is solid; much is weak, with improvement needed in how data are collected, the kinds of data to collect, and how to handle data. Questioned were some of the concepts basic to determination of range conditions." A number of other range subjects were critically in need of much more basic research. Having the greatest immediate practical implications, the biological response of rangelands to various management systems is simply not understood very well. Moreover, the BLM faces "other technical problems . . . [including]the increasingly important need to have available sound methods for measuring resource interactions and tradeoffs. As competition increases for the many resources of the range, knowing how to predict the effects of multi–resource management systems upon each of the components of the ecosystem will become increasingly important."[58] With a weak foundation of basic knowledge, any new management systems proposed in grazing EISs are likely to show a hit–and–miss pattern of success.

The lack of a solid scientific foundation meant that, despite all the expenditures for inventories and analysis, BLM was unable to provide definitive estimates of such critical matters as the carrying capacity of the range or the likely forage response to new grazing system.

The pouring of money into grazing EISs on top of a shaky scientific foundation of knowledge about rangelands and with a resource of modest value at stake has raised some eyebrows. One range scientist considered that the EISs were "pure busywork carried out in the name of decision making, but serving only to divert energy, attention and effort from management functions to useless paper work."[59] A commenter on the Challis draft supplemental EIS wondered if "much more could be accomplished if less were spent in on EISs and more were spent on the ground improvements."[60] An economic consultant called in to examine the economic analysis in three grazing EISs concluded that this analysis was not bad. The real economic problem was the expense of preparing EISs themselves.

The entire ES preparation and review process is appallingly tedious and costly. The public and courts should be made aware of the incredible costs of research, printing, coordination, review, correcting, reprinting, and even monitoring, so that these costs can be placed in perspective with the real impacts of the BLM's proposed actions. *The government is spending millions of dollars to prepare impact statements for projects with economic impacts which can hardly be measured.*

The government not only spends money on the ES process that is disproportionate with the potential disruptions to people and the environment, but ultimately this kind of governmental functioning is discouraging to the public as well as to those in government employment.[61]

An alternative to the grazing EISs would have been to spend a large share of the money instead on basic research into rangeland scientific questions. Before the grazing EISs were begun, no one seems to have examined closely the question of whether the state of scientific knowledge was adequate for the aims of the EISs. The alternative of a basic research effort would have required a more national focus, a direction precluded by the court's requirement to prepare site specific EISs.

The New BLM

Traditionally, the BLM was seen as an agency dominated by livestock interests. However, the growth of recreational use of public rangelands and the emergence of the environmental movement in the 1960s led to the development of strong counterpressures. In a recent study, Paul Culhane sees this development in an optimistic light; the BLM has now been freed to pursue policies in the public interest, because it has interest groups pressing from all directions. As Culhane puts it, "The Forest Service and BLM thus find themselves in a very powerful position. . . .The agencies, whose commitment to multiple use demands a balanced course of action, can play their more extreme constituents off against each other to reinforce the agencies' preferred middle course. By using both extreme elements in their constituencies, the bureau and service generate a multiple clientele for their multiple–use mission."[62]

To be sure, Culhane seems to assume that multiple–use management constitutes a form of professional management in the public interest. More plausibly, multiple–use management actually means interest group competition and accommodation in the public land area—a special public land term for the "interest group liberalism" of American government described by Lowi. Indeed, the lack of objective standards in multipleuse management leads to protracted political bargaining before any decisions can be reached. The court decision mandating grazing EISs set the framework for the 1970s bargaining over future rangeland use. However, the resulting bargaining process has not yielded many decisions. Rather, the most obvious consequences has been to greatly lengthen the time and expense for negotiation. The new BLM has had to give far more attention to the process of mediating conflicts among multiple interest groups. Inventories, planning, environmental impact statements, and public participation efforts have all been central instruments in this costly bargaining process.

As a result of developments of the 1970s, a new breed of BLMer more suited to emerging tasks began to show up in increasing numbers. The old typical BLMer was trained as a professional range manager at a western state university. The new BLM employee is more likely to have been trained in some other social or physical science field. Of those working on the early grazing EISs, 4 percent were economists, 3 percent outdoor recreation planners, 4 percent archeologists, 7 percent fisheries and wildlife biologists, 3 percent hydrologists and geologists, and 3 percent soil scientists. Range conservationists, the mainstay of the old BLM, were 20 percent of the EIS preparers.

The change in BLM tasks has also required the BLM to bring in new people at high levels. A 1980 article described the divisions emerging between "western managers who had come up through the ranks of the agency, starting at low levels in various mud–on–their–boots field jobs, and the Department of Interior Washington hierarchy, many of whose members have backgrounds as professionals, staff advisors, academics and lobbyists rather than as managers with field experience." One old–time BLMer complained that "those guys know pencil work a lot better than dirt work" and that this "preference was starting to inundate us out in the field."[63]

Environmentalists, however, saw these developments in a much more favorable light: "If a government agency can have a renaissance, the BLM is becoming one of the best examples. Starting in the 1960s with a rise in environmentalism the agency began to discard its lethargy. Now revitalized by FLPMA, the agency is enjoying larger budgets and new responsibilities. While its programs, and many of its line managers, reflect the cowboy and miner orientation of the agency, many of its new wildlife recreation and wilderness specialists are challenging it from within."[64]

For the older BLMers, accustomed to managing by professional judgment and oriented to on–the–ground results, the large resources devoted to grazing EISs became an unfortunate symbol of the new BLM. More surprisingly, many of the social scientists and other recent BLM employees thought little better of the grazing EISs. They found it difficult to connect the huge amount of work going into these EISs with any direct policy consequences and wondered whether there was a real purpose. The BLM study of grazing EISs mentioned earlier reported on the views of EIS team participants: "11 percent categorically state that they believe that the EIS they were working on would not be of significant value in BLM resource management," a further "63 percent expressed doubt that it would be of value or felt that it would only be of limited value," and only "26 percent stated that it would [be of value]."[65]

The study concluded that "the range EIS program is suffering serious morale problems." Only 2 percent of BLMers working on a range EIS described their experiences as "greatly satisfying" and 17 percent as "usually satisfying." On the other hand, 43 percent described the experience as "sometimes frustrating" and 16 percent as "totally frustrating." Twenty–two percent said it was "O.K."[66]

The limitations of agency management through the courts are evident in considering some possible responses that might be made to a wide unhappiness among BLM employees and other indications of problems with the grazing EISs. Normally, a manager facing such a situation would review the program and take steps to deal with the problems. He might eliminate the program altogether, cut it back severely, reorganize it, or perhaps decide that the problem lies mostly in execution and put in new people. None of these options is feasible, however, when the court set the basic direction for the program. If the true problem is that the program is a bad idea, the court must be persuaded of this. However, since court intervention was, formally at least, based on fundamental legal principles, the court is hardly likely to decide that these principles no longer apply. Even getting a judge to take up new evidence and reconsider his original decision is likely to be difficult. Bad programs ordered by a judge are therefore likely to go on without much feedback.

If the court was basically right, but the problem is bad execution, the situation is no better. The judge may believe that an agency is observing the form but not the spirit of his decision. But unlike a corporate president, cabinet secretary, or other top manager, the judge does not have the authority to put in a new management team for the program; there simply is not all that much he can do to see that his decision is implemented as he wants.

Economic Dilemmas in Rangeland Management

A striking feature of the public rangelands is their low economic productivity. There are some BLM rangelands where a square mile (640 acres) is necessary to provide sufficient forage for one cow for one year. More typically, an area larger than 100 acres is required; the average BLM rangeland will support one cow for one month on about 15 acres.

The total values at stake in public rangeland grazing thus are not large. The total capital value of all grazing on BLM rangelands is very likely no more than $1 billion. Grazing resulted on most of these rangelands not by choice, but because grazing turned to be the only land use feasible. The expectations of Congress in passing the Homestead Act of 1862 were unrealistic in the arid— often desert—climate of the West. Although in some areas of the Northern Great Plains homesteading lasted for a while, considerable land eventually shifted back from farming to grazing.[67] Farming in most parts of the West is sustainable over the long term only with construction of expensive irrigation projects.

In 1981 the BLM spent about $70 million for direct rangeland outlays. This did not include rangeland outlays with a specific wildlife or recreation purpose. Expenditures in this latter category constituted about $55 million, most of which also involved rangelands. Thus, total direct rangeland expenditures in 1981 were on the order of $125 million.[68] Moreover, the BLM, like most organizations, must maintain a large overhead for the support of its direct program activities. This overhead includes the BLM cost of operating area, district, state, and Washington offices, as well as the Denver Service Center. Adding in the overhead brings the total 1981 cost of BLM rangeland management to around $230 million per year—about what would be expected for an agency with a total budget in 1981 of $454 million, and for which rangeland management takes up much of its attention.

It is clear that annual expenditures of $230 million for rangeland management cannot be justified by the economics of livestock grazing on public rangelands. The government itself collects only about $20 to $30 million per year in grazing fees, plus perhaps another $1 million per year in recreation and wildlife fees. To be sure, the social value of grazing as well as of recreation and wildlife activities on public rangelands both greatly exceed the amounts collected by the federal government. But even if livestock grazing is worth four times the fees that the government collects, the total annual value would still only be around $100 million. The public benefits of recreation and wildlife activities on the public rangelands are considerably larger. The BLM estimates that total recreation visitor days in 1981 equalled 64 million for all BLM lands. Valuing these days conservatively at $5 per day, the total value of recreation visits would have been $320 million in 1981, easily exceeding the level of BLM rangelands expenditures. However, most of the recreation on public rangelands would have occurred whether or not the BLM made any expenditures on behalf of recreation. Indeed, only about $10 million was budgeted in 1982 for direct recreational purposes.

A significant part of the BLM expenditures for wildlife are justified by the need to reconcile domestic livestock grazing with wildlife demand for forage and habitat. If the domestic livestock were not present, very likely far less would be spent on wildlife by the federal government. For example, wild horse and burro management becomes contentious and expensive because of the direct conflict with forage availability for livestock. In the absence of livestock grazing on federal rangelands, the federal government might well turn wildlife programs over to the western states. In most of the United States, management of wildlife habitat is a state responsibility.

In short, the expenditures made on BLM rangelands mainly involve domestic livestock issues, even when they ostensibly involve other resources. Yet, as noted above, the economic values derived from livestock grazing fall far short of the public expenditures for the rangelands. Politically, it would, of course, be impossible to eliminate livestock grazing. But if the recent levels of rangelands costs were to be taken as given, economically, elimination of grazing would be the rational course.

On the other hand, rather than eliminate grazing, the problem could also be resolved by cutting the costs of public rangeland management. Another possibility would be to redirect some of the expenditures for rangeland management to be of greater real benefit to wildlife and recreational users of public rangelands. The recreation and wildlife values obtained from public rangelands might be able to sustain much greater expenses than grazing use can justify.

The costs for rangeland management have been so high because of the scientific management philosophy that was emphasized in the 1970s. Scientific rangeland management is expensive for several reasons. First, range conditions and management are site specific. One good study is not enough for all rangelands; rather, many separate expensive studies for varying sites and conditions are necessary. Second, as noted above, range science often does not produce definitive management prescriptions on which a broad professional consensus is easily achieved. Lacking an objective scientific answer, rangeland decisions easily become political. Then, instead of accepting government decisions, each party is likely to demand further analysis in hopes that the next round of studies will lend greater support to its case. And, third, scientific management has tended to require centralized decisionmaking, adding large overhead costs to the already substantial field expenditures for inventories, planning and other rangeland study elements.

An alternative to scientific management is trial and error. Another way to manage the public rangelands would be to test new management systems and approaches on a limited number of experimental sites. On an agencywide basis, smaller changes in management practices could be instituted. If subsequent experience showed that the new systems and methods were working well, then they could be adopted on a wider basis. But unlike past efforts, there would be no attempt to decide through a onetime study the future management of the public rangelands.

A trial–and–error strategy is also an incremental approach. Students of government have found that designs for comprehensive change seldom succeed. The history of comprehensive land use planning in the United States is generally one of high hopes and few results. In range management it may also be desirable to move a few steps at a time, taking care to evaluate each step before moving to the next.

Nevertheless, it is by no means certain that an incremental strategy of trial and error would reduce rangeland management costs to a level in line with the low economic values created by livestock grazing. The current rangeland system may simply make incompatible demands. Good management is considered to require large amounts of data and analysis; yet, good management also requires that data and analysis be limited so as to bear a reasonable relationship to the values at stake. If this dilemma proves to be unresolvable, consideration of further options involving more basic institutional change may be necessary.[69]

To be sure, one possibility is simply to accept a permanent condition of management costs much greater than any identifiable values from rangeland use. The rangelands may have an intrinsic value to the nation that justifies substantial expenditures to maintain them in high quality— without regard to identifiable benefits. Maintenance of ranching lifestyles may also have an intrinsic value to the nation; society may be willing to provide a permanent government subsidy for this purpose. The ranching life might, as it were, be considered a "rare and endangered" existence, having an historical and symbolic significance to the whole nation.

But if major changes are to occur, one option would be to transfer greater management responsibility to the states.[70] State governments have a large interest in the management of rangelands in each state, greater really than the federal government for most rangelands of ordinary interest. The federal government might contract with states to manage the lands; states might first submit an implementation plan that would have to be federally approved. Alternatively, the direct ownership of rangelands could be transferred to the states.[71]

Another main option would be to expand the role of the private sector. Congress might authorize longer terms for grazing leases and permits—perhaps fifty years.[72] Under the lease terms, ranchers would have wider management flexibility and would bear the cost of range improvements and other management actions. Western state governments would be encouraged to maintain wildlife programs, much as eastern states with mainly private lands do at present. Rather than long–term leases, another possibility would be outright sale of rangelands to ranchers.[73] In order to make this feasible, the federal government would probably have to offer special financing and below–market land prices to ranchers who currently use the lands.

Prospects for Economic Analysis

Assuming public rangelands remain under federal management, the prospects for greater use of economic analysis are uncertain. The number of economic calculations performed was significantly greater in the 1970s, but it is hard to know just how influential they were in rangeland decisionmaking. One has the sense that the precise numbers obtained often were not very important. However, the pressures to undertake economic analysis may have served a "consciousness-raising" purpose. The very fact of new agency requirements for economic assessments sent a signal to agency employees to pay more attention to the costs of their proposals. It also indicated a perceived need to think more carefully about the actual benefits achieved. Of course, since economic analysis can be expensive, one might question whether frequent requirements for economic assessments are a cost–efficient way to raise an agency consciousness. Perhaps a single strongly worded memorandum from the BLM director would do, or expanded agency training programs.

The use of economics cannot be divorced from the bureaucratic and institutional context where such use is sought. One cannot simply say that economic efficiency is a desirable goal, and therefore economic methods should be widely adopted by the agencies. It is necessary to ask, who will do the economic calculations, why will they do them, and why and how will these calculations be followed in agency decisionmaking? In short, what are the incentives for use of economics?

The interest groups that benefit from use of public lands typically are skeptical of economic analysis—at least as far as their own use is involved. Instead, they prefer to maximize user benefits without much regard to costs. Thus, there is little pressure from outside interests for economics. The one main exception is environmental groups opposed to the development of projects that they believe will not withstand economic scrutiny.

The chief outside pressure on BLM for economic analysis of rangeland projects thus far has come from budget offices in the Interior Department and OMB. Economics offers an objective standard for setting priorities among the numerous claimants for limited federal funds. Yet, while budget officers can require that formal economic assessments be made, they cannot require that the substance of agency decisionmaking closely reflects the results of these assessments. It would be impossible for a few budget reviewers to police agency conduct of economic analyses to ensure that they were not manipulated to reach predetermined conclusions.

In the end, the use of economics is likely to be determined by the importance given to economics in the belief system of a public agency. Recent students of American business have emphasized the importance of a "corporate culture" in explaining differing business results. The same is true of public agencies. The belief system or governing ideology of public land management is scientific management, as derived from the conservationism of the Progressive era. The BLM understanding of scientific management is professional range management, not the professional expertise of economists. Until an economics ethos becomes ingrained in public rangeland management, the role of economics is likely to remain minor. Perhaps the only way to achieve a new agency culture of economic efficiency advocates would be to make wholesale changes in personnel. As noted above, there is no strong interest group pressure to make such changes.

The 1970s willingness of the BLM to spend excessive sums on rangeland planning illustrates the difficulties of achieving economic efficiency in the absence of an ingrained tradition of cost consciousness. In the mid-1970s, there were strong political pressures to "do something" about rangeland conditions. The public land agencies responded by sharply increasing their expenditures on the rangelands—if mostly for inventories and planning. The question of whether this level of expenditure was economically rational may never have come up; in any case, it certainly was a minor consideration.

Recent Events

The focus of this paper has been on the 1970s.[74] However, the arrival of the Reagan Administration in 1981 brought new officials skeptical of much that had gone on the previous decade. Moreover, the general effort to cut back sharply on federal expenditures has also left its mark on rangeland programs, much as it has affected many other areas of government activity. Since 1981 there have been a number of significant changes or shifts in emphasis in rangeland policy.

Rather than implement immediate reductions (or increases) in domestic livestock grazing, as recommended by grazing EISs, the Reagan Administration has adopted a program of limited initial changes followed by monitoring of the results. If the monitoring then reveals a need for further changes, they will be made at a later time. In another new procedure, all grazing allotments will be assigned to one of three categories: custodial, need to improve, or maintain. In the allocation of scarce BLM funds, the allotments in the need to improve category will tend to have a higher priority. Allotments in the custodial category would have few prospects for effective management actions, while allotments in the maintain category would already be in satisfactory condition. Individual ranchers proposing new range investments for allotments in the custodial or maintain categories would be largely responsible for the cost of financing these investments.[75]

Partly reflecting strong general pressures to cut federal spending, the Reagan Administration has reduced sharply the expenditures for rangeland inventories and for land use planning. For example, planning funding fell from $16 million in 1981 to $9 million in the 1982 budget.[76] These reductions provoked strong criticism from environmental organizations. They perceived cuts in planning and rangeland studies as an abandonment of the 1970s attempt to assert greater government control over the use of public rangelands. There has also been an effort to give a greater role in management to ranchers and to shift responsibility to the field generally. Under a new cooperative management program announced in 1983, a few specially selected ranchers will be given wider latitude to manage livestock grazing. For these ranchers, tenure will be made more secure by renewing the 10-year permit every 5 years on a rolling basis. If the program works out in a limited trial, it could be expanded to a larger number of ranchers.[77] Environmentalists have also criticized these efforts as an undesirable transfer to private ranchers of greater responsibility for management of public rangeland.

The BLM has reviewed and revised its procedures for conducting economic analysis, and reaffirmed a commitment to economics as part of its planning. A third generation of guidelines for benefit–cost studies of range investments was issued in 1982.[78] This time, the benefit–cost guidelines were worked out in close consultation with western university economists who in effect gave their blessing. Finally, the asset management program of the Reagan Administration has raised the issue of whether some lands could be managed more economically in the private sector. The strong political opposition to the asset management program has also shown the degree to which any change in the current system will be difficult. There are many beneficiaries from current arrangements, and they tend to resist change, whether or not it would enhance the economic efficiency of public land management.

Notes

1. For a review of such criticism, see Robert H. Nelson, ""The Public Lands," in Paul R. Portney, *Current Issues in Natural Resource Policy* (Washington, D.C.: Resources for the Future—distributed by the Johns Hopkins University Press, 1982); see also Marion Clawson, *The Economics of National Forest Management* (Washington, D.C.: Resources for the Future, 1976).

2. See John V. Krutilla and John A. Haigh, "An Integrated Approach to National Forest Management," *Environmental Law* (Winter 1978); also Richard M. Alston, *Forest-Goals and Decision Making in the Forest Service* (Ogden, Utah: Intermountain Forest and Range Experiment Station, U.S. Forest Service, September 1972).

3. Samuel P. Hays, *Conservation and The Gospel of Efficiency: The Progressive Conservation Movement, 1890–1920* (Cambridge, Mass.: Harvard University Press, 1959).

4. See Phillip N. Truluck, ed., *Private Rights and Public Lands* (Washington, D.C.: The Heritage Foundation, 1983); and articles on "Land Use and Resource Development" in *The Cato Journal* (Winter 1982).

5. See Joseph Sax, "For Sale: A Sign of the Times on The Public Domain," to be published in a forthcoming volume from Resources for the Future, Sterling Brubaker, ed., based on a September 1982 Conference on "Rethinking The Federal Lands."

6. See the works of Charles E. Lindblom, such as Lindblom, "The Science of 'Muddling Through,' " *Public Administration Review* (Spring 1959); see also R.W. Behan, "The Privatization Alternative for the Future of the Federal Public Lands: A Penultimate Comment," Presented to the annual meeting of The Western Political Science Association, Seattle, Washington, March 25, 1983; and Christopher K. Leman, "The Revolution of the Saints: The Ideology of Privatization and Its Consequences for the Public Lands" prepared for a panel on "The Political Economy of Privatization: Two Points of View," at a National Symposium on Selling the Federal Forests, University of Washington, Seattle, Washington, April 22–23, 1983.

7. For the history of public rangelands see E. Louise Peffer, *The Closing of The Public Domain: Disposal and Reservation Policies, 1900–50* (Stanford, Calif.: Stanford University Press, 1951); Phillip O. Foss, *Politics and Grass: The Administration of Grazing on The Public Domain* (Seattle: University of Washington Press, 1960); and Wesley Calef, *Private Grazing and Public Lands* (Chicago: University of Chicago Press, 1960).

8. See Robert H. Nelson, *The New Range Wars: Environmentalists versus Cattlemen for the Public Rangelands* (Draft), Office of Policy Analysis, U.S. Department of the Interior, 1980; also Paul J. Culhane, *Public Lands Politics: Interest Group Influence on The Forest Service and the Bureau of Land Management* (Baltimore: Johns Hopkins University Press for Resources for The Future, 1981).

9. See Christopher K. Leman and Robert H. Nelson, "Ten Commandments for Policy Economists," *Journal of Policy Analysis and Management.*

10. See Gifford Pinchot, *Breaking New Ground* (New York: Harcourt Brace, 1947).

11. Laurence A. Stoddart and Arthur D. Smith, *Range Management* (New York: McGraw-Hill, 1943), p. 2.

12. Laurence A. Stoddart, Arthur D. Smith and Thadis W. Box, *Range Management* (New York: McGraw-Hill, 1975), p. 90.

13. New Mexico Allotment Management Plan Handbook, BLM Manual Supplement, Bureau of Land Management, New Mexico State Office, October 10, 1969, p. 14.

14. Bureau of Land Management, U.S. Department of the Interior, *Livestock Grazing Management on National Resource Lands: Final Environmental Statement* (Washington, D.C.: December 1974), p. I–1.

15. Public Law 94–579, Sec. 102(a)7.

16. Public Law 95–514, Sec. 2(b)(2).

17. See Culhane, *Public Lands Politics.*

18. See John Kenneth Galbraith, *American Capitalism: The Concept of Countervailing Power* (Boston: Houghton Mifflin, 1956).

19. Theodore J. Lowi, *The End of Liberalism: Ideology, Policy and the Crisis of Public Authority* (New York: W.W. Norton, 1969), p. 71.

20. Ibid., p. 287.

21. See Krutilla and Haigh, "An Integrated Approach to National Forest Management."

22. Memorandum from Office of Management and Budget to Interior Deputy Assistant Secretary for Policy Development and Budget, August 29, 1975.

23. See Earl E. Houseman, et al., "Special Report on Grazing Fee Survey," Statistical Reporting Service, U.S. Department of Agriculture (Washington, D.C.; November 29, 1968).

24. See *Study of Fees for Grazing Livestock on Federal Lands,* A Report from the Secretary of the Interior and the Secretary of Agriculture, October 21, 1977.

25. U.S. Forest Service, Department of Agriculture, *The National Range Resources—A Forest-Range Environmental Study,* by the Forest-Range Task Force, Forest Resource Report No. 19 (Washington, D.C.: Government Printing Office, December 1972).

26. Ibid., p. 79.

27. Ibid., p. 82.

28. U.S. Department of Agriculture, *Opportunities to Increase Red Meat Production from Ranges of the USA (nonResearch), Phase I,* prepared by the Department of Agriculture Work Group on Range Production (Washington, D.C.: June 1974).

29. Ibid., p. 51.

30. See Robert H. Nelson, "Benefit-Cost Analysis of Public Range Investments: A Case Study" (unpublished paper, 1976).

31. Comments of Max Lieurance, Chief of BLM Range Division, at BLM Economics Workshop, Salt Lake City, May 22 and 23, 1979.

32. "Program Decision Option Document on AMP Feasibility Analysis," Bureau of Land Management, November 6, 1975.

33. Bureau of Land Management, U.S. Department of the Interior, *Range Condition Report,* prepared for The Senate Committee on Appropriations, January 1975.

34. "Forage and Livestock Valuation Methods," Committee Report from the Range Economics Workshop of the Bureau of Land Management, Tucson, Arizona, March 19–23, 1978.

35. BLM Instruction Memorandum No. 76–455, from Associate Director (George Turcott) on "Allotment Management Plan (AMP) Economic Analysis," August 26, 1976.

36. Bureau of Land Management, U.S. Department of the Interior, *Final Environmental Impact Statement on the Proposal Rio Puerco Livestock Grazing Management Program* (May 1978), p. IX–8.

37. OMB internal paper transmitted from Chief, Interior Branch, Natural Resources Division (Rodney Weiher) to Director, Bureau of Land Management (Frank Gregg) and Director, Office of Policy Analysis (Lester Silverman) on "Range Investments on Public Lands," May 23, 1979.

38. Secretary's Memorandum No. 1999, ""Statement of Range Policy," Secretary of Agriculture, October 25, 1979, p. 3.

39. "The Sense of the Symposium on Rangeland Policies for the Future," a report to the Assistant Secretary of Agriculture for Conservation, Research and Education, the Assistant Secretary of the Interior for Land and Water Resources and the Council on Environmental Quality, by the Interagency Ad–Hoc Committee on the Sense of the Symposium, June 1, 1979, p. 7.

40. *Managing the Public Rangelands*, Public Review Draft, Bureau of Land Management, November 1979, p. 1.

41. Arthur H. Smith and William E. Martin, "Socioeconomic Behavior of Cattle Ranchers, with Implications for Rural Community Development in the West," *American Journal of Agricultural Economics* (May 1972), p. 217.

42. See *Study of Fees for Grazing Livestock on Federal Lands*.

43. Memorandum from Roy Allen (BLM Wyoming State economist) to Bob Browne, "Methodology Used for Assessing the Economic Impacts in the Sandy Grazing Environmental Statement," August 29, 1978, p. 2.

44. See John F. Dwyer, John R. Kelly, and Michael D. Bowes, *Improved Procedures for Valuation of the Contribution of Recreation to National Economic Development* prepared for the Office of Water Research and Technology, U.S. Department of the Interior (September 1977).

45. *Natural Resources Defense Council v. Morton*, 388 F. Suppl. at 841 (1974).

46. See Christopher K. Leman, "Formal Versus De Facto Systems of Multiple Use Planning in the Bureau of Land Management: Integrating Comprehensive and Focussed Approaches," National Resource Council, *Developing Strategies for Rangeland Management* (Boulder, Co.: Westview Press, forthcoming).

47. For an examination of the history of comprehensive planning as a basis for zoning, see Robert H. Nelson, *Zoning and Property Rights: An Analysis of the American System of Land Use Regulation* (Cambridge, Mass.: MIT Press, 1977).

48. *Grazing Environmental Statement Review Report*, Bureau of Land Management, April 30, 1979.

49. Budget Issue paper on "BLM Inventory, Planning and Environmental Analysis Expenditures for the Public Rangelands," transmitted by deputy Assistant Secretary for Policy, Budget and Administration to Assistant Secretary for Land and Water Resources, June 22, 1977.

50. "Plaintiffs' (NRDC) Summary of the Evidence in Opposition to the Federal Defendant's Notice of Proposed Deviation," in *NRDC* v. *Andrus* (formerly Morton) February 28, 1978, p. 2, 7.

51. Ibid., p. 4.

52. Ibid., p. 20.

53. NRDC comments on the Challis Draft Supplemental EIS, reprinted in Bureau of Land Management, U.S. Department of the Interior, *Final Supplemental Environmental Statement on a Revised Range Management Program for the Challis Planning Unit* (November 1978), p. A–232.

54. Special Evaluation, Social Economic Analysis in Bureau Decision Making, Bureau of Land Management, Office of Program Evaluation, 1979, p. 6.

55. Ibid., p. 15.

56. Thadis W. Box, Don D. Dwyer and Frederic H. Wagner, "The Public Range and its Management," A Report to the Council on Environmental Quality, March 19, 1970, p. 43.

57. Ibid., pp. 32, 42.

58. "The Sense of the Symposium on Rangeland Policies for the Future," p. 6.

59. Boysie E. Day, "Range Management, An Ecological Art," in *Rangeland Policies for the Future*, proceedings of a symposium, January 28–31, 1979, Tucson, Arizona (Washington, D.C.: Government Printing Office, 1979), p.92.

60. Comment of the Idaho Farm Bureau Federation on the Challis Draft Supplemental EIS, reprinted in *Final Supplemental Environmental Statement on a Revised Range Management Program for the Challis Planning Unit*, p. A–216.

61. "An Evaluation of the Economic Analysis Contained in Three BLM Grazing

Environmental Statements," prepared for the Montana Public Lands Council and Old West Rangeland Monitoring Project by T.A.P., Inc., April 2, 1979, p. 18.

62. Culhane, *Public Lands Politics*, p. 336.

63. Doug Gill, "BLM Exodus Highlights Schism," *The Denver Post*, April 6, 1980, p. 66.

64. Bernard Shanks, "BLM, Back in the Spotlight after Years of Neglect," *High Country News*, January 26, 1979, p. 5.

65. *Grazing Environmental Statement Review Report*, p. 32.

66. Ibid., pp. x, xi.

67. See Walter Prescott Webb, *The Great Plains* (Boston: Ginn and Co., 1931).

68. Robert H. Nelson and Gabriel Joseph, "An Analysis of Revenues and Costs of Public Land Management by the Interior Department in 13 Western States—Update to 1981," Office of Policy Analysis, U.S. Department of the Interior (September 1982).

69. See Nelson, "The Public Lands," and Robert H. Nelson, "Ideology and Public Land Policy: The Current Crisis," to be published in a forthcoming volume from Resources for the Future, Sterling Brubaker, ed., based on a September 1982 conference on "Rethinking the Federal Lands."

70. See Robert H. Nelson, "Making Sense of the Sagebrush Rebellion: A Long Term Strategy for the Public Lands," paper prepared for presentation at the Third Annual Conference of the Association for Public Policy Analysis and Management, Washington, D.C., October 23–25, 1981; a shortened version is found in Robert H. Nelson, "A Long Term Strategy for the Public Lands," in Richard Ganzel, ed., *Resource Conflicts in the West* (Reno: Nevada Public Affairs Institute—University of Nevada, March 1983).

71. For examination of this issue with respect to another public land resource, see Robert H. Nelson, *The Making of Federal Coal Policy* (Durham, N.C.: Duke University Press, 1983).

72. For new leasing proposals, see Marion Clawson, *The Federal Lands Revisited* (Washington, D.C.: Resources for the Future, forthcoming); also Marion Clawson, "Major Alternatives for Future Management of Federal Lands," in the forthcoming Resources for the Future volume on "Rethinking The Federal Lands," Sterling Brubaker, ed.

73. See Truluck, ed., *Private Rights and Public Lands*; "Privating Public Lands: The Ecological and Economic Case for Private Ownership of Federal Lands," in *Manhattan Report* (May 1982); Gary D. Libecap, *Locking Up the Range: Federal Land Controls and Grazing* (Cambridge, Mass.: Ballinger Publishing Company for the Pacific Institute for Public Policy Research, 1981); and B. Delworth Gardner, "The Case for Divestiture of the Federal Lands," in the forthcoming Resources for the Future volume on "Rethinking The Federal Lands," Sterling Brubaker, ed.

74. For further assessment of the events of the 1970s, see *Developing Strategies for Rangeland Management*, a report prepared by the Committee on Developing Strategies for Rangeland Management—Assembled by the National Resource Council of the National Academy of Sciences (Washington, D.C., 1981).

75. See Bureau of Land Management, "Rangeland Improvement Program Policy," attachment to Instruction Memorandum No. 83–27, October 15, 1982.

76. See Bureau of Land Management, U.S. Department of the Interior, *Managing the Nation's Public Lands*, a Program Report Prepared Pursuant to Requirements of the Federal Land Policy and Management Act of 1970, (January 31, 1983).

77. See Bureau of Land Management, "Policy and Procedures for Implementing Cooperative Agreements," Instruction Memorandum No. 83–485 (April 22, 1983); also Leonard U. Wilson and Frank Lundberg, "Cooperative Management on the Public Rangelands," in Ganzel, ed., *Resource Conflicts in the West*.

78. See Bureau of Land Management, "Evaluating, Ranking and Budgeting Implementation of Rangeland Improvements," attachment to Instruction Memorandum No. 83–27 (October 15, 1982).

3

Beyond the Sagebrush Rebellion: The BLM as Neighbor and Manager in the Western States

SALLY K. FAIRFAX

Introduction

The recent Sagebrush Rebellion invites us to look closely at trends in American federalism. At the same time, it may tempt us to view those trends in the potentially deceptive context of a federal–state conflict. Although research in the field has not been dominated by "dual federalism" for several decades, the lawyerly preoccupation with preemption analysis never passed nor has public acceptance of newer theories threatened the traditional civics book notion.[1] The whole concept of conflicting sovereigns seems to have taken on renewed vigor in recent years, and the trend is potentiated as lawyers emerge as major players in policy analysis.[2] There are certainly grounds for arguing, across the board and in specific policy arenas, that the states are increasingly effective rivals of the federal sovereign. Nevertheless, casting the analysis in those terms obscures both important influences on decisionmaking and significant developments in our federal system. It also inappropriately narrows the search for effective institutional arrangements. This is particularly troubling in the public resources field where the lure of the Rebellion is strong and the property clause is arguably the decisive arbiter of ultimate authority.[3]

The purpose of this paper is to use recent history in public domain policy to highlight inadequacies in the renascent federal–state conflict model, both generally and as it is applied in the specific area of federal land management. Part I will describe the general context in which the faulty model is rejuvenated in the Sagebrush Rebellion, emphasizing the evolution of the major and most controversial (perhaps the most troubled) western landholder—the Bureau of Land Management (BLM). Part II will deal with selected recent events in federal land management. The major point is to argue that federal–state relations, more specifically federal– state conflict, is a very poor conceptual guide for understanding contemporary federal land management.

The Bureau is not institutionally in a position to assert federal authority "without limitations" over the resources in its charge.[4] Lacking adequate funding, data, and personnel, it must perforce take a cooperative approach to planning and management.[5] Moreover, the BLM is required by statute to deal cooperatively with states and localities, and that mandate may alter the nature of federal management.[6] Finally, perhaps most significantly, the relevant advocates are increasingly engaged in forum shopping. Special interest groups, each seeking the most favorable arenas in which to air grievances and pursue their goals, are functioning as a major force in institutional diversification and reorientation of "our federalism."[7]

One can draw a state revival out of the Sagebrush Rebellion and simply argue that the worm has turned in traditional federal–state turf battles. Certainly, the ubiquitous but ill–defined pendulum swings, and it would appear that it is headed in the states' direction. To stop at that gross level and fail to analyze the pattern on an issue–by–issue basis would, however, be misleading. In the public lands field, there is ample reason to suggest that the state is neither a consistent nor the only salient awakening actor and that the BLM's management policies will be increasingly implemented through imaginative, area–specific arrangements devised by a diverse array of neighbors, government bodies, and affected interests.

I. Prologues to the Sagebrush Rebellion

At first blush, even at second blush, the Sagebrush Rebellion appeared to be fairly straightforward, simple, and familiar. Particularly from popular media and fund–raising broadsides of various persuasions, the idea emerged that livestock operators—after years of growing federal interference with their activities—righteously or otherwise banded together to beat back the rapidly tightening tentacles (again, righteously or not depending on your preferences) of the federal government. Asserting ancient and highly questionable entitlements, the rebels sought the "return" to the state governments of lands increasingly regulated by the federal government.

Neither the BLM nor its predecessor agencies have ever been popular with the ranching industry except as an easily manipulated token that occupied a space on the board and thereby blocked imposition of more onerous burdens. The recent fracas is part of a nearly century–long series of disputes in which livestock operators have pressed to solidify their advantages on the public domain. Although security of land tenure in the form of title transfer has nominally been the traditional goal, stock operators have largely achieved their purposes without the transfer (and concomitant financial burdens associated with ownership) by simply intimidating the agency. The conventional wisdom about BLM managers being subservient vassals for powerful western ranchers does not, however, tell the whole story.[8]

Significantly, the Bureau was established by an executive reorganization of the Grazing Service at the beginning of a 1940s rendition of the Sagebrush Rebellion. In the waning days of Nevada Senator Pat McCarran's investigation of the

Grazing Service, the stock operators, already primed for a fight, began a movement to take title to their allotments. The lands were, under the terms of the Taylor Grazing Act, being held "pending final disposition;" and the livestock industry was, by the end of World War II, financially in a position to own the lands. Conservationist response to the proposed "land grab" was intense and the industry effort failed. However the prolonged assertion of national public interest in the public domain lands was unusual. Perhaps emboldened by this apparent public support, the Bureau began, albeit with great difficulty and frequent defeats, to alter the direction of its programs. During the 1950s, it engaged in a series of battles to bring the levels of grazing use, liberally overallocated in the 1930s, into line with obviously overextended resource capabilities. The basic components of the early BLM effort seem relatively simple: determine the grazing capacity of the land, determine the patterns of use, and allocate a prorated share of the capacity to the users.

Nevertheless, the agency continually lost conflicts with the livestock industry. Philip Foss and Wesley Calef, in a series of classic case studies of this troubled period of range adjudication, have chronicled activities designed to keep the agency weak and largely ineffectual. Although Nevada's Senator McCarran was by every account the most awesome and devoted defender of the large cattle ranchers, there has never been a shortage of westerners in the Senate and elsewhere who could harass the BLM whenever it suited the goals of the livestock industry.

The notion that the BLM works for the livestock industry was almost literally correct: in the wake of extensive 1940s budget cuts, the Grazing Advisory Boards actually paid agency employees out of "their share" of the grazing fees, range improvements funds. However, while the tendency to emphasize the defeats and the problems of the BLM is understandable, it may lead to serious error. It overlooks the fact that, in spite of the handicaps under which the Bureau labored, ranching interests consistently found it necessary to bully the district managers in order to contain and forestall the Bureau's meager conservation efforts. Notwithstanding all the advantages to the industry in determining grazing policy, clear evidence began to accrete in the early 1960s that the Bureau was developing the ability and the ambition to chart an independent course toward public range regulation.

Current preoccupation with the Federal Land Policy and Management Act (FLPMA) as the start of a new era for the Bureau results in two major errors. First, it fails to recognize that FLPMA was too late to be formative. The Bureau not only survived in a uniquely hostile bureaucratic environment long before FLPMA was enacted: it had gradually transformed itself into a multiple–use management agency. Much of what appears to be "new and improved" in FLPMA is simply formal congressional acquiescence in programs and concepts the Bureau had been following for a decade or more. The success of this pre–FLPMA effort is an important indicator of the Bureau's herculean efforts. Far from being the servile captive of industry, it has bent without breaking and has stood for something. Its resilience and dedication is the BLM's major resource for future development.

Although it had not met with overwhelming success in its 1950s efforts to readjudicate the range and reduce allotments, the Bureau continued in the 1960s to evince standard patterns of bureaucratic ambition and self–aggrandizement. Emulating Forest Service approaches, the Bureau instigated a number of legislative proposals drawing heavily on Multiple Use Sustained Yield Act concepts, and it instituted a major land management planning program centered on a new and controversial concept called rest–rotation grazing.

The maturation of the increasingly professional BLM staff is evident in a 1960 Bureau publication, "Project 2012." The document can be understood in the context of Mission 66 and Operation Outdoors, massive multi–year public relations and budget request programs from the Park Service and the Forest Service, respectively.[9] During the 1950s, both of those agencies launched major long–range planning efforts to cajole extra dollars out of Congress for sorely needed postwar refurbishments. Belatedly, the Bureau followed in their path. The weak point in the BLM effort, especially in comparison to the other two, is its failure to tie program descriptions to outputs or cost estimates. Nevertheless, the fact that the Bureau was, in 1960, capable of producing such an unexceptional document continues to surprise people. In fact, the document was but one part of an extensive legislative offensive undertaken by the Bureau. The Bureau strategy was to eliminate sequentially the agency's serious real estate management problems (pertaining to acquisitions, sales, classifications, rights–of–way, leases, permits, withdrawals) through a series of separate bills, each attacking a severable portion of the problems.[10]

The legislative initiative of the Eisenhower and Kennedy years was stalled by Colorado Congressman Wayne Aspinall's insistence that, before any partial steps were taken, a commission to review all the public land laws should be established. Finally, as part of a complex legislative compromise, a statute establishing the Public Land Law Review Commission (PLLRC) emerged from Congress simultaneously with the Wilderness Act of 1964 and two relatively obscure public domain statutes, the Public Land Sales Act and the Classification and Multiple Use Act (CMU).

One of these, the CMU, was critical to expansion of BLM activities. Three major activities were authorized by the statute. First, the Secretary was directed to develop criteria for determining which of the lands administered by BLM ought to be disposed of and which should be retained for multiple–use management, at least until the PLLRC reported. Second, the BLM was to review the public domain lands "as soon as possible" and classify them for retention or disposition using the criteria. Third, the Bureau was to manage the retained lands for multiple uses, again pending the PLLRC report. Through the CMU Act the BLM was granted authority and an affirmative duty to inventory and assess land for a broad range of multiple uses. The fact that this general administrative authority to inventory and analyze a full spectrum of range resources was not conferred until 1964 explains part of the present inadequacies in the Bureau's data base. Because the CMU Act authority was not accompanied by massive infusions of dollars, the statutory change did not precipitate the major improvement in Bureau information systems that one might suppose.

The CMU Act was also a clear indication that public domain lands held three decades in the "pending final disposition" limbo were to be retained and managed. This led to the most significant ramification of the CMU Act: the BLM's evolving attitudes toward itself and its mission. The classification authority, notes former Bureau Director Marion Clawson, "gave the Bureau a psychological lift that has led to its taking the intitiative more and more often."[11]

This initiative can be seen in the immediate and long–term implementation of the act. The Bureau seized upon the classification process as a major public involvement and agency assertiveness program. The normal regulation–writing procedures were modified to allow for extensive public involvement in drafting classification criteria, and the classification process was carried out with an emphasis on public involvement that was unprecedented in mid–1960s land management. The Bureau used the opportunity of the CMU Act in an effort to build its public profile and recruit a constituency for its efforts to become a multiple–use, land management agency.[12]

Even more notable is the Bureau's effort to parlay temporary classification and management authority into a multiple–use mandate. The statutory classification and management authorities were to expire six months after the PLLRC submitted its report. As originally published, the classification regulations contemplated that the classifications—Bureau decisions regarding retention of lands for specific purposes and exclusion of incompatible entry or disposition applications—would lose their effect when the authority to classify expired. However, the Bureau subsequently changed its mind and concluded that "classifications for retention, including their segregative effect and the authority to manage classified lands, continue indefinitely."[13] The Bureau continued to operate on this tenuous assertion of authority until 1976.

Emboldened by this heady fabrication, the Bureau undertook many activities on public domain lands. The Bureau began to designate "primitive" areas and research natural areas. It even dedicated several wildlife preservation areas, such as the Birds of Prey Natural Area near Boise, Idaho. Most startling, perhaps, were the BLM's efforts, beginning in 1965, to move toward a more intensive range management and rehabilitation program. The 1950s efforts simply to reduce livestock numbers were transposed into a major planning and investment program based on concepts of rest–rotation grazing. This in turn led to the development of "allotment management plans" (AMPs) with cooperating ranchers. The Bureau was attempting to move beyond a mere gatekeeper function and create a reliable, dependent constituency by developing a program of investments, subsidies, and land management.[14]

The upheavals of the late 1970s appear to have originated in several events that dramatically altered the ranchers' ability to direct federal grazing programs in preferred channels. In the "NRDC grazing case," the court found the environmental impact statement on the BLM's grazing program to be inadequate.[15] The BLM was tied by the settlement in the case to an unrealistic schedule for preparation of over 200 EISs. Management activities have been delayed while the Bureau struggled to push the proper papers through the unwieldy, underfunded, and contentious EIS process. The second event, the 1976 passage of the Federal Land Policy and Management Act, further interrupted the livestock industry's sense of control by diversifying and strengthening the BLM's mission and authority.

The 1934 Taylor Grazing Act was, as the name implies, a dominant–use statute. After it was enacted, wildly inflated estimates of available forage were allocated exclusively to established domestic livestock producers, that is, those with a private property ranching base. Western water law, which approves appropriations for beneficial uses, reinforced the dominant role of ranchers in rural areas by providing them with water for irrigation of supplemental hay production. FLPMA's multiple–use mission required that these forage resources be reallocated to many uses including livestock, watershed, wildlife, wilderness, and recreation. The statute required that the reallocation take place as part of a massive and ill– defined inventory and planning program. The combined effect of the NRDC lawsuit and the legislation was to suggest enormous cuts in livestock grazing allotments while establishing a planning process so expensive and inconclusive that the costs of devising the plans vastly exceeded the benefits of implementing them.[16] The ranchers' immediate needs and goals have been submerged in a sea of nettlesome forms, procedures, and regulations, and their long–term position on the federal lands has been seriously eroded.

The antifederal component of the stock operators' position was sharpened into an apparent federal–state conflict by a third event of the "pre–Rebellion" period. In a major interpretation of the property clause, the Supreme Court apparently extended federal authority over federal lands to the point that traditionally exercised state authorities were severely diminished. In sustaining a federal statute that preempted long–established concepts of state authority over wildlife, the court held that federal authority over federal lands was "without limit." Although the phrasing was taken from numerous older cases, the previous assertions of unlimited authority all pertained to the federal authority to *dispose* of federal lands.[17] Although not totally unexpected, the decision is a long way from the original notion that, in holding lands, the government acted as a mere proprietor and like any other body was subject to regulation by the state sovereign. The fact that *Kleppe* v. *New Mexico* resulted in the displacement of state estray law, and the solidification of protection for the much–lamented wild horses and burros, undoubtedly sharpened the ranching community's attention to these finer points of constitutional law and the erosion of state prerogatives.

If the first impression of the Sagebrush Rebellion—that which reflects the livestock industry's concern—suggests the validity of the federal-state conflict approach, a broader, slightly more analytical, view also seems to confirm it. Few commentators have failed to note that the Sagebrush Rebellion was not limited to aggrieved ranchers but includes a broad range of protestants whose voices added to the audibility and durability of the Rebellion.[18] Even in this broader perspective, however, there is some evidence to suggest the utility of the federal-state conflict model.

The western states have long argued that federal lands diminish their ability to control their own economic development and their own resources. This complaint must always be balanced against the benefits, in terms of subsidies, development, open space, recreation, et cetera, that federal landholdings generate. Not unexpectedly, the beneficiaries of federal largess within the western states are less apt than the losers to complain of federal usurpations. Nevertheless, there are clear patterns of federal encroachment that have general import and credibility.

One obvious bone of federal–state contention over the public lands that has been exhumed, refurbished, and unfortunately beclouded by Sagebrush Rebellion rhetoric arises over selection and management of state lands. Although it was frequently alleged by anti–rebels that federal lands were national resources to be managed in the national interest, this contention is simply not true in all cases. Many states have uncontested claims to enormous acreage dating from the state land selections granted as a precondition to their joining the union. As a successor agency to the General Land Ofice, the BLM is charged with administering those grants; it has been slow indeed in processing state applications.[19] Moreover, BLM management programs—most obviously wilderness designations—interfere with the ongoing process by effectively withdrawing land from state selection, thus diminishing the value of the state's right.

More ubiquitous are the difficulties states encounter in trying to manage the "school lands" (sections 16 and 36 in every township.) Not infrequently, those dispersed state holdings are completely surrounded by federal lands. For good or ill, the federal government is not reliably cooperative in granting the states access to their lands. Finally, the states are frequently unable to manage their lands according to their own priorities when they conflict with the BLM plans for federal lands. For example, 85 percent of the 6 million acres granted to Utah at statehood are held in isolated single sections "which are all but immune from any logical state management strategy."[20] Exchanges to "block in" state lands and eliminate the sources of conflicts are time–consuming, arduous undertakings, and infrequently consummated.[21] Utah's interesting land consolidation effort, "Project Bold," suggests that Secretary Watt's "Good Neighbor" policy, which emphasizes granting public lands to local governments for public purposes, has not been fruitless. Although Good Neighbor *grants* have become entangled in privatization *sales*, the emphasis on title adjustments has generated aggressive state action in at least one quarter.

Western governors have been actively asserting a number of additional claims that are only loosely, if at all, related to the partisan pleas of commodity users of the federal lands. For example, faced with the prospect of the long pending MX missile site in their states, officials in Utah and Nevada alleged that their position as equals of other states is inherently eroded by the extent of federal landownership. The fact that the federal government could plop such an albatross into their midst without much more than a ceremonial "by your leave" made them unavoidably diminished sovereigns when compared to states where federal initiatives are curtailed by lack of free access to federal land.[22] Other western governors have been similarly aroused when federal landholdings in their states have appeared to be targeted for controversial public projects, nuclear power plants, toxic dumps, and the like which, they argued, should be more evenly distributed among the sovereign states.[23]

Less episodically, western governments lament, to some avail, the loss of control over development and potential revenues incurred when resources are extracted from federal lands within their states. There is growing disparity between investments and returns on public lands managed by the BLM. The BLM budget is low and declining, while public land use and receipts are increasing rapidly. Shortfalls in funding mean backlogs in handling applications for use of and access to public lands, inadequate inventory and planning activities, and limited monitoring and protection programs. Moreover, the disparity translates into a flow of resources out of the states. Leaving aside redistribution of reclamation funds and the high–yielding "O & C" revenues to counties containing BLM Oregon and California revested lands, approximately 42 percent of the BLM receipts are returned to the states annually.[24] Very little of this money is returned for mitigation, management, or maintenance in the affected areas. Generally speaking, grazing revenues and a percentage of geothermal receipts are "returned" to the county of origin. The rest of the monies are placed in the general school fund. The fact that the state "as a whole" gets a major share of the revenues bears no necessary and direct relationship to the local fiscal, economic, and social impacts of public landholdings. Although there are grounds for arguing that adverse impacts should be mitigated and productive capacities maintained, some areas would be rich indeed, if the revenues from minerals extraction were returned to the counties. Hence while neither the just nor the pro–state resolution to this problem is clear, the opportunity for state hostility to federal land management is obvious.

Not all conflicts are so limited, of course. Although the Sagebrush Rebellion was widely and appropriately viewed as a western event, it was in part a western branch of a continuing national movement. This section could be expanded considerably by even a partial recitation of the degree to which federal mandates in diverse areas have defined directions, options, priorities, and expenditures in state programs. In areas as diverse as aid to the disabled, air pollution, education, prison reform, and a host of others, the states have complained bitterly of a federal influence that has become, in the words of the Advisory Commission on Intergovernmental Relations (ACIR), increasingly "pervasive, more intrusive, more unmanageable, more ineffective, more costly, and, above all, more unaccountable."[25] The ACIR also notes that the states have become increasingly active and effective as "resisters of federal regulatory activities."[26]

The resistance works. Witness the success of western governors in forcing President Carter's much touted "Opinion of the Solicitor on Non–Indian Federal Reserved Water Rights" into the shredder.[27] Perhaps more startling is the recent Supreme Court decision upholding California's finely crafted nuclear power plant siting act, in the face of facially unassailable claims that the field requires a uniform federal standard.[28] Data gathered for a recent ACIR report indicate that the last two decades have seen an unprecedented growth in capacity at the state level.

> Those who looked at the states in the 1930s or even in the early 1960s and decided that they lacked the capability to perform their roles in the federal system because they operated under outdated constitutions, fragmented executive structures, hamstrung governors, poorly equipped and unrepresentative legislatures, and numerous other handicaps should take another look at the states today. The transformation....has no parallel in American history.[29]

The enhanced state respectability and capability in this role are dramatized by yet another major federalism case, *National League of Cities* v. *Usery*, in which the Supreme Court found limits on the reach of the federal government's authority under the Commerce Clause for the first time in forty years.[30] It held that although the general question of regulation of wages and work hours was well within congressional power, federal regulation of wages and hours of state and local government employees was not constitutionally permissible; the court reasoned vaguely that it would infringe upon the power of "states as states."

The decision has not been significantly clarified or relied upon in subsequent decisions. Yet, the bold assertion of expansive federal authority in *Kleppe* combined with the startling if ambiguous restriction on the commerce power in

Usery have suggested to many that the federalstate conflict is the dominant characteristic of nascent trends in American federalism. One might formulate a hypothetical case of *Kleppe v. Usery* as a guide to approaching Sagebrush Rebellion–type public lands issues.[31] My purpose is to argue that to do so would be understandable, but erroneous.

II. The BLM: Unconvincing Goliath

Federal–state relations are rarely purely conflictual because, except in very rare circumstances, the players are neither willing nor able to force the issue into a conflict situation. In quite exotic cases it is perfectly plausible to envision gladiators, such as a Governor and an Interior Secretary, locked in dubious battle and asserting the prerogatives of their offices and levels of government. Normally, that just does not happen. First, *the* federal position does not emerge with clarity and finality from individuals or some office in Washington to fall like a maze or strait jacket on the struggling states. Federal authority is not asserted apart from the day–to–day implementation of federal programs. It is inevitably an increment in a crowded policy arena with all the earmarks of data gaps, conflicting purposes, and institutional capacities. Only rarely is a specific federal agency in a position to assert "federal" authority fully, and the BLM is particularly unlikely to be loosed as a spear carrier for some federal cause. Second, the sovereign may not try to assert its rights. In this connection FLPMA is, *Kleppe* notwithstanding, a fascinating study of federal deference to state and local priorities. Third, there are other actors, including but not limited to the localities, that impinge upon potential federal–state conflicts. Finally, augmenting the previous three points, there are idiosyncratic aspects of any policy arena or decision which make general models problematic.

In the present context it is sufficient to reiterate that the BLM is not in much of a position boldly to assert federal authority "without limit" or to follow through on such an assertion made in its behalf by some other party. Although the problem of inadequate legal authority for the Bureau was apparently solved by the passage of FLPMA, the agency continues to lack institutional credibility and professional authority. In an imaginative effort to assess the power of seven major executive agencies in the resources field, Nienaber and McCool devised a ranking system based on federal resource agency executives' perceptions of sister agencies' efficacy on four variables. In dealing with Congress, the executive branch, and interest groups, as well as responding to NEPA, bureaucratic peers ranked the BLM significantly lower than the other six agencies.[32]

The perceived inability of the Bureau to respond to NEPA reflects a second and very much related problem for the Bureau: it does not have a data base on which to rely in making and defending its management decisions. For example, current BLM grazing impact statements are typically based on data gathered in a single season simply because the Bureau does not have condition, trend, or use data over time. Because range productivity varies widely depending on the weather, among other things, the Bureau's analyses are unconvincing and its management decisions are vulnerable, politically if not legally.

The Bureau's position as a federal standard bearer is further eroded by its related inability to defend itself in executive agency infighting. Part of the problem stems from the unsurprising fact that the federal government does not always speak with one voice. Any federal agency may be limited in its ability to press its authority to or near the theoretical limits—or even to assert its authority at all—by the fact that there are numerous other federal agencies with different, frequently conflicting, interests operating in the same arena. The Bureau has more to contend with, however, than this standard jostling. For reasons probably not unrelated to Nienaber's data, it usually loses in in-house dustups. The Alaska land settlement, in which over 100 million acres of "national interest" lands were identified for special management, was such an enormous defeat for the Bureau that many people failed to see the BLM as a contender. It never had a prayer as its responsibility for much of the State of Alaska was reallocated to various competitors. Similarly, the Bureau took a severe drubbing, spiritually and in the press, as a particularly feckless Secretary of the Interior tried yet again to regain the long–lost Forest Service for his department. Cecil Andrus argued long and hard that the BLM needed a "center of excellence" and bargained away significant portions of the Bureau's authority over coal leasing in an effort to have the Forest Service reorganized into his domain. He failed but the Bureau was not well served by the process.

Finally, it is worth noting that, in terms of asserting federal authority over federal lands, the Bureau is in a very awkward position because of its lands. Popular concepts about "federal" resources appear to have developed in the context of the National Park Service and the National Forest Service lands. Although those two agencies have isolated difficulties due to "inholdings" (occasional privately owned parcels within their administrative boundaries), the BLM is in just the opposite position. It has occasional areas of contiguous or "blocked in" ownership, but the more typical pattern is that BLM lands are fragmented parcels admixed with state, private, and other federal lands in a crazy quilt of landowner goals, resources, and constraints. Moreover, and again unlike the more familiar federal lands of the National Parks and Forests, the lands under BLM authority are not unencumbered. Generally they are leased to private users under a system of grazing permits and many permittees regard the allotments as their own. Although the permittees' view is clearly not true from a legal standpoint, the issue is sufficiently controversial to create problems for the Bureau.

The fact that a federal agency is not necessarily in a position to assert authority fully within an administration or in terms of its resources for dealing with external adversaries must be added to another equally obvious fact: the sovereign does not always *choose* to exercise the available authority. *Kleppe* is a particularly bad muse in this connection, for in FLPMA Congress has explicitly abjured authority without limit and instructed the Bureau to manage the federal lands in cooperative consultation with all affected interests and government bodies.

This directive has two components, neither of which approximate defining identifiable boundaries on the federal role. The first and most ambiguous is the emphasis given to public involvement in the statute. Dithyrambs on this topic are the ubiquitous feature of the 1970s congressional enactments. Nevertheless, the FLPMA provisions appear to have gone a significant step beyond the standard in this genre by calling on the Bureau to involve the public not only in the planning but also *management* of federal lands. Specific Bureau efforts, prior to the passage of FLPMA, to remove those words failed, probably adding to their weight albeit without clarifying their meaning. It does suggest, however, that Congress specifically declined to impose its will, in preference to that of interested citizens, on the public lands.[33]

Even clearer congressional deference—this time to states and localities—is found in a bundle of provisions, each stating in a slightly different way that BLM plans should be mindful of, coordinated with, considerate of, or consistent with state and local plans and regulations. The question of whether or not a BLM plan is consistent with state and local plans and regulations is ultimately resolved by the Secretary. That fact has caused many to look with dismay on the "consistency" requirements. It is true that they appear to give the Secretary final say in a litigated dispute, but it only *appears* that way; more importantly, very few day–to–day land management decisions get litigated. Although power under present court interpretation of the property clause may be boundless in theory, in practice Congress has essentially demurred and relied explicitly on a locally oriented negotiations process called land–use planning.

Congress was, it seems reasonable to suggest, reluctant to go out on a limb by defining clear programs and priorities in a contentious policy arena notably lacking in consensus. For much the same reasons, one can reasonably suppose that BLM personnel will not always consider bold assertion of federal prerogatives to be their top priority. Nor, in fact, are BLM field personnel likely to conceive of the issues with which they deal in that light. More apt to be trained in natural sciences than in things political, BLM personnel tend to talk in terms of cooperative approaches to joint management problems rather than protection of federal prerogatives.[34] Turf guarding is readily apparent, but it is not as clearly patterned as the federal–state conflict model would suggest. BLM personnel are not infrequently allied with nonfederal or private groups in pursuit of a management goal. Similarly, the Bureau's perceived rivals are frequently other federal agencies with whom the Bureau is unlikely to ally in defense of some vague notion of federal supremacy.

All of this suggests a final and critical inadequacy in the federalstate conflict model. There *are* other actors who do not readily fit into one camp or the other. The localities are critical and obvious; under FLPMA's consistency and public involvement provisions, they have an importantly enhanced and protected role. It is not entirely clear, in fact it is thoroughly unclear, what the policies and regulations are with which the BLM must be consistent. However, in areas where county, regional, or other bodies have well–developed planning and environmental protection programs, the potential for local leadership is great.

Less obvious and more interesting is the role that private interest groups are increasingly playing in the shifting of patterns of authority. Whereas conservation and environmental groups have in the past looked primarily to Congress and the courts for vindication of their interest, they have become increasingly sophisticated in seeking alternative forums and are frequently attentive participants in BLM planning. Similarly, stock operators have begun to diversify their reliance on local advisory boards and congressional delegations from the western states; they have gained considerable ground in the courts of late. This forum shopping, long familiar in the realm of litigation, requires that analysts keep a keen eye on specific issues. As abilities and ambitions of state and local officials have increased, so has the tendency for advocates opportunistically to seek the best forum.

Conclusion

Appearances to the contrary notwithstanding, the Sagebrush Rebellion and its aftermath should not be seized upon as a signal that a federalstate conflict is about to become the dominant theme in western resource management. The conflict will be significantly more diverse and inclusive than the model suggests, and patterns of interaction will be determined more by opportunism than by any preconceived notions of what different levels of government can or should undertake. Because the Bureau is in a very poor position to push federal authorities to some theoretical limit, it should take advantage of this period of flux in federalism to develop new institutional approaches to its considerable management problems.

This might include a recognition of the fact that since the Taylor Grazing Act in 1934 the congressional approach to BLM lands has been different from prevailing dogma surrounding the National Forests. Unlike the U.S. Forest Service, which essentially manages "its" lands as a proprietor, the BLM has been directed by Congress and required by the nature of the lands and its limited budget and personnel to achieve its goals working through and with local interests and allotment holders. Throughout its institutional history, the BLM has orchestrated an effort to blur those key distinctions as it sought to gain the resources, respect, and authorities associated with Forest Service multiple–use management. Changing expectations about the nature of federal authority as compared with state and local priorities suggest that this tactic may have decreasing utility in the future. Events such as the Sagebrush Rebellion also indicate that the Bureau ought to be exploring a full range of cooperative mechanisms— including but definitely not limited to lease–type arrangements with private livestock operators—to achieve its conservation goals. In a period when Big Brother has eroding appeal, the BLM would be well advised to develop its capacities as neighbor and friend.

Notes

1. Harry N. Scheiber, "Federalism and the Legal Process: Historical and Contemporary Analysis of the American System," 14 *Law and Society Review* 633 (1980).

2. See, for example, Clayton, "The Sagebrush Rebellion: Who Should Control the Public Lands?" *Utah Law Review* 505 (1980); Haslam, "Federal and State Cooperation in the Management of Public Lands," 5 *J. Contemporary Law* 149 (1978); and Touton, "The Property Power, Federalism, and the Equal Footing Doctrine," 80 *Columbia Law Review* 817 (1980).

3. Touton, op. cit., at 817–33.

4. *Kleppe* v. *New Mexico*, 426 U.S. 529, 539 (1976).

5. Fairfax, "Coming of Age in the Bureau of Land Management: Range Management in Search of a Gospel," National Academy of Sciences/NRC, Committee on the Development of Strategies for Rangelands Management, Washington, D.C., 1981.

6. C. Yale, S. Fairfax, and R. Twiss, "Federalism and the Public Lands: The State's Role in Managing BLM Lands in California," prepared for California State Lands Commission, November 1981. A portion of this analysis appeared in Fairfax, "Old Recipes for New Federalism," 12 *Environmental Law* 945, 968–78 (1982).

7. Scheiber, op. cit., at 660, citing Justice Black.

8. This account draws upon Fairfax, *supra* note 5.

9. S. Dana and S. Fairfax, *Forest and Range Policy* (McGraw–Hill, Second Edition, 1980), 192–94.

10. Senzle, "Genesis of a Law: I and II," 80 *American Forests*, January and February 1978.

11. M. Clawson, *The Bureau of Land Management* (Praeger, 1971), 50.

12. Harvey, "Public Land Management Under the Classification and Multiple Use Act," 2 *Natural Resources Lawyer* 238 (1969).

13. Id. at 245.

14. R. Nelson, *The New Range Wars: Environmentalists versus Cattlemen for the Public Rangelands* (Department of the Interior, Office of Policy Analysis, 1980).

15. *Natural Resources Defense Council* v. *Morton*, 388 F. Supp. 840 (1974).

16. R. Nelson, *Basic Issues of Public Range Management in the 1980s* (Department of the Interior, Office of Policy Analysis, 1980), 114–17.

17. Touton, *supra* note 2, at 824 (footnote 53 and material cited).

18. Fairfax, "Riding Into a Different Sunset," 79 *Journal of Forestry* 516 (1981).

19. Yale, Fairfax, and Twiss, *supra* note 6.

20. State of Utah, "Proposal to Exchange and Consolidate State and Federal Land Holdings" (Project Bold), Salt Lake City (n.d.), p. 6 (mimeographed).

21. Webb, "Utah's Big Old Land Deal," 13 *Environmental Action* 17 (1982).

22. Holland and Benedict, "The Great Basin States and the MX," in Richard Ganzel, ed., *Resource Conflicts in the West* (Nevada Public Affairs Institute, 1983), 60–79.

23. Kearney and Garey, "American Federalism and the Management of Radioactive Wastes," 42 *Public Administration Review* 14 (1982); and Tyson, "The Intergovernmental Cleanup at Love Canal: A First Crack at 'The Sleeping Giant of the Decade,'" 10 *Publius* 101 (1980).

24. Yale, Fairfax, and Twiss, "The Federal Land Policy and Management Act and the Sagebrush Rebellion: Cause or Cure?" in Ganzel, *supra* note 22, at 40–41.

25. Advisory Commission on Intergovernmental Relations, *In Brief: State and Local Roles in the Federal System* (B–6), (Washington, D.C., November 1981), p. 2, citing ACIR, *The Federal Role in the Federal System* (B-4), December 1980.

26. Id. at 12.

27. Tarlock and Fairfax, "Federal Proprietary Rights for Western Energy Development: Analysis of a Red Herring," 3 *Journal of Energy Law and Policy* 1 (1982).

28. *State Energy Resources Conservation and Development Commission* v. *Pacific Legal Foundation* (1981).

29. ACIR, In Brief, *supra* note 19 at 3.

30. Fairfax, "Old Recipes....," supra note 6.

31. Id.

32. J. Nienaber and D. McColl, "Agency Power: Staking Out Terrain in Natural Resources Policy," Department of Political Science, University of Arizona (December 1980), Chapter 5 (draft).

33. Achterman and Fairfax, "The Public Participation Requirements of the Federal Land Policy and Management Act," 21 *Arizona Law Review* 501 (1979).

34. Wilson and Lundberg, "Cooperative Management on the Public Rangelands," in Ganzel, *supra* note 22 at 94–106.

4

Cross-Jurisdictional Conflicts: An Analysis of Legitimate State Interests on Federal and Indian Lands

CHARLES F. WILKINSON

Introduction

During the last fifteen years the stakes have steadily gone up in the combat among the state, tribal, and federal governments on federal and Indian lands. Tensions have been aggravated by the familiar exertions of too many bureaucrats from various governments seeking to shoehorn their jurisdiction into every available vacant hollow. The conflict in the American West has also involved increasingly big money. Leaving aside the private fortunes to be made, so shallow a phrase as "jurisdiction" sets the ground rules for the annual distribution of billions of dollars of tax revenues and in–lieu payments.[1]

But the emotional stakes may be higher yet. We have seen it in the eyes of those westerners watching the helicopter lifts of wild horses and burros out of sage and juniper canyons—in the eyes of people like Wild Horse Annie, who have expended so much of their souls to provide some measure of benevolence to these animals, and in the eyes of ranchers who view the same animals as pests that steal valuable forage from cows and sheep and that erode away hillsides with their clumsy, plate–sized hoofs.[2] We have heard the essence of the conflict in the blaring horns of hard-working California North Coast loggers who organized noisy caravans of log trucks to drown out equally impassioned pleas of

Ellen Guerin and Valerie Lind Hedquist made major contributions toward the publication of this piece. I dedicate this chapter to Prof. Ralph W. Johnson of the University of Washington School of Law, who has as good a mind and spirit as the public and Indian lands know, or are likely to know. This chapter first appeared in the *UCLA Journal of Environmental Law and Policy*, Vol. 2, Spring 1982, Number 2.

environmentalists at congressional hearings on an expanded Redwood Park.[3] The California State Director of the Bureau of Land Management (BLM) is quite confident that he both saw and heard the millennial conflict when he stood on a sand dune in the Southern California desert during the hottest days of the off–road vehicle conflict and saw the members of the Desert Lily Lovers Society coming up one side of his sand dune and, coming up the other, the stalwarts of the Barstow Bombers.[4]

The level of emotional intensity has been at least as high over Indian issues, though there are fewer people and acres involved: while the United States owns over 30 percent of all land in the country, Indians own about two and one–half percent of all the nation's land, or 52 million acres.[5] The recent but already historic fishing rights dispute of the Pacific Northwest brought us vivid memories of good fishers, Indian and non–Indian, who had much in common but who saw themselves left with no alternative but to wage a seemingly ceaseless campaign of sit–ins, fish–ins, and resistances to arrest in order to preserve their respective livelihoods, traditions, and, in the case of the Indians, religions.[6] Ironically, they fought each other while knowing full well that the depletion of the fish runs was caused by decades of poor logging practices, overdevelopment of the watersheds, and collections of dams, which taken together clouded, warmed, and throttled most of the great salmon and steelhead streams from the Klamath River in California north to the Canadian border.[7] The Indian fish wars leave us with even more disturbing memories of state officials who sought figuratively to stand astride Puget Sound and the Columbia River in a manner all too reminiscent of the way in which Orville Faubus literally stood astride a schoolhouse threshold in Arkansas two decades earlier.[8]

Indian–state conflicts raise questions that may be even more elusive of resolution than the resource dilemmas. Exactly who, for example, should decide adoption and guardianship of young Indian children whose parents have by any standard gone awry—progressive, well–educated state social workers, typically with ample compassion for Indian people, or new tribal judges and newer–yet tribal social workers, who understand the traditions and needs of their people but typically lack formal education in the weighty task of assigning young children to new homes and new parents?

One can quickly see, then, that bland words like "jurisdiction", "regulation," and "governmental authority" fail to alert us to the real nature of the struggles on federal and Indian lands. These phrases all obscure a deeper well–spring: whether one group will be allowed to impose its ideology on others. Wild Horse Annie, the western rancher, the long– time Redwood Park hiker, the admirer of the mariposa lily, the Barstow Bomber, the Indian fisher, the steelheader, the tweed–jacketed professional social worker, the tribal elder, and scores of other groups and individuals have each found that a particular government—state, federal, or tribal—holds out the best hope for furthering their own deeply held ideologies. To all of them the fulfillment or frustration of their ideologies is a deadly serious business. The Sagebrush Rebellion is one, but not the only, result.

In this article, I will first summarize the legal and policy structure that allocates jurisdictional prerogatives and limitations—the balance of ideology— on and near the federal and Indian lands in the West. I will then assess the legitimate interests of the states on federal and Indian lands. The question of legitimate interests is pivotal: the fulfillment of those interests should be, and I think will be, at the center of the continuing debate over whether, and how, to alter the allocation of finances and ideology in this field.

I. Federal and Indian Jurisdiction over Adjacent Private Lands

Both the federal and Indian governments have authority to exercise jurisdiction on adjacent private lands as an incident of their power to regulate their own lands. The United States has infrequently regulated off the federal lands to further public lands policy but, where it has, federal power has been upheld.[9] Such authority seems plainly to exist under the Property Clause, which gives Congress broad powers to legislate over and protect federal lands.[10] The limits, as opposed to the existence, of Property Clause power over adjacent lands, may in the future pose more difficult questions. We may reasonably wonder, for example, what Congress and federal land managers can do to protect migratory wildlife whose habitat is primarily on federal lands or to preserve the peace and quiet on public lands in the face of noise, traffic, signs, and other activities on adjacent nonfederal lands.[11] But such questions have just begun to come to the fore in the past few years and are not likely to be asserted by the current administration, which seems little interested in those kinds of activist management practices.

Indian tribal governments have similar powers over private and holdings within their reservations. In March 1981 the Supreme Court, while recognizing that such tribal authority exists, seemed to impose fairly tight strictures on the power: activity on nontribal lands must directly affect some significant tribal interest or the tribe is without power.[12] The tough cases—those involving pollution and migratory animals that use tribal lands as part of their habitat—have not yet reached the Supreme Court. But they soon will, since tribal governments, unlike federal officials now in power, are avid to establish control over nearby lands when activities there affect tribal interests. In my judgment it is still too early to define the contours of those tribal interests that will be sufficient to allow tribal regulation of non–Indians on non–Indian land within reservation boundaries.

Indian child custody, adoption and similar proceedings present a dramatic departure from the principles just discussed. Tribes have often resolved those kinds of issues for on–reservation children. With the express sanction of Congress in the Indian Child Welfare Act of 1978, most tribes now have jurisdiction over these critical issues even though the children may reside off the reservations.[13] The 1978 act is surely constitutional even though it reaches into a sensitive subject matter historically committed to state jurisdiction.[14] To date, it is perhaps Congress' furthest reach into state authority under the Indian Commerce Clause and demonstrates the sweeping power of Congress to vest regulatory authority in tribal governments, even outside of Indian country.

Indian reserved hunting and fishing rights, Indian reserved water rights, and, to a much lesser extent, federal reserved water rights also affect private lands, though they do not normally involve regulation per se.[15] These rights are extraterritorial to Indian and federal lands, because persons on private land may be required to allow anadromous fish or water to remain in the stream even though they would be allowed to take the fish or capture the water under state law were it not for the reserved right. *United States* v. *New Mexico*[16] makes it clear that most federal lands carry only minimal reserved water rights.[17] But the probable limited impact of federal reserved rights does not apply to Indian resource rights. The Indian fishing cases, of course, have made their mark during the 1970s on non–Indian commercial and sport fishing and on the public consciousness.[18] A major resource issue in the 1980s and 1990s is likely to be attempts by tribes to exercise their superior reserved water rights in the American West, a region characterized by its geographic aridity and by its contemporary water crisis.[19]

II. State Jurisdiction on Federal and Indian Lands

In spite of the potential significance of federal and tribal jurisdiction on private lands, most of the emotions alluded to earlier have been vented over the reverse situation, state jurisdiction on federal and Indian lands. The subject has also been characterized by a number of misconceptions. We need to divide the inquiry into three separate kinds of lands: federal enclaves, federal resource lands, and Indian lands.

A. Federal Enclaves

Most federal enclaves were established in earlier eras when a need was perceived to create jurisdictional islands under exclusive federal control so that specified federal activities could be conducted without interference from the states. "Federal enclave" too often is wrongly used as a generic term to describe all public lands and even Indian lands. In fact, enclaves comprise only about 6 percent of all federal lands. They were created by agreements between states and the federal government in which the state in question expressly transferred its jurisdiction to the United States.[20] Enclaves include most post offices and federal office buildings, all military bases, some national parks, and other miscellaneous holdings. Almost without exception, resource development on federal enclaves is not an issue.

They are not making many federal enclaves today,[21] but the ones in existence continue to operate under so–called exclusive federal jurisdiction. It is an outmoded kind of installation. Recent cases have shown that the United States has ample protection from the states by virtue of its sovereign immunity, its superior sovereignty, and its ability to pass specialized laws to oust state jurisdiction when necessary.[22] Further, "pure" federal enclaves could often result in clumsy and ambiguous arrangements in the many situations where no federal law is applicable and state law could not be invoked due to the exclusivity of federal law. Congress has sought to ameliorate the inconvenience by voluntarily limiting federal exclusive jurisdiction to allow the operation of specified state regulatory laws and tax laws.[23] This "assimilation" of state law into enclaves is good lawmaking because the federal government is not equipped to exercise the broad range of authority we call the police power: health, safety, wildlife, criminal, civil, and commercial laws are all the business of the states, which have comprehensive codes to regulate such matters. In addition, cash payments are made to the states in lieu of taxation of federal enclave lands.[24] Thus, even in federal enclaves, states exert considerable legal influence and receive substantial revenues.

B. Resource Lands

For lack of a better term, all federal lands other than federal enclaves can be lumped under the category of resource lands. Unlike federal enclaves, states generally need no invitation or permission to extend their laws onto the approximately 700 million acres of resource lands. True, the states cannot directly tax or zone federal land, but that intergovernmental immunity of the United States is narrow.[25] True, the United States can preempt (or override) state laws, but preemption requires affirmative action by Congress.[26] In the meantime, state law governs.[27]

The result, contrary to the popular perception, is that state laws have extensive application on federal resource lands. That is exactly as it should be, for—as noted above—the states' business is to promulgate and enforce a comprehensive police power.

States have influence over, and receive benefits from, federal resource lands even in those areas where their laws do not control. As is the case with federal enclaves, Congress has voluntarily made adjustments to accommodate the states. Numerous statutes require the federal government to consult extensively with the affected states over resource development and land use planning matters.[28] And finances have hardly been ignored: recognizing the states' inability to impose a property tax on federal lands, Congress makes voluntary payments to the states. These payments totalled over $2 billion in 1979, the last year for which figures are available.[29]

But there is more, and we now move toward some mighty irony. July 2, 1981 may prove to have been a fateful day in the history of the American West. On that date the Supreme Court handed down its opinion in *Commonwealth Edison Co.* v. *Montana*,[30] upholding Montana's 30 percent severance tax on coal extracted in the state, including coal on the federal resource lands. This enormous (that is Justice Blackmun's phrase)[31] source of revenue may generate as much as $20 billion for Montana alone through the year 2010.[32] I say there is powerful irony here because the Sagebrush Rebels' complaints over the burdens western states must endure because of the presence of the public lands now begin to ring a bit hollow. Indeed, there are many who have begun calling the western states the "haves" and the remaining states the "have nots" in the critical area of energy, due to the concentration of easily accessible, low–sulphur federal coal and other fuel minerals in the West.[33]

C. Indian Lands

The law regarding state jurisdiction on Indian lands is markedly different because state jurisdiction is much more circumscribed, especially when Indians themselves are involved. For most, but not all, purposes state law does not operate within the boundaries of Indian reservations because of the long–standing federal commitment to protect tribal self–government.[34] The courts have been fairly quick to find that federal laws have occupied the field and that state laws have been excluded, especially when non–Indians are doing business with Indians.[35] When no specific federal law is controlling, various results in favor of or against state jurisdiction have been reached depending on whether the application of state law would "infringe upon tribal self–government."[36]

Where only Indians are involved, the courts have been extremely reluctant to allow any state or local jurisdiction because the application of a state law to a tribe or reservation Indian normally has some impact on tribal self–govern-ment.[37] The only state laws that have been upheld against Indians in Indian country during the modern era are requirements that Indians collect state taxes when sales of cigarettes are made to non–Indians; sales to Indians, however, are exempt from taxation.[38] States receive financial revenues from Indian reserva-tions in a variety of ways.[39]

The restrictive nature of state jurisdiction in Indian country is due to a number of legal and historical factors. The central reason is that Indian treaties were, at their essence, not just a guarantee of a tribal land base but also a promise of a tribal jurisdictional base. The treaties were intended to make jurisdictional islands of Indian reservations where the tribes could govern themselves and be free of local non–Indian pressures.[40] Those notions have been eroded somewhat and are not absolute today, but they continue to explain why the courts and Congress have generally been stingy in allowing state jurisdiction in Indian country.

III. Legitimate Interests of the States

The preceding discussion is a summary of the law as it is. I would like to spend some time now evaluating it in very general terms. The ultimate question is whether the existing structure substantially recognizes and fulfills the legitimate interests of the states.

To begin the discussion I must allude to the Sagebrush Rebellion. This movement is not new. It is a continuation of sentiments that were heartfelt when the first settlers cut over the Appalachians into the Ohio Valley: administrators in far–away Washington should not control settlers on the furthest reaches of the continent. That concern— variously phrased as local rights or state rights— raged in Alabama and Missouri in the 1830s, in Texas and Wisconsin in the 1840s, in California and Oregon in the 1850s, in Nevada and Colorado in the 1860s, and so on.[41] And it rages today. But that general concern for local control needs to be parsed. Some elements of that broad concern are legitimate and some are not.

The Sagebrush Rebellion is bankrupt on an essential issue. The most expan-sive claim of the Sagebrush Rebels is that the states are entitled as a matter of law to have the public lands in the West, or most of them, transferred to the western states. The Rebels usually argue that the original thirteen states do not have large blocs of federal lands within their boundaries and that the western states, as a matter of constitutional law and fairness, should be similarly situated.[42] This is misleading as a matter of history and law.

The original thirteen states owned—under settled principles of American property law—most of the land within their boundaries after the American Revolution. When they entered the union they refused to relinquish that ownership to the new federal government.[43] That was their prerogative, as property owners.

Then the United States acquired vast areas of land. The young nation negotiated the Louisiana Purchase with France, the Northwest Compromise with Great Britain, the Treaty of Guadalupe Hildalgo and the Gadsden Purchase with Mexico, the Alaska Treaty with Russia, and others.⁴⁴ The United States acquired that land in fee simple absolute—as complete owner—subject only to a small number of perfected private titles and to the rights of Indians to reside upon the land.⁴⁵ There were no states to claim ownership.

The United States began to create territories, then states, out of its public domain. Whenever a state was to be carved out of a territory, the United States made a bargain with the representatives of the territory. The bargaining for land was always fierce. In addition to smaller grants for specific purposes, the early states received grants for schools of one section in each township, roughly 1/36th of the land within the state. Later states normally were granted two school sections in each township. Utah, Arizona, and New Mexico received more and Alaska, in 1959, received more yet.⁴⁶ But it was always a bargain with explicit rules crafted by American property law;⁴⁷ the United States owned the land and states obtained whatever they were able to bargain for. After each statehood transaction most land in the region remained in federal ownership, just as it had been before the transaction. Most western states expressly agreed to "forever disclaim all right and title to the unappropriated public lands" within the state and to leave federal lands "at the sole and entire disposition of the United States."⁴⁸

A few cases during the mid–nineteenth century suggested, always in dictum, that the United States retained its land with the idea that it would later be transferred away to the states or private parties.⁴⁹ The Supreme Court was paraphrasing—with considerable accuracy—the nature of congressional policy at that time. Congressional policy has since changed, just as any property owner has the right to consider selling its property and then to recant when circumstances change. The general language in those two or three court opinions has never been a rule of law, has not been followed, and has been discredited in a modern context.⁵⁰

The United States owns the land. The states have no legitimate claim to ownership. But the existence of federal ownership does not settle policy issues of how much control the western states should have over land owned by the federal government within their boundaries.

Plainly state control cannot be justified simply by saying that federal land should be subject to state regulation because private land within state boundaries is subject to state law. This analysis first ignores the fact that in our federal system the United States is a sovereign superior to local, county, and state governments—a superiority born of necessity after the quantum of local power under the Articles of Confederation had proved too cumbersome.⁵¹ That essential supremacy is rooted in the Constitution and subject to no serious debate today. The unacceptable scattering effect of fifty superior governments is evident. Second, claims to state control must recognize that the ultimate owners of the public lands are literally all of the citizens of the United States. United States citizens in the West may be specially affected by the public land—for good and for bad—but all citizens have a stake in the western public lands and their resources: energy, hard rock minerals, timber, beef, water, recreation, and wilderness.

Nevertheless, citizens of western states can claim a special interest in public lands management. Because they live there, they may use, or be affected by uses on, the public lands to a disproportionate degree. Public lands policy more often and more directly affects their businesses and jobs; their hiking trails and snowmobile runs; their air, water, and vistas; and their tax rates.

Recognizing, then, a somewhat more diffuse national interest and a somewhat more direct western interest, how can we articulate the legitimate interests of the state governments on the federal lands? First, there is, in my view, an absolute right to economic equity. Federal installations cause local and state governments economic burdens (the costs of some roads, police and court systems, water and sewer service, and other services) and also economic benefits (federally constructed roads, fire protection, landing strips, and others). State and local governments should be made whole for the net costs of these financial burdens. This is no easy task. It is a tremendously technical and complex matter. As noted, the financial burdens imposed on the states by public lands policy are addressed by an extensive, shifting matrix of statutes.[52] Federal policy and law in this area should be continually re-evaluated as conditions change to be certain that the states are receiving economic equity.

What of the right asserted by the Sagebrush Rebels to control activities on public lands? There is sound basis for a state interest in regulating some conduct by private persons on the public lands; activities there may have impacts on private lands and, as noted, states have functioning systems to exercise the full range of police powers. But it flows irresistably from the fact of federal ownership and from the superior constitutional federal sovereignty that federal programs should not be subject to interference by the states. Thus states have a second broad legitimate interest, the right to exercise the police power over activities by private persons on the public lands up to the point that an ongoing federal program is thwarted. To give an example in resource development, it seems to me that a state should legitimately be able to regulate and control the environmental impacts of hardrock mining or mineral leasing but that it should not be able to zone federal land so as to prohibit that kind of activity outright.

Even in those areas of federal policy not directly subject to state control, the western states continue to have a legitimate interest in being heard and responded to when federal concerns do not rationally outweigh local concerns.[53] Though the eastern states' voice in public land policy can be adequately met by representation in Congress, that level of participation is insufficient for western states: access to Congress alone would not recognize the special western interest in the public lands and would not be sufficiently site–specific. It would not insure sufficient influence over on–the–ground decisions. The right of affected western governments to be heard must be expressly recognized by statutes and must be implemented in the land–management agencies. As noted, there is in place a range of statutory provisions that require federal land managers to consult with state and local interests and, where practical, to conform federal programs to state and local requirements.[54] Like the western interest in economic equity, the right to be heard should be continually evaluated and improved so that those western interests specially affected have a truly effective statutory right to be heard—early and at length. This would allow those often–conflicting state and local views to be meshed with more general (and also often–conflicting) national considerations.

Thus I would catalogue three legitimate interests of the states on the federal lands: the right to economic equity; the right to impose the police power where it is not inconsistent with federal programs; and the right, even when programs are outside the police power, to have special notice and an opportunity to be heard.

The legitimate interests of state and local governments are different with regard to Indian land. First, the federal role is not the same as with federal lands: Indian lands are held by the United States pursuant to a special trust relationship, not for the general public, but for Indian tribes and individuals.[55] Therefore, Indian lands are not public lands.[56] Federal policy, properly reacting to views of Indians expressed during treaty negotiations and ever since, has been to protect Indian tribes from state and local interests—to provide a buffer between Indians and racism, economic sharp dealing, and, more abstractly but no less real, ethnocentric views of non–Indians on how Indians should behave.[57] Further, the presence of tribal governments makes the states' responsibilities more modest in Indian country, where there are operating tribal legislatures, courts, police, natural resource agencies, social service bureaus, and often schools. Just as the burdens on the state in Indian country are markedly less, so are their powers. That diminished role flows directly from the special place of Indian lands in our history, law, and policy.[58]

The states plainly have an absolute right to economic equity in Indian country, as they do on the public lands. But on Indian lands the states cannot lay claim to being a special constituency. The special constituency there is the Indian people. Thus, on Indian lands, the legitimate state interests in participating in Indian policy is akin to the role of non–Westerners in public lands policy: state officials should have representation in the making of Indian policy but access is most appropriate in Congress, where broad decisions are made.

Day–to–day decisions on Indian lands should be made primarily by tribal officials and secondarily by federal officials, with only the narrowest state encroachment allowed when activities on Indian lands have extraordinary impacts on essential state interests.[59] Tribal governments are small and young; traditional governments were smothered by federal policy during the nineteenth century and were not allowed to breathe again until the 1960s.[60] Like any small or young governments–whether they be developing nations, federal territories and states in the old American West, or cities and counties—Indian tribes will have their growth stunted if their responsibilities are borne by others. These concerns, the keynote of both old treaties and new statutes, burn brighter and hotter than the interests of the states.

IV. Evaluation of Current Policy

And how do current policy and law comport with these legitimate interests of state and local governments? My own conclusion is that the public lands legislation of the 1970s, easily the most intense period of congressional scrutiny in this field in the nation's history, fairly reflects these legitimate interests.[61] The laws of the modern era are characterized, above all, by careful compromise among the many interests.[62] The states are accorded broad police power over private conduct in federal territory when the United States has not asserted its supremacy.[63] In several instances state authority extends to resource development.[64] As they should be, the states are denied landownership and policy control over those areas, mostly involving resource development, where the

United States has legislated pursuant to its paramount authority. But in spite of superior federal constitutional power, the system now in place is struggling mightily to assure the western states financial equity[65] and policy influence[66] on a highly preferential basis. The existing corpus of legislation, much of it recently enacted, needs to be refined, a process that will be a continuing one. But wholesale changes would be wrong; the present system, which respects local views while being premised on national supremacy, fairly reflects the overriding reality that these resources are national, not local, possessions.

Similarly, I would judge that the states' legitimate interests are being substantially achieved on Indian lands. Where no interests of the tribes or individual Indians are at stake, state police power can be exercised.[67] If there is some nexus with Indian interests, state law has a narrow ambit and tribal or federal law is likely to control. Unlike the states' broader role on the public lands, the states have no general statutory right to be consulted on decisions on Indian lands. This limited theater for the states befits a situation where tribal governments can fill any governmental void: "The tradition in Indian country is tribal, not state, police power."[68]

In broadest terms, then, the comprehensive scheme of public lands statutes allows substantial financial returns to the states, a considerable residuum of state police power over private activities where the federal government has not preempted state law, and broad–based consultation in those areas where the United States has legislated. This might be called a "mixed police power– intensive consultation" model. The statutes and court cases in the Indian field allow significantly less state participation, a "tribal dominant" model. These schemes are preferable to the total exclusion of state law because such an approach would fail to meet the states' legitimate needs. Other models, "state ownership" (title to land in the states) and "state supremacy" (title to public land in the United States and title to Indian lands with Indian tribes, but paramount legislative authority in the states) are in turn inappropriate because state interests are outweighed by clearly established supremacy, the need for a relatively uniform national development and conservation policy, and special Indian rights.

Yet another approach, tilted toward state control, is the "state law subject to federal veto" model found in the Coastal Zone Management Act.[69] That act allows states, with federal funding, to develop plans for the coastal zone that go into effect upon approval by the Secretary of Commerce. But coastal zone planning, which is entirely appropriate in its own context, is fundamentally distinguishable because most lands in the coastal zone are not in federal ownership so that the proper starting point is traditional state police power. Indeed, federal lands in the coastal zone are excepted from coverage under the act.[70]

Calls for revamping the current system to allow for increased state control come from many sides—Sagebrush Rebels pressing to expand development opportunities,[71] non–Indian groups aiming to limit tribal powers,[72] even environmentalists seeking to curtail the authority of a development–oriented administration in favor of western states that show signs of becoming increasingly conservation–minded.[73] Policymakers should take a larger view. Change should be made on principle, not in response to the personalities on the stage at a particular moment in time. Progressive, newly formulated policies have been conceived during a decade of intensive reflection and formulation. They should be given time to mature in an atmosphere substantially devoid of the disruptive effects of wholesale legislative restructuring.

Neither public land policy nor Indian policy, then, needs still more statutes and land transfers. The main outlines of the present system preserve vital prerogatives of the national public and Indian tribes while assuring fairness to the western states on both federal and Indian lands.

Conclusion

Ultimately, we have learned from recent federal land and Indian policy that not all great experiments in the laboratory that is federalism occur at the state and local level. There are many examples of creative federal contributions, but a leading one is so bold and idealistic as to amount to an attempt to disprove the premise of the Turner thesis.[74] After the passage of the Alaska Lands Act in late 1980, we—who were the first nation in the world to experiment with legislatively protected wilderness—now have some 80 million acres of land in wilderness.[75] That is 4 percent of all land in the country. You can add to it roadless land of a like quantity in national parks, forests, and wildlife refuges. This is not a frontier to live in, which is the point that Turner made. But it is a frontier to be in, so powerful a wild frontier as to enflame the minds of many of us though we be hundreds or thousands of miles from it.

Our Indian land policy, with all of its horror, is still considered the most progressive of any nation toward its aboriginal people.[76] Today there are still the Ye-be-chei dances on the Navajo Reservation, the Sun Dance in South Dakota, and the Grey Horse dances in Oklahoma. And there are still discrete groups to challenge the premises of the majority society, to stand as bastions against a world that is too fast, too rude, and too materialistic. This geographic and cultural diversity is one of the proudest elements of public land and Indian policy.

So, in sum, we risk a good many dangers when we fail fully and fairly to protect the legitimate interests of the state and local governments in the West. But we take risks, too, when we accord those governments more than their legitimate interests. No, the public and Indian lands are not a burden on the American West. They are the hallmark of the American West. They are perhaps the most distinctive and positive and glorious elements of the way of life in the West. Without them, the American West really would pass on.

Perhaps our descendants in misty generations hence will have to answer tragic questions: How much was it worth to have lost the frontier? How much was it worth to have lost the culture of another people? But I continue to hope and believe that these and similar questions are ones to which no American will ever be held to answer.

Notes

1. See, e.g., Advisory Comm. on Intergovernmental Rel., Payments In Lieu of Taxes on Federal Real Property (1981); U.S. General Accounting Office, Report to Congress: Assessing the Impact of Federal and State Taxes on the Domestic Minerals Industry

(1981); Advisory Comm. on Intergovernmental Rel., The Adequacy of Federal Compensation to Local Governments for Tax Exempt Federal Lands (1978).

2. Trueblood, "Disaster on the Western Range," *Field and Stream,* January 22, 1975, at 14; Weiskopf, "Wild West Showdown," *Sports Illustrated,* May 5, 1975, at 87; see generally Wild Free–Roaming Horses and Burros Act, 16 U.S.C. §§ 1331–1340 (1976); Note, *Constitutionality of the Free Roaming Wild Horses and Burros Act: The Ecosystem and the Property Clause in Kleppe v. New Mexico,* 7 Envtl. L. 137 (1976).

3. See generally Fraker & Lubenow, "Redwood Protest: Loggers' Demonstration Against Major Expansion," *Newsweek,* Apr. 25, 1977, at 30; "Giant Battle in Redwood Country," *Bus. Week,* April 25, 1977, at 30; "Logger Outcry Obscures Slow Death of Ancient Park Redwoods," *Nat'l Parks & Conservation Mag.,* June, 1977, at 22. For the Redwood Park litigation, see *Sierra Club* v. *Department of the Interior,* 424 F. Supp. 172 (N.D. Cal. 1976); *Sierra Club* v. *Department of the Interior,* 398 F. Supp. 284 (N.D. Cal. 1975); *Sierra Club* v. *Department of the Interior,* 376 F. Supp. 90 (N.D. Cal. 1974); Hudson, *Sierra Club* v. *Department of Interior: The Fight to Preserve Redwood National Park,* 7 Ecology L.Q. 781 (1979).

4. Jim Ruch (California Director for the Bureau of Land Management), untitled address, reprinted in *The Public Trust Doctrine in Natural Resources Law and Management: Conference Proceedings* 204 (H.C. Dunning ed. 1981).

5. For Indian land statistics, see *U.S. Dep't of Commerce, Federal and State Indian Reservations* (1975). For public land holdings, see *Bureau of Land Management, U.S. Dep't of the Interior, Public Land Statistics* 10 (1977).

6. "American Friends Service Committee." *Uncommon Controversy* 108–13 (1970).

7. See, e.g., A. Netboy, *The Columbia River Salmon and Steelhead Trout: Their Fight For Survival* 142–147 (1980).

8. This comparison was expressly made by the Ninth Circuit Court of Appeals in language approved by the Supreme Court. Quoting from *Puget Sound Gillnetters Ass'n v. United States Dist. Court,* 573 F.2d 1123,1126 (9th Cir. 1978), the Court made this observation concerning the conduct of Washington state officials:

> The state's extraordinary machinations in resisting the (1974) decree have forced the district court to take over a large share of the management of the state's fishery in order to enforce its decrees. Except for some desegregation cases, the district court has faced the most concerted official and private efforts to frustrate a decree of a federal court witnessed in this century. The challenged orders in this appeal must be reviewed by this court in the context of events forced by litigants who offered the court no reasonable choice.

Washington v. *Washington State Commercial Passenger Fishing Vessel Ass'n,* 443 U.S. 658, 696 n.36 (1979) (citations omitted). See also *United States* v.*Washington,* 520 F.2d 676 (9th Cir. 1975), *cert. denied,* 423 U.S. 1086 (1976), where the concurring opinion stated:

> The record in this case, and the history set forth in the *Puyallup* and *Antoine* cases, among others, make it crystal clear that it has been recalcitrance of Washington State officials (and their vocal non-Indian commercial and sports fishing allies) which produced the denial of Indian rights requiring intervention by the district court. This responsibility should neither escape notice nor be forgotten.

502 F.2d at 693 (Burns, J., concurring).

9. E.g., *Camfield* v. *United States,* 167 U.S. 518 (1897) (upholding a statutory prohibition against fences on private lands that limit access to public lands). Federal control over nonfederal lands may also be achieved by administrative regulations, rather than by express action in a statute, if the regulations are adopted pursuant to delegated authority from Congress. *United States* v. *Brown,* 552 F.2d 817 (8th Cir.), *cert. denied,* 431 U.S. 949 (1977) (prohibition of hunting by Park Service); *United States* v. *Lindsey,* 595 F.2d 5 (9th Cir. 1979) (prohibition of fires by Forest Service).

Private activities can also be affected—though not, technically, regulated—by the establishment of federal reserved water rights. *Cappaert* v.*United States,* 426 U.S. 128 (1976).

Federal reserved rights for purposes other than Indian reservations generally appear to be limited in scope. See *United States* v. *New Mexico*, 438 U.S. 696 (1978), and the authorities cited infra in notes 16 and 17.

The most recent court case involved a 1978 statute prohibiting motorized vehicles on some nonfederal holdings within the Boundary Waters Canoe Area Wilderness in Minnesota, *Minnesota* v. *Block*, 660 F.2d 1240 (8th Cir. 1981), *cert. denied*, 102 S. Ct. 1645 (1982) (upholding congressional prohibition of motorboats and snowmobiles within designated areas of the Boundary Waters Canoe Area Wilderness). See Generally Gaetke, *The Boundary Waters Canoe Area Wilderness Act of 1978: Regulating Non--Federal Property Under the Property Clause*, 60 Or. L. Rev. 157 (1981).

10. The Property Clause of the Constitution provides that "Congress shall have Power to dispose of and make all needful Rules and Regulations respecting the Territory or other Property belonging to the United States. . . ." U.S. Const. art. IV, § 3, cl. 2. See *Kleppe* v.*New Mexico*, 426 U.S. 529 (1976) (upholding the Wild and Free Roaming Horses and Burros Act on the ground that the Property Clause authorizes Congress to protect wildlife living on public lands).

11. For a discussion of the power of Congress and, more specifically, the Park Service, to regulate private lands adjacent to the national parks, see Sax, *Helpless Giants: The National Parks and the Regulation of Private Lands*, 75 Mich. L. Rev. 239 (1976). See also supra the Redwood Park opinions, note 3. The Park Service has attempted, although not always successfully, to eliminate "aesthetic nuisances" adjacent to the parks. See, e.g., *United States* v. *Arlington County*, [12 Decisions] Env't Rep. (BNA) 1817 (1979) (Arlington Tower); cf. *New Jersey Builders Ass'n* v. *Dep't of Envl. Protection* [13 Decisions] Env't Rep. (BNA) 1541 (1979) (Pine Barrens). See generally Gaetke, *Congressional Discretion Under the Property Clause*, 33 Hastings L.J. 381 (1981).

12. *Montana* v. *United States*, 450 U.S. 544, 565–66 (1981) (no tribal authority to regulate fishing by non–Indians on non–Indian land where the tribe was not historically a fishing tribe and where the tribe had long acquiesced to state regulations of fishing by non–Indians). A subsequent case stands for considerably broader tribal powers than suggested by *Montana*, although tribal regulation over non–Indians lands was not directly involved. *Merrion* v. *Jicarilla Apache Tribe*, 102 S. Ct. 894 (1982). Four circuit court opinions since *Montana* have upheld tribal authority over non–Indians. *Colville Confederated Tribes* v. *Walton*, 647 F.2d 42 (9th Cir. 1981), *cert. denied*, 102 S. Ct. 657 (1981); *Confederated Salish & Kootenai Tribes of the Flathead Reservation* v. *Namen*, 665 F.2d 951 (9th Cir. 1982); *Cardin* v. *De La Cruz*, 671 F.2d 363 (9th Cir. 1982); *Knight* v. *Shoshone & Arapahoe Indian Tribes*, 670 F.2d 900 (10th Cir. 1982).

13. Indian Child Welfare Act of 1978, 25 U.S.C. §§ 1901–1963 (Supp. I 1981). See generally Guerrero, *Indian Child Welfare Act of 1978*, 7 Am. Indian L. Rev. 51 (1979).

14. E.g., *In the Matter of the Guardianship of D.L.L. and C.L.L.*, 291 N.W.2d 278, 281 (S.D. 1980). Congressional power under the Indian Commerce Clause, U.S. Const. art. I, § 8, cl. 3, is very broad. See, e.g., *United States* v.*John*, 437 U.S. 634 (1978); *United States* v. *Antelope*, 430 U.S. 641 (1977); *United States* v. *Kagama*, 118 U.S. 375 (1886). The Supreme Court has never found a congressional action to be beyond the reach of congressional power under the Indian Commerce Clause.

15. Some tribes, however, have sought to directly regulate non–Indian water rights on private land. See, e.g., *Colville Confederated Tribes* v. *Walton*, 647 F.2d 42 (9th Cir. 1981), *cert. denied*, 102 S. Ct. 657 (1982) (upholding tribal and federal authority).

16. 438 U.S. 696 (1978) (holding that the federal reservation of the Gila National Forest did not impliedly reserve a minimum instream flow for "aesthetic, environmental, recreational, or "fish' purposes"). However, the majority opinion stated in dictum: "Where water is necessary to fulfill the very purposes for which a federal reservation was created, it is reasonable to conclude, even in the face of Congress' express deference to state water law in other areas, that the United States intended to reserve the necessary water." Id. at 702.

17. Exceptions include some low–elevation wildlife refuges established early in this century, which may force cutbacks in irrigation on adjacent private lands. Federal Water Rights of the National Park Service, Fish & Wildlife Service, Bureau of Reclamation & the Bureau of Land Management, 86 Interior Dec. 553, 602–04 (1979) (Solic. Op.).

18. See, e.g., Jones, "Clamor Along the Klamath," *Sports Illustrated*, June 4, 1979, at 30; Nash, "Chippewas Want Their Rights: Disputes Over Hunting and Fishing Regulations on Reservations in Michigan and Minnesota," *Time*, Nov. 26, 1979, at 54; Starnes, "New Indian Ripoff," *Outdoor Life*, Oct. 1979, at 15.

19. See generally Back & Taylor, *Navajo Water Rights: Pulling the Plug on the Colorado River?*, 20 Nat. Resources J. 71 (1980); DuMars and Ingram, *Congressional Quantification of Indian Reserved Water Rights: A Definitive Solution or a Mirage?*, 20 Nat. Resources J. 17 (1980); Note, *Adjudication of Indian Water Rights: Implementation of the 1979 Amendments to the Montana Water Use Act*, 41 Mont. L. Rev. 73 (1980); Note, *Indian Reserved Water Rights: the Winters of Our Discontent*, 88 Yale L.J. 1689 (1979); Pelcyger, *The Winters Doctrine and the Greening of the Reservations*," 4 J. Contemp. L. 19 (1977).

Popular attention has also focused on the issue of Indian water rights. See, e.g., Gebhart, "Who Owns the Missouri?," *Progressive* 44 (1980); *N.Y. Times*, Dec. 17, 1980, at 15, col. 1; see also, Boslough, "Rationing a River," *Science '81*, June 1981, at 26.

20. On federal enclaves, see generally G. Coggins & C. Wilkinson, *Federal Public Land and Resources Law* 144–60 (1981).

21. In 1940 Congress curtailed the establishment of federal enclaves:

> [T]he flood of transfers of legislative jurisdiction was stayed, by an amendment to section 355 of the Revised Statutes of the United States which eliminated the presumption of Federal acceptance [40 U.S.C. § 255 (1977)]....This ended a period of 100 years during which the Federal Government, with relatively minor exceptions, acquired legislative jurisdiction over substantially all of its land acquisitions within the States.

U.S. Dep't of Justice, Federal Legislative Jurisdiction 49–50 (1969).

22. E.g., infra notes 25 & 26.

23. Enclaves are said to be under "exclusive" federal jurisdiction but in fact state laws have considerable effect in enclaves. Federal jurisdiction in an enclave can be less than exclusive if the state reserved regulatory jurisdiction when the installation was created. *Collins* v. *Yosemite Park & Curry Co.*, 304 U.S. 518 (1938). Criminal cases arising in enclaves go to federal court, but state substantive criminal law applies if no federal statute is on point. 18 U.S.C. § 13 (1976); *United States* v. *Sharpnack*, 355 U.S. 286 (1958). National parks often incorporate state fishing laws and state fishing licenses are required. The Buck Act allows states to collect income, gasoline, sales, and use taxes in federal enclaves. 4 U.S.C. §§ 104–110 (1976). Congress also has provided that state unemployment and workers' compensation laws apply. 26 U.S.C. § 3305(d) (1976); 40 U.S.C. § 290 (1976). A wide range of other laws apply, including divorce, motor vehicle, child custody, and voting provisions. Impact Aid is provided to some local school systems. See generally *Evans* v. *Cornman*, 398 U.S. 419 (1970).

24. See the reports of the Advisory Commission on Intergovernmental Relations, supra note 1.

25. E.g., *United States* v. *County of Fresno*, 429 U.S. 452 (1977) (upholding California tax on federal employees on their possessory interests in housing owned and supplied to them by the federal government as part of the employees' compensation).

26. E.g., *Kleppe* v. *New Mexico*, 426 U.S. 529 (1976) (upholding federal preemption of state laws relating to wild horses and burros).

27. See generally Wilkinson, *The Field of Public Land Law: Some Connecting Threads and Future Directions*, 1 Pub Land L. Rev. 1, 19–23 (1980).

28. E.g., Federal Land Policy and Management Act of 1976, 43 U.S.C. §§ 1720, 1721(c), 1733(d), 1747, 1765(a) (1976); National Forest Management Act of 1976, 16 U.S.C. § 1612 (1976); Fish and Wildlife Coordination Act, 16 U.S.C. § 661 (1976); Intergovernmental Cooperation Act of 1968, 42 U.S.C. §§ 4231–4233 (1976). State wildlife laws have pervasive application on federal resource lands. Coggins, *The Law of Wildlife Management on the Federal Public Lands*, 60 Or. L. Rev. 59 (1981).

29. Advisory Comm. on Intergovernmental Rel., Payments in Lieu of Taxes on Federal Real Property, supra note 1, at 47.

30. 453 U.S. 609 (1981).

31. Id. at 641 (dissenting opinion of Justice Blackmun).

32. Id.

33. E.g., Science and Public Policy Program, University of Oklahoma, *Energy from the West: A Technology Assessment of Western Resource Development* (1981).

34. See *White Mountain Apache Tribe* v. *Bracker*, 448 U.S. 136 (1980); *McClanahan* v. *Arizona State Tax Comm'n*, 411 U.S. 164 (1973). See generally F. Cohen, *Handbook of Federal Indian Law* 270–79 (1982 ed.).

Under "Public Law 280," states have civil and criminal judicial, but not regulatory, jurisdiction over designated reservations. See Act of Aug. 15, 1953, ch. 505, § 7, 67 Stat. 588 (codified as amended at 18 U.S.C. § 1162, 25 U.S.C. §§ 1321–1326, 28 U.S.C. §§ 1360, 1360 note). For a detailed analysis, see Goldberg, Public Law 280: *The Limits of State Jurisdiction Over Reservation Indians*, 22 UCLA L. Rev. 535 (1975). The Act allows state judicial jurisdiction over private causes of action on affected reservations but does not extend state tax or regulatory laws into Indian county. *Bryan* v. *Itasca County*, 426 U.S. 373 (1976).

There are special complex, statutory provisions for criminal cases on reservations not covered by "Public Law 280." See generally Clinton, *Criminal Jurisdiction Over Indian Lands: A Journey Through a Jurisdictional Maze*, 18 Ariz. L. Rev. 503 (1976); F. Cohen, *Handbook of Federal Indian Law* 286–308 (1982 ed.).

35. *Central Machinery Co.* v. *Arizona State Tax Comm'n*, 448 U.S. 160 (1980); *Warren Trading Post* v. *Arizona State Tax Comm'n*, 380 U.S. 685 (1965).

36. *White Mountain Apache Tribe* v. *Bracker*, 448 U.S. 136 (1980) (state not allowed to tax non-Indian logging company doing business with tribe); *Washington* v. *Confederated Tribes of the Colville Reservation* , 447 U.S. 134 (1980) (state allowed to tax sales of cigarettes by Indians to non–Indians); *Kennerly* v. *District Court*, 400 U.S. 423 (1971) (state court not allowed to assert jurisdiction over contract action by non–Indian against Indian); *Williams* v. *Lee*, 358 U.S. 217 (1959) (same). Cf. *Montana* v. *United States*, 440 U.S. 544 (1981) (state allowed to regulate fishing by non–Indians on facts of case).

37. *Washington* v. *Washington State Commercial Passenger Fishing Vessel Ass'n*, 443 U.S. 658 (1979); *Fisher* v. *District Court*, 424 U.S. 382 (1976); *McClanahan* v. *Arizona State Tax Comm'n*, 411 U.S. 164 (1973).

38. *Washington* v. *Confederated Tribes of the Colville Indian Reservation*, 447 U.S. 134 (1980); *Moe* v. *Confederated Salish and Koutenai Tribes*, 425 U.S. 463 (1976). Outside of Indian country, state law presumptively applies to Indians. E.g., *Mescalero Apache Tribe* v. *Jones*, 411 U.S. 145 (1973).

39. F. Cohen, *Handbook of Indian Law* 676–77 (1982 ed.).

40. *Worcester* v. *Georgia*, 31 U.S. (6 Pet.) 515 (1832); *Merrion* v. *Jicarilla Apache Tribe*, 102 S. Ct. 894 (1982).

41. See generally Leshy, *Unraveling the Sagebrush Rebellion: Law, Politics and Federal Lands*, 14 U.C.D. L. Rev. 317, 317–329 (1980).

42. The Rebels' position is summarized in Note, *The Sagebrush Rebellion: Who Should Control the Public Lands?*, 1980 Utah L. Rev. 505, 516–25. See also Note, *The Property Power, Federalism and the Equal Footing Doctrine*, 80 Colum. L. Rev. 817 (1980).

43. Although the original states retained title to lands within their borders, they agreed to cede to the United States their claims to lands beyond their western boundaries. P. Gates, *History of Public Land Law Development* 49–57 (1968).

44. Id. at 75–86.

45. In treaties providing for acquisition of land from foreign nations, the United States agreed that grantees of land from those nations would retain title to their property. B. Hibbard, *A History of the Public Land Policies* 23–25 (1965). On the rights of the United States vis–à–vis the Indian tribes, see *Johnson* v. *M'Intosh*, 21 U.S. (8 Wheat.) 543 (1823).

46. P. Gates, *History of Public Land Law Development* 285–318 (1968).

47. The Supreme Court has analogized statehood transactions to contracts between private parties. E.g., *Andrus* v. *Utah*, 446 U.S. 500, 507 (1980); *Stearns* v. *Minnesota*, 179 U.S. 223, 249–50 (1900).

48. Act of Mar. 3, 1975, 18 Stat. 474, 475 (Colorado Enabling Act). See also, e.g., Act of Mar. 21, 1864, 13 Stat. 30, 31 (Nevada Enabling Act); Act of July 16, 1894, 28 Stat. 107,

108 (Utah Enabling Act). Earlier statehood acts had disclaimed state ownership over public lands with different language, but with the same legal result. See, e.g., Act of Sept. 9, 1850, 9 Stat. 452, 452–53 (1850) (California Statehood Act); Act of Feb. 14, 1859, 11 Stat. 383, 384 (Oregon Statehood Act). See generally Leshy, *Unraveling the Sagebrush Rebellion: Law, Politics and Federal Lands*, 14 U.C.D. L. Rev. 317, 324–25 (1980).

Of the western states, only Texas and Hawaii depart from the pattern outlined in the text. Texas, as an independent Republic at the time of statehood, retained title to lands within its state boundaries. P. Gates, *History of Public Land Law Development* 80–83 (1968). Therefore, there was never public domain land in Texas and all public lands in that state have been acquired by the United States since statehood. In Hawaii there was no unalienated public domain land remaining by the time of statehood in 1959. Id. at 316.

49. *Pollard* v. *Hagan*, 44 U.S. (3 How.) 212, 221–24 (1845). See also *Dred Scott* v. *Sandford*, 60 U.S. (19 How.) 393 (1856); *Martin* v. *Waddell*, 41 U.S. (16 Pet.) 367 (1842).

50. A unanimous Supreme Court has squarely held that the federal government exerts sovereign authority over the public lands. *Kleppe* v. *New Mexico*, 426 U.S. 529 (1976). See generally Wilkinson, *The Public Trust Doctrine in Public Land Law*, 14 U.C.D. L. Rev. 269, 278–80, 305-06 (1980) and the authorities cited there.

51. E.g., L. Tribe, *American Constitutional Law* 321–24 (1978).

52. See supra notes 23, 23, 28 & 29.

53. The Public Land Law Review Commission, in acknowledging that state and local interests have special concerns in public lands management, identified six categories of interests that should be recognized in making public lands policy: (1) the national public; (2) the regional public; (3) the Federal Government as sovereign; (4) the Federal Government as proprietor; (5) state and local governments; (6) the users of public lands and resources. Public Land Law Review Comm., *One Third of the Nation's Land: A Report to the President and to the Congress* 6 (1970).

54. See supra note 28.

55. F g , *United States* v. *Sioux Nation of Indians*, 448 U.S. 371 (1980); *Delaware Tribal Business Comm.* v. *Weeks*, 430 U.S. 73 (1977); *Seminole Nation* v. *United States*, 316 U.S. 286 (1942).

56. F. Cohen, *Handbook of Federal Indian Law* 209 n.19 (1982 ed.).

57. Early federal policy imposed a wide range of restrictions, most of which remain in effect today, on persons engaging in trade with Indians. See generally F. Prucha, *American Indian Policy in the Formative Years: Indian Trade and Intercourse Act, 1790-1834* (1962). In the mid–1800's, reservations were set aside for tribes. S. Tyler, *A History of Indian Policy* 71–88 (1973). Large blocs of land were lost to Indians as a result of the allotment policy of the late 19th century, D. Otis, *The Dawes Act and the Allotment of Indian Lands* (F. Prucha ed., 1973), and the termination policy of the mid-20th century, Wilkinson & Biggs, *The Evolution of the Termination Policy*, 5 Am. Indian L. Rev. 139 (1977). Since the early 1960's the Indian land base has actually expanded and Congress has adopted a policy of "self–determination" for Indian tribes. Israel, *The reemergence of Tribal Nationalism and Its Impact on Reservation Resource Development*, 47 Colo. L. Rev. 617 (1976); *Bryan* v. *Itasca County*, 426 U.S. 373, 388 n.14 (1976) ("Present federal policy appears to be returning to a focus upon strengthening tribal self–government ."").

Two progressive pieces of recent legislation go further than protecting Indian resources and expressly seek to preserve Indian culture and religion. See American Indian Religious Freedom Act of 1978, 42 U.S.C.A. § 1996 (West Supp. 1981); Indian Child Welfare Act of 1978, 25 U.S.C. §§ 1901–1963 (Supp. III 1979).

58. A leading explication of the special legal status of Indian tribes is *Morton* v. *Mancari*, 417 U.S. 535 (1974), upholding a hiring preference for Indians in the Bureau of Indian Affairs. The court recognized that numerous federal laws "single out for special treatment a constituency of tribal Indians living on or near reservations." Id. at 554. "This unique legal status is long standing, see *Cherokee Nation* v. *Georgia*, 5 Pet. 1, 8 L. Ed. 25 (1831); *Worcester* v. *Georgia*, 6 Pet. 515, 8 L. Ed. 483 (1832), and its sources are diverse." Id. at 555.

59. The courts have recognized a state right to regulate special Indian rights in rare and extreme circumstances. Thus state regulation of Indian fishing is allowed for "conservation purposes." E.g., *Department of Game* v. *Puyallup Tribe*, 414 U.S. 44, 49 (1973);

Washington v. *Passenger Fishing Vessel Ass'n*, 443 U.S. 658, 682 (1979). This state right is exceedingly narrow and is apparently triggered only on a showing that the tribes themselves are incapable of exerting sufficient governmental control and that the state has first prohibited all fishing by non–Indians. *United States* v. *Michigan*, 471 F. Supp. 192, 280–81 (W.D. Mich. 1980), aff'd, 653 F.2d 277 (6th Cir. 1981); *United States* v. *Washington*, 520 F.2d 676, 686 (9th Cir. 1975), cert. denied, 423 U.S. 1086 (1976).

60. See F. Cohen, *Handbook of Federal Indian Law*, ch. 2 (1982 ed.).

61. In the recent legislation, see generally S. Dana & S. Fairfax, *Forest and Range Policy* (2d ed. 1980); G. Coggins & C. Wilkinson, *Federal Public Land and Resources Law* (1981).

62. The two dominant statutes are the National Forest Management Act of 1976, Pub. L. No. 94–588, 90 Stat. 2949 (codified at 16 U.S.C. §§ 1600–1614 (1976) and scattered sections of 16 U.S.C.) (NFMA) and the Federal Land Policy and Management Act of 1976 (codified at 43 U.S.C. §§ 1701–1782 (1976) and in scattered sections of 7, 16, 30 and 40 U.S.C.) (FLPMA). Both are umbrella laws that attempt to bring most of the basic authority of the Forest Service and the BLM, respectively, within their ambits. Lobbying was intense and both laws struck a middle course. On the NFMA, see the several articles in the symposium in 8 Envtl. L. 239 (1978). A symposium on FLPMA is at 21 Ariz. L. Rev. 267 (1979). One can gain a sense of the six-year flow of trade–offs in FLPMA by reviewing the summary of its legislative history. Senate Comm. on Energy and Natural Resources, Legislative History of the Federal Land Policy and Management Act of 1976, 95th Cong., 2d Sess. (Comm. Print 1978). See also, e.g., *Rocky Mountain Oil & Gas Ass'n* v. *Morton*, 500 F. Supp. 1338, 1340 (D. Wyo. 1980), appeal pending ("FLPMA's policy directives clearly attempt to strike a balance between the development of mineral resources and environmental concerns."); *Utah* v. *Andrus*, 486 F. Supp. 995, 1002 (D. Utah 1979) ("FLPMA represents an attempt on the part of Congress to balance a variety of competing interests . . .").

63. See supra notes 23 & 27 and accompanying text.

64. State regulatory authority is especially broad over resource activities on the public lands in regard to water law, e.g., *California* v. *United States*, 438 U.S. 645 (1978), *California--Oregon Power Co.* v. *Beaver Portland Cement Co.*, 295 U.S. 142 (1935); and hardrock mining, e.g., 30 U.S.C. §§ 24, 26, 28, 28b, 38 and 43 (1976), *State ex rel. Cox* v. *Hibbard*, 31 Or. App. 269, 570 P.2d 1190 (1977). There is also room for state law to operate in other areas, such as mineral leasing, e.g., *Texas Oil & Gas Corp.* v. *Phillips Petroleum Co.*, 277 F. Supp. 366 (W.D. Okla. 1967), aff'd, 406 F.2d 1303 (10th Cir. 1969), but federal preemption of state and local regulatory activities is common, e.g. *Ventura County* v. *Gulf Oil Corp.*, 601 F.2d 1080 (9th Cir. 1979), aff'd 445 U.S. 947 (1980).

65. See supra notes 23, 24, & 29–33 and accompanying texts.

66. See supra note 28.

67. See supra notes 34–40 & 55–60 and accompanying texts.

68. F. Cohen, *Handbook of Federal Indian Law* 270 (1982 ed.).

69. 16 U.S.C. §§ 1451–1464 (1976 & Supp. IV 1980).

70. "Excluded from the coastal zone are lands the use of which is by law subject solely to the discretion of or which is held in trust by the Federal Government, its officers or agents." 16 U.S.C. § 1453(1) (1976). See generally Note, *Coastal Zone Management and Excluded Federal Lands: The Viability of Continued Federalism in the Management of Federal Coastlands*, 7 Ecology L.Q. 1011 (1979).

Although excluded from the Coastal Zone Management Act's definition of the coastal zone, activities on federal lands may be subject to the Act's consistency provisions, requiring that federal actions directly affecting the coastal zone be consistent with the state's federally– approved program to the maximum extent practicable, 16 U.S.C. § 1456(c), (d) (1976). See generally Deller, *Federalism and Offshore Oil and Gas Leasing: Must Federal Tract Selections and Lease Stipulations Be Consistent with State Coastal Zone Management Programs?*, 14 U.C.D. L. Rev. 105 (1980); *California ex rel. Brown* v. *Watt*, 683 F.2d 1253 (9th Cir. 1982). In some circumstances the Secretary of Commerce can override the consistency requirement upon a finding that the proposed federal action is consistent with the Act's objectives or otherwise necessary in the interests of national security. 16 U.S.C. § 1456(d) (1976).

The consistency provisions of the Coastal Zone Management Act have their counterparts in existing laws relating to the public lands. See supra note 28; *Columbia Basin Land Protection Ass'n* v. *Schlesinger*, 643 F.2d 585, 602–06 (9th Cir. 1981).

71. E.g., 125 Cong. Rec. S11665 (daily ed. Aug. 8, 1979) (remarks of Sen. Orrin G. Hatch).

72. Clinton, *Isolated in Their Own Country: A Defense of Federal Protection of Indian Autonomy and Self-Government*, 33 Stan. L. Rev. 979, 980–82 (1981) (discussion of opposition to special separate status of Indian tribes).

73. Governor Bruce Babbitt of Arizona, who opposes the Sagebrush Rebellion and many of the resource development policies of the current administration, is studying various legislative approaches that would limit federal authority and increase state authority on the public lands. Though not committed to such an approach, Babbitt believes that the Coastal Zone Management Act, discussed supra in note 70, may provide a model preferable to current law. Babbitt's premise is that state preservation practices may well be superior to federal policies. Symposium, Northwestern School of Law, Lewis & Clark School of Law, Portland, Or. (Feb. 5, 1982) (unpublished address).

Babbitt's views are a reminder that it is overly simplistic to view the western states as being more development–oriented than the federal government during the modern era. In fact, there are numerous examples of attempts by states to halt or limit federal development activities on conservation grounds. See, e.g., *Washington Dept. of Game* v. *FPC*, 207 F.2d 391 (9th Cir. 1953), cert. denied, 347 U.S. 936 (1954) (opposition by Washington to federal dam); *Federal Power Comm'n* v. *Oregon*, 349 U.S. 435 (1955) (opposition by Oregon to federal dam); *South Dakota* v. *Andrus*, 614 F.2d 1190 (8th Cir. 1980) (opposition by South Dakota to issuance of mining patents); *California* v. *Bergland*, 483 F. Supp. 465 (E.D. Cal. 1980) (opposition by California to development on roadless areas in national forests); *California ex rel. Brown* v. *Watt*, 683 F.2d 1253 (9th Cir. 1982) (opposition by California to off–shore oil leases).

74. Turner based his analysis of the importance of the frontier in the nation's past, and its demise in the late nineteenth century, on the statement in the Eleventh Census that, as of 1890, "the unsettled area [of the United States] has been so broken into by isolated bodies of settlement that there can hardly be said to be a frontier line." F. Turner, *The Frontier in American History 1* (1920)

75. On the United States' pioneering efforts in establishing official wilderness areas, see J. Hendee, G. Stankey & R. Lucas, *Wilderness Management* 27–59 (1978). See generally R. Nash, *Wilderness and the American Mind* (rev. ed. 1973); G. Coggins & C. Wilkinson, *Federal Public Lands and Resources Law* 766–839 (1981). After the addition of approximately fifty–seven million acres to the Wilderness Preservation System by means of the Alaska National Interest Lands Conservation Act of 1980, Pub. L. No. 96–487, 94 Stat. 2371 (codified in scattered sections of 16 U.S.C.), the system today includes approximately 79.8 million acres. Telephone conservation with Richard Joy, Wilderness Office, United States Forest Service, Washington, D.C. (Oct. 19, 1981).

76. See generally Cohen, *Original Indian Title*, 32 Minn. L. Rev. 28 (1947) and the several articles beginning at 27 Buffalo L. Rev. 617 (1978).

5

How the Privatization Revolution Failed, and Why Public Land Management Needs Reform Anyway

CHRISTOPHER K. LEMAN

A. Benefits and Costs of Ideology

At least sixteen symposia were held between 1980 and 1983 on the subject of whether to keep or sell the public lands.[1] Witnessing seven of these, I observed some patterns. The critics, many of whom were economists, unreservedly attacked the public lands agencies and their performance. They strongly invoked historical traditions and economic theory, provoking defenders of the lands to respond in kind. The result was polarization, with little room for middle positions, especially for reforms in policy and management within the existing pattern of ownership. The attack on the public lands intertwined factual assertions about the superior performance of private over public organizations, with moral absolutes about the superiority of the private model, and thus reflected an ideology—an ideology of privatization. All of us hold views on how the world works and on how civil life should be carried on, but an ideology combines such views in an unusually structured form whose coherence is constantly renewed by its combativeness toward existing institutions and its relentless quest for a utopian alternative. The public lands were singled out for

Research and writing for this paper have been aided by fellowships from Resources for the Future and by grants from the Land Use Project of the University Consortium for Research on North America, Harvard University, with support from the Lincoln Land Institute. I regret that space does not allow me to thank the dozens of people whose ideas or assistance were essential in the preparation of this paper.

particular attack because public ownership clashes with some peoples' deepest beliefs.

Ideologies can be very powerful things, as Michael Walzer argued in *The Revolution of the Saints*, a classic study of the origins of Britain's 17th century Puritan uprising, the world's first modern revolution.[2] Every revolution since has depended in part on the radical hostility to the established order of an ideological vanguard like Walzer's Puritans. Today, some of the most deeply ideological people are those who are not happy with the spread of equality and the growth of government. Conservative theorist Kevin Phillips points out that some parts of the conservative movement reflect such disenchantment with American institutions that they really are no longer conservative; their proposals represent radical change. He calls this element "revolutionary conservatism," a term I will use here.[3]

Ideologies have both strengths and weaknesses. As the distinguished sociologist Edward Shils has observed:

> Ideologies hostile to the prevailing outlook and the central institutional system of a society have not infrequently contained truthful propositions about important particular features of the existing order, and have drawn attention to particular variables which were either not perceived or not acknowledged by scholars or thinkers who took a more affirmative, or at least a less alienated, attitude toward the existing order. On the other hand, ideologies have no less frequently been in fundamental error about very important aspects of social structure, especially about the working of the central institutional system, about which they have had so many hostile fantasies. [4]

This statement sums up very well the dilemma posed by the ideology of privatization. As a perception of the public lands, it has some serious flaws. At the same time, however, the critics have noticed some things that the rest of us overlooked and would be unwise to ignore—even though the temptation is to let the ideology of privatization fall of its own weight.

To weigh the costs and benefits of the ideology of privatization, this chapter initially explores the movement's serious misreadings of American history and political practice, which led it to overestimate existing support and alienate potential supporters. The essay then shows that many proponents of privatization have enlisted the support of economic theory beyond what it can sustain, and that they have overemphasized the importance of who owns land while neglecting the specifics of how it is to be managed. To remedy the less constructive aspects of the recent debate, the chapter ends with a series of recommendations for research and reform in improving the public lands.

B. Misreading American Traditions

The strongest critics of the public lands have claimed that history and economics come down entirely on their side of the argument, but the fact is that these fields are consistent with a range of alternative institutional arrangements.[5] To choose among these views, we must look not to the social scientists, but within ourselves. In the current controversy, the experts will have their hands full just trying to refute the excessive claims made for social science.

One common, and erroneous, contention has been that the public land system is somehow inconsistent with American political theory and historical practice. Actually, a strong government role in landownership and in other matters is deeply rooted in American history. Far from being united against a government role, the Founding Fathers disagreed mainly about whether that role should be extended. In his valuable 1978 book, *The Governmental Habit*, conservative economist Jonathan Hughes observes that "economists have been largely inept in their understanding of the place of economic control in American society."[6] State and local governments in particular were always heavily involved. Far from being miracles of free enterprise, the railroads were the greatest 19th century instance of business-government partnership. In some states, government officials even served on their boards of directors. The federal government, too, got involved in the economy through tariff policy, internal improvements, and land grants, initially to states and then to companies and individuals. In fact, the American use of federal land grants still stands as the greatest governmental intervention in a national economy in the Western world. Hughes observes that governmental controls on the economy "have existed so long in peaceful conjunction with Anglo–American ideas about property rights that their long–lived existence must be considered by the historians as somehow 'natural' within the reality of our laws and customs."[7]

Despite this history, some critics have tried to argue that the consideration of federal control and the diminution of land sales after 1890 arose from a foreign ideology. This is hard to square with the fact that the most influential proponent of public land management ever was Theodore Roosevelt, about as nationalistic a president as we have had; federal management was also strengthened under such other Republican presidents as Rutherford Hayes and Calvin Coolidge. Nevertheless, the critics are upset that a few public figures studied abroad (at a time when there were no forestry schools in the United States, Gifford Pinchot studied forestry for thirteen months in France; imagine that!) or, worse, yet, that a few were not even born in this country.

Consider, for example, the campaign that accused Carl Schurz, one of the greatest Secretaries of the Interior, of being a German socialist.[8] In fact, Schurz was a founder of the Republican Party, a monetarist, a Civil War general, and a Missouri senator before he arrived at the Interior Department.[9] Upon appointment as Secretary, Schurz announced the intention to "conduct this department upon business principles," and one of his first reforms was to institute a system of competitive bidding for government contracts.[10] Some socialist! Carl Schurz's effort to consolidate federal control on the public lands was motivated by concern for the rule of law, not by hostility to private ownership. In fact, in his last year as Secretary, the government disposed of three times as much land as it had in his first year. Like many immigrants, Schurz had come to America not to remake it but to respect the values it represented—which he took as his own. In his efforts to restore order and equity on the public lands, he was honoring values of the Founding Fathers, better in fact than were many native–born Americans of that era.

Once the accusation of foreignness has been raised, very few political ideas can escape it. The only truly indigenous American political theory, that of the American Indian, firmly supported public ownership! In contrast, it is very difficult to find *any* American prior to the 20th century who seriously argued, as do today's revolutionary conservatives, that a system of private property rights could replace government entirely. The intellectual roots of that movement must be traced to such foreign "subversives" as Britain's Adam Smith, the French physiocrats, and the Austrian free market economists. The term "laissez–faire" is hardly a product of the American frontier.

Another polemical ploy has been to apply the label of "socialism" to the public land agencies. Those who do so choose to overlook the far greater government role in other sectors. For example, the Defense Department employs about half of all federal personnel and owns about three–quarters of all federal building space, including many factories and retail stores. The critics seem to save the term "socialism" only for those governmental tasks that they do not like.

Since Americans have long argued over whether to make government more active or to limit its role, it is not surprising that the actual outcome has almost always been a mixed one that does not depend solely on either the public or private sector, but rather on both. Contrary to being some kind of anomaly, the public lands typify this mixing. Socialism is a system in which government owns and manages the means for producing and distributing goods. Our public lands system falls far short of this arrangement. A more accurate—if polemically less satisfying—description of the role of the Forest Service and the Bureau of Land Management is as a *rentier*: that is, a landowner who in effect leases out the land for development by others. Although the resources on these lands are publicly held, the means for producing and distributing them are largely in private hands, and in the case of commodities, virtually entirely so. As Marion Clawson has pointed out, the arrangements that convey public land resources to others are sometimes called contracts or permits, and sometimes there is no charge for them, but in effect they are all forms of leasing.[11] The most common examples are in timber, grazing, oil and gas, coal, ski areas, and park concessions. Even campers and picnickers enjoy a form of lease.

C. Loss of Already Low Political Support

Given this long American tradition of mixing public and private roles in the economy and specifically on the public lands, revolutionary conservatism has made little political headway. Kevin Phillips judges, and I agree with him, that for better or worse, we are not presently in a revolutionary situation insofar as the current critique of government is concerned. On the specific issue of privatizing the public lands, virtually no support emerged beyond the advocacy of a group of committed intellectuals. Although many were economists, they were a clear minority even in their own profession. Western state and local governments went on record against the idea. Environmental, conservation, and recreation groups—including wilderness advocates, hunters, and off–road vehicle enthusiasts—achieved an unprecedented unity in opposition. Not a single major commodity sector supported privatization: livestock, mining, oil and gas, coal, or timber. That included not only businesses that use the public lands, but also those that do not. Small business and nonprofit groups had little reason to welcome privatization; given the current diversity of private delegations of authority on the public lands, wholesale turnover would likely diminish their role. And after more than a year of debate, an April 1983 poll by ABC News and the *Washington Post* found only 11 percent of the American public supporting the sale of *any* National Forest land.[12]

Members of Congress became increasingly outspoken against privatization, particularly in the Republican Senate. Senator Charles Percy (R–Ill.) backed off on his 1981 resolution favoring disposal as a means to raise revenue. Senator John Chafee (R–R.I.) cosponsored legislation that would strictly control land sales. And strong criticism of the administration's plans came even from the conservative chairs of the committees most concerned with the public lands, Senators James McClure (R–Id.) and Jesse Helms (R–N.C.). Both were up for reelection in 1984 and privatization did not sit well with voters in Idaho and North Carolina, nor indeed in other states Republicans depended on.

The political problems that faced any proposal to sell off any significant proportion of the public lands were exacerbated by the Reagan Administration's mishandling of its specific initiative, the Asset Management Program. As presented in 1982, the program was extreme and poorly thought–out, and the administration had to retract and correct it gradually.[13] The initiative was cooked up primarily within the White House. Steve Hanke, then a presidential staff member and a prime architect of the plan, recalls that "Great care was taken to avoid sabotage by bureaucrats and political appointees who were located outside the Executive Office of the President."[14] Input from those who could have made the program more workable and more politically palatable was limited. Secretary Watt and Assistant Secretary Crowell may have been in a position similar to that faced by Interior Secretary Cecil Andrus in 1977 when the White House staff imposed on him a "hit list" of water projects in key western states. The privatization predicament was probably even worse because Carter eventually withdrew his hit list, while the Reagan Administration was unable to let go of its Asset Management Program. As Daniel Poole of the Wildlife Management Institute—and a sportsman who knows about these things—observed: "The President's people have shot themselves in their collective foot."[15]

Given the poor sense for political strategy reflected in how the privatization proposal was presented, it is not surprising that very little thought was given to how such a program would work if it could be enacted. The hard lesson of past revolutions has been that they usually proceed differently from what the radicals initially envision.[16] Recent writings on American policymaking, too, have shown that the implementation of any new policy usually involves many twists and turns that can frustrate its original purpose.[17] In fact, it has been argued that almost every major public lands policy of the past—the Homestead Act, the Taylor Grazing Act, and so on—has turned out differently from what was originally intended.[18] There is strong reason to expect that a similar fate would befall the effort to privatize the public lands.

Historical experience suggests that the land sales process is fraught with problems, such as depressed prices from a glut on the market, strategic purchasing by some buyers to obtain a virtual monopoly on other lands, outright corruption within government and fraud outside of it.[19] In addition, pressures would be strong to help various groups that might lose out if the lands were really sold on the open market—state and local governments, nonprofit groups, ranchers, and so on. They would be helped through land sales below market value, covenants requiring the new owners of land to deal with them, and outright federal subsidies. Such compromises would likely be necessary to pave the way for privatization, but in the process they would undermine the efficiency case for it and could even produce a result less efficient than we have right now.

D. Overextending Economic Theory

Economists were prominent among the advocates of privatization. A great contribution they made was to gain more visibility for allocative efficiency as one criterion by which to judge the public lands system. However, in doing so they also managed to sour many people by insisting on an often sloppy and one–sided interpretation of what efficiency recommends.

The economic case for privatization made excessive claims for the conclusiveness of efficiency as a guide for policy. Consider Steve Hanke's proposal that: "All lands that do not yield a real rate of return of 10 percent in public ownership should be classified as surplus and privatized."[20] Obviously, this view ignores the nonmonetary values of these lands, which can be quite substantial. By his standards, even the Washington Monument does not produce a sufficient rate of return, and should be sold off immediately.[21] But even on its own terms, Hanke's version of asset management is not efficient, because his guideline for 10 percent return on private sector investment is an average. Many private sector activities—for example, most private ranches, and many forest products and mining companies at the time that Hanke wrote—earn less than a 10 percent real rate of return. Thus Hanke imposes a stricter test on the public lands than he does on the private sector. Obviously, we will have to look further for an efficient approach to managing public assets.

Certainly it is essential to recognize the opportunity cost of the capital represented in the public lands. But it is equally important to look at the indirect, hidden costs of actually unloading these lands. Looking back, it seems clear that the large–scale land disposals of the 18th and 19th centuries gave insufficient consideration to the option value of the land. So much land was disposed of in the East and South that governments and corporations have found it difficult to assemble land in the large blocks needed for recreation, transportation, and so on. Public recreational land there still falls short of demand, despite costly efforts by governments and private groups to reassemble these lands. Another case where land disposal probably went too far and too fast was the privatization of some farm and forest lands that really could not sustain continued farming, logging, or other commercial activities. This policy encouraged the emergence of a marginal economic sector that has been ever since a drain on the Federal Treasury, a drain that has at times been alleviated when some of these lands eventually find their way back into federal ownership because no one wants to keep them.

An even more fundamental problem in many economic arguments for dismantling the public lands has been to ignore questions of equity. In applying economic theory, questions of distribution are unavoidable. The frequency with which inequities have been cited as an objection to privatization suggests that equity is no more avoidable in political practice than it is in economic theory. No set of institutions, however attractive in principle, can be truly efficient if the citizenry is not satisfied with it. A theorist who refuses to recognize that fact is an enemy not just of equity, but of efficiency itself.

Given the radical nature and thus the political infeasibility of their proposals, supporters of privatization need to be more open to compromises than they have been so far. In adapting to social values other than efficiency, they do risk undermining the efficiency argument for privatization. But if the efficiency case is in fact so vulnerable to other important concerns, how strong is it really? All other things being equal, perhaps privatization schemes will achieve a more efficient result; but the question is, how much more efficient? If the difference is not great, then is a change worth all of the work and conflict necessary to obtain it? Arguably the efforts of researchers and political leaders would return more in efficiency if they were directed not into an ideological crusade against public landownership, but rather into substantive improvement in the current management of these lands.

E. Policy and Management, Not Ownership, the Key

Many of the complaints about public land management turn out to be rooted in policy and management—in what our society has directed these agencies to do, and in how they have carried out this mandate—rather than in inherent weaknesses of government. It is unfair to berate the Bureau of Land Management for not having done more with some choice urban sites it has in Las Vegas and elsewhere, because until recently, revenue was not a major priority, and even now, the Federal Land Policy and Management Act makes it national policy that disposal not be emphasized. Similarly, the ban on export of federal logs has been singled out for criticism, but that is a national policy, not a case of Forest Service rigidity. As a matter of fact, it is quite likely that the log export ban would be carried over into any system of privatization, as a covenant imposed on the new landowners.

A key complaint of property rights theorists who have been prominent among the privatizers has been that those who manage public land are not the immediate owners, and thus do not have the incentive to do a good job that a private owner–manager would.[22] Certainly the management of a resource should be carried out in accordance with the interests of the owners, but that does not mean that they must do the managing. Institutional arrangements short of outright ownership by managers may be quite sufficient to secure the desired behavior. And these arrangements had better be sufficient, because the owner–operator approach is surprisingly rare in today's private sector, a fact to which many proponents of the property rights view seem oblivious. Perhaps the most prominent case is in corporate management where Berle and Means in their classic 1935 work heralded the emergent separation between ownership and control of capital. Edward S. Herman's recent sequel documents the widening of this separation.[23]

Private lands in urban areas are subject to a dazzling array of owner–manager arrangements. Proponents of privatization have sometimes cited the Empire State Building as an example of good management in the private sector; they do not seem to realize that the building's ownership situation undermines their argument.[24] The Prudential Life Insurance Company owns the land under the Empire State Building. A 114–year master lease conveys the land to Empire State Building Associates, who own the building and have in turn, by another 114–year lease, delegated its operation to a partnership, Empire State Building Company. This partnership subleases space on a 10–year basis to the hundreds of businesses actually located in the building. These businesses in turn sublease some of their space in the building to other businesses. When I asked whether this complex system didn't violate the principle that ownership and management should be synonymous, one of the managers replied: "Who cares? It works."[25]

Even American agriculture, whose productivity is often contrasted favorably with that on the public lands, in fact depends far less on the farmer being both owner and operator than is usually supposed. The fact is that only one–third of all American farmers own every acre that they farm; the other two–thirds must lease some land from others. Fully 30 percent of the private land now under cultivation is farmed not by its owner, but rather by a lessee.[26] The ubiquity of leasing arrangements is even apart from the fact that much farm land is heavily mortgaged to creditors, who include some government entities. Certainly it is only in a very incomplete sense that most farmers "own" their land.

How lands are managed, rather than who owns them, is the key to efficiency. To the extent that they do not recognize this fact, then efficiency is too important to be left to the economists, or to others of a similar mind. Despite avowals of wanting public land management to be more businesslike, the Reagan Administration made surprisingly little progress toward that end, perhaps reflecting a skepticism as to whether government should even be involved in land management. In contrast, Public Buildings Commissioner Richard O. Haase, under a March 1983 presidential order, led a nationwide effort by the General Services Administration to decrease federal reliance on leasing of private facilities and to depend more on government ownership.[27] Secretary of Transportation Elizabeth Dole led an effort to redevelop Washington's Union Station—a past example of government bungling—into a retail and office complex like Boston's Faneuil Hall Market Place.[28] Thus, ironically, while revolutionary conservatives inside and outside the Reagan Administration were trying to dismantle the public lands on the grounds that government cannot work, the real conservatives were showing that by managing it better, government can be made to work after all.

F. Research and Reform to Revitalize Public Land Management

The ideology advocating privatization of the public lands seems to have had more costs than benefits for public debate.[29] But there were benefits, perhaps the most important being to force many noneconomists to consider allocative efficiency as one standard by which to judge how these lands are managed. The tragedy of the privatizers was that their polemicism, political naivete, and resistance to compromise undermined their professed objective of promoting efficient land management. In the short run perhaps their criticisms got more attention as coming from the outside than if the criticisms had been made primarily internally. But in pursuing an almost exclusively outside strategy, the critics made blanket opposition to change seem to many people preferable to a course of internal reform. And vastly complicating intellectual debates was the fact that the sales plans were already grinding their way through the bureaucracy and were under challenge in Congress and the courts. Whereas recent advances in deregulation were based on decades of scholarly debate, the Reagan Administration moved to privatize the federal lands before the idea had been studied or even discussed publicly, and thus before a respectable case had been made.

As the intellectual movement against government becomes institutionalized, it could make the public lands system permanently unworkable, without setting forth any viable alternative. Joseph Schumpeter made a famous argument that capitalism fostered within itself a new class of people whose ideological opposition would prove its undoing.[30] Schumpeter's argument does help explain ongoing increase in the role of government in economic life. But the rise of the public sector has occasioned a reverse phenomenon, in which today's mixed economy fosters an even newer class, just as ideological, who scorn government, not capitalism. The effect of the critique could be to hamper the effectiveness of government programs, and to discredit them with potential supporters.[31] Government *can* work, but it will not if subjected to constant negativism, with few constructive suggestions or politically feasible reforms to improve its functioning.

Yet blanket defense of the status quo would in my view be just as mistaken as blanket opposition to it. There is power in ideas, and the substantive objections raised by the critics are serious enough that if improvements do not come, defense of the public lands against future campaigns for privatization will be increasingly difficult. The effort to make government more efficient faces difficulties, because although the constituency for efficiency is a broad one, it is diffuse and thus vulnerable to opposition from an odd coalition. Whereas many conservatives oppose the effort because they think it is doomed, many liberals oppose it because they find the objective all too feasible and they do not like it. The proponents of wholesale privatization have too easily had conceded to them the assertion that allocative efficiency is inherently impossible on the public lands. I think this view is a mistaken one that does not consider the potential for retailing economic thinking within government.[32] It is unfortunate that proponents of public lands have been ambivalent and even resistant toward efficiency as one objective for management. Those who deny to government certain techniques that business finds indispensable will unwittingly stack the deck against government as a viable alternative to the market.

Many current policies deserve some rethinking, such as current timber policies in which harvest is probably too high in those regions with less productive stands and too low in those where timber grows fastest. Another debatable policy is that states and localities receive such a large share of timber and mineral revenues from the public lands, often in larger quantities than the taxes the land would yield if it were in private ownership. Many other dubious policies need also to be examined.

Reforms in management offer as great an opportunity for improving the public lands system as do reforms in policy, yet it is far more difficult to obtain discussion of them. Opponents of the public land agencies obviously have little incentive to suggest improvements. But the friends of the land agencies have also tended to resist organizational improvements. For example, policy questions such as wilderness have absorbed the energies of the environmental groups, distracting them from management. As an official of one national group told me, "management questions are not politically sexy."[33] It is essential to the survival of the public lands that the agencies and their friends confront the need to improve management. Issues that deserve attention include land adjustment, delegation by leasing and other approaches, user fees, interorganizational relations and public participation, budgeting, personnel, and leadership.

Land Adjustment

In the current debate, changes in federal landownership patterns have been advanced for such ambitious purposes as reducing the national debt and dismantling the governmental presence. The controversy has overshadowed and even thwarted more modest objectives for land adjustment, such as making way for urban growth and improving the management of existing public lands. Many western towns are near public land, and some are entirely surrounded by it. It is reasonable for the federal government to dispose of land to localities or to private owners in such cases, and various laws and procedures exist for this purpose. However, the demand is not limitless. Having promised to deal with the problem, Secretary Watt in 1981 invited western governors to nominate BLM lands needed for local growth. The total area singled out was 959,000 acres— a large amount, to be sure, but it represented only four–tenths of one percent of total BLM acreage.[34]

Another significant but limited purpose of disposal is to make existing management more efficient and effective. Federal lands that are remote from the offices that manage them, or are mixed in with lands owned by others, impose special costs in building and maintaining roads, laying out land lines, issuing special–use permits, obtaining rights of way, or administering generally. Where other policy considerations do not recommend otherwise, a good case can be made that the disposal of such land would improve the management of the public lands that remain. Thus even in the current polarized climate, Congress passed and President Reagan in 1983 signed the Small Tracts Act, authorizing the Forest Service to sell small quantities of land in certain specified cases; a total of up to 120,000 acres could eventually be sold under this act in the next decade.[35] Of course, certain lands that are difficult or expensive to manage may be retained in public ownership for the sound reason that they are particularly valuable for public purposes. In fact any national program for land adjustment should contemplate selective acquisition of lands needed to fill out the federal lands portfolio. Some acquisitions would actually improve management efficiency by reducing problems of mixed ownership.

If it is recognized that acquisition as well as disposal may be desirable in certain cases, exchange is a reasonable instrument for land adjustment. In recent years, land exchange has been more widespread than outright disposal. For example, between 1970 and 1982, BLM sold only about 137,742 acres, but it exchanged another 1,244,864 acres.[36] Marion Clawson has calculated that even at this rate, the Forest Service and BLM would take more than 500 years to accomplish by exchange the elimination of all nonfederal lands encompassed within their exterior boundaries (a goal that of course, need not be reached so long as the more serious instances of mixed ownership are eliminated.)[37] Even so, very large transactions are possible, as in Project BOLD, the State of Utah's proposal to exchange about 700,000 acres of scattered state lands for a similar area of federal lands.

The Asset Management Program does not welcome land exchange, partly in opposition to the addition of more lands to the public domain, and partly because such transactions do not go through normal budgetary channels. However, exchange can be the most practical means of disposing of certain lands, and treating it as two separate transactions—a sale and a purchase—may complicate matters for both parties. Land exchange is common not only in the federal government; forest products corporations routinely trade lands with each other. In fact they do so much more easily than the public land agencies, which face various procedural and political hurdles. Rather than complicate public land exchange further, as by the freeze on future exchanges instituted by the Reagan Administration, new ways need to be found to smooth the way. One promising proposal is Clawson's for a Land Exchange Board.[38] An *ad hoc* version of such a board is currently in use in the State of Utah's handling of its BOLD proposals.

There is no question that the public land agencies have a long way to go in managing the land adjustment process better. They have given critics too many grounds for indignation. For example, once having decided to dispose of a piece of federal land, and to do so at less than market value, the agencies have often taken too few precautions to assure that state or local governments or other beneficiaries will use the land as originally intended.[39] And to the extent that revenue–raising becomes an explicit national policy in certain land sales, these agencies are duty bound to carry this policy out effectively just as they would any other policy.

Delegation by Leasing and Other Approaches

Selling public land is not the only way to address problems of mixed ownership, or for that matter to raise revenues or reduce the federal role. The federal government can delegate almost any public lands function to others; in fact, as Arizona Governor Bruce Babbitt has pointed out, the only function that cannot be discharged without actually owning the land is the act of selling it.[40] As land disposal is final indeed, it makes sense to consider keeping the lands in public ownership but delegating to others the responsibility for resource development and resource protection. If used selectively, such an approach allows resort to state and local governments, nonprofit groups, and businesses where they would most promote public purposes, while retaining the public lands under direct federal control where that would be the most desirable approach.

The various forms of leasing are the most widespread form of delegation now used by the federal land agencies. Examples include the sale of oil, gas, coal, timber, and grazing rights.[41] Less well–known cases of leasing include that of cropland and wildlife refuges. Critics have rhetorically asked, "Why should the government grow trees when it does not grow wheat, corn, oats or apples?"[42] They have not taken the time to learn the real answer: the government does grow wheat, corn, oats, apples, and lots of other crops. It does so as a lessor. More than 5 million acres of cropland—not even including pasture, range, and forest land—are now leased to farmers by governments.[43] That is an area the size of New Jersey. Some amenity resources are also managed by lease. The Minnesota Department of Natural Resources in 1940 signed a lease with the U.S. Department of Agriculture to manage the 81,000-acre Beltrami Wildlife Management Area.[44] Leasing of public land is even more widespread at the state level and, in Canada, at the provincial level.[45] Too little experimentation with leasing managements has taken place on the U.S. federal lands, and we have drawn too few lessons from the experiences of other levels of government and other countries.

Forms of delegation other than leasing have greatly increased in recent years. Tasks contracted out to businesses, nonprofit groups, or governments include recreation management (trails, visitor centers, and so on), land inventory, land treatment (especially reforestation), fire protection, and range management.[46]

The federal land agencies also are coming to depend more on the proper management of land that is not federally owned. Leasing and related approaches can be applied on state, local, and private land to secure practices that are compatible with those on adjacent federal lands, thus reducing problems of mixed ownership without need for disposal or acquisition of federal lands. The federal land agencies have long secured rights of way and easements on private lands, and more recently they have begun also to use cooperative agreements and other mechanisms.[47] Another widespread practice is for nonprofit groups to purchase land or easements near federal lands such as along the Appalachian Trail, thus protecting the area without need for extension of federal landownership.[48]

Not all delegations of federal authority to others are desirable, of course, and some have had severe consequences for efficiency or the environment, and sometimes for both. Under the 1977 Surface Mining Control and Reclamation Act, surface owners have the right to veto the development of federal subsurface mineral rights, severely constraining the federal coal management program.[49] In another example, a December 1981 order by the Director of BLM disclaimed federal rights to stock watering from new developments on BLM land, allowing grazing permittees to file as the exclusive or shared holders of these new water rights, even if the BLM had entirely paid for these improvements. Since water determines the utility of arid lands, a few ranchers could now veto BLM actions, and they obtained this right without any compensation to the public.[50] These and other undesirable delegations of federal authority have received as little attention as have the many desirable delegations. Since delegation of federal authority on the public lands is on the increase, it is imperative that researchers and managers thoroughly explore which delegations are most appropriate and how they are best managed.[51]

User Fees

When public land resources are consumed, management is usually most effective if the value of these resources is recaptured as fully as possible in fees charged to the users. The case for user fees is not only one of equity to the public and to competing suppliers on private land (thus avoiding a giveaway of public resources); it is also one of efficiency, in that the user's internalization of the full cost of the resource discourages overuse and encourages conservation. Current instances of underpricing of public resources include hard rock minerals, grazing rights, and outdoor recreation.

Policy or management considerations may of course recommend against recovering from users the full cost of a resource. The traditional justification for giving away free the minerals removed from the public lands was to encourage domestic mineral production. However, when the Public Land Law Review Commission in 1970 recommended the imposition of continuing royalties in this case, the American Mining Congress accepted the idea in principle.[52] Given the marginal economics of today's grazing operations, many ranchers would be hurt if federal grazing fees were raised to their true value. However with the fee now at scarcely one– quarter of what forage would cost on private lands, and having been dropping as a result of 1978 legislation, a significant increase seems reasonable.[53]

User fees are most likely to aid management if the administering agency is able to keep at least a portion of the revenues and reinvest them in its operations much as would a private company. However, this linkage of income and outgo has long been resisted by the Office of Management and Budget, which seeks maximum control over agency spending. In my view, OMB's concern is misplaced, as the agencies would likely manage their affairs most efficiently if they could recapture some of the revenue generated. This favorable result can be found in those minor cases where OMB's wishes have been overridden, as in BLM's 5100 account for the processing and monitoring of right–of–way applications and environmental statements.

Both the millions of people who use the public lands for outdoor recreation, and the resulting problems of overuse, equity, and efficiency provide arguments for charging user fees on these lands.[54] Currently the National Park Service has the greatest authority to charge for entry on public lands, but the other land agencies are at least able to charge for campgrounds and other facilities. Various objections have been raised as to the equity and the practicality of such charges, but these objections can generally be handled. Unanswerable, however, is the strong political opposition to instituting anything resembling an entrance fee or a federal hunting or fishing license on the public lands and waters. The Reagan Administration in February 1982 submitted to Congress a bill authorizing such charges, and the resulting storm of bipartisan protest led it to withdraw the bill only four days later.[55] This incident symbolizes the political problems that have long plagued efforts to charge the users of public lands, not only for outdoor recreation, but for other resources. However, the budget crunch and other unprecedented changes have created new incentives for the adoption of user fees. And although the federal lands have been slow to respond, user fees in the past few years have spread in most other federal programs. State and local parks, too, have greatly increased their fees.[56] It is time again for a similar effort on the federal lands.

Interorganizational Relations and Public Participants

Bureau management was never a self–contained process, and recent trends have made external relations even more important to successful management. Traditional intergovernmental relations, themselves intensifying, between federal land managers and other units of government are now increasingly supplemented by the need to deal with many other organizations and individuals who have ideas about how the federal lands should be managed and who in some cases have been delegated direct responsibilities on these lands.

Examining writings on the public lands since 1950, Sally Fairfax notes that among the topics generally missed by scholars was the new significance of intergovernmental relations—even though the land agencies were at that time developing many new instruments for cooperation with state and local governments.[57] Even if these agencies legally may defy the lower levels of government, they cannot as a practical matter do so because of the strong clout of these areas in Congress and the administration.[58] The techniques now used include formal intergovernmental bodies such as the Alaska Land Use Council; federally–sponsored cooperative arrangements such the BLM's regional coal teams; state–sponsored efforts like the Colorado Joint Review Process; and various other ongoing instances of cooperation, some based on memoranda of understanding, and some more informal.[59]

Effective management of the public lands also requires good relations with many nongovernmental organizations. Land managers must constantly deal with organizations that have been delegated direct management responsibilities. In addition, consensus–building requires—as do the Administrative Procedure Act, the National Environmental Policy Act, and many other laws—that a wide range of groups and individuals be included in the policy and management process. Much has been written about new processes for public involvement.[60] However, researchers have generally not examined interorganizational and public participation efforts in their full contexts, where bureau managers often juggle relations with their own employees, with a galaxy of other federal agencies, with state and local government, with organizations delegated to manage land, and with other groups and individuals. Such cooperative efforts often involve mixed ownerships, such as the Forest Service's Laguna Morena Demonstration Area in California. On a much larger scale are consultative efforts on development projects involving particularly widespread environmental and economic impacts, as in the intensive federal–state–local– nonprofit–business involvement in examining Colorado's Mt. Emmons project and Wyoming's Riley Ridge. Properly managing a process involving so many different organizations is a fundamentally different task than directly administering a self–contained program. Because the resulting consensus and cooperation are usually achievable in no other way, it is essential for the bureaus to make a determined effort to deal with this complex new reality. Researchers have been particularly slow to recognize and draw lessons from recent experience.

Budgeting

Comparisons between corporations and public agencies have shown budgeting to be an unusually important instrument for management in the public sector. Every president has used the budget to get at bureau behavior. The Reagan Administration gave new visibility to budgeting, but it tended to use budget cuts mainly as a blunt instrument, as in draconian cuts that were imposed on the planning system of BLM, rather than as a careful tool of management. In fact, Secretary Watt began to dismantle the managerially effective budgeting system that had been used throughout the Interior Department in the last decade. For the first year since 1972, in 1983 the bureaus were not required to file with the Department a Program Strategy Paper outlining and justifying their budget proposals and preparing special analyses as requested by the Secretary.[61]

The bureaus have long known the importance of budgeting to their internal management. The failure of the governmentwide Programming-Planning-Budgeting reforms of the 1960s did not deter the Forest Service from adapting the idea to its needs, and the other federal land agencies have to some extent followed suit.[62] As a result of these changes, the quality of information on public land decisions has greatly improved. Unfortunately, in many cases the information has not influenced decision making, as with the Forest Service data showing significant differences within and between National Forests in the unit costs of road building and timber production. And in some cases, the data have been used to obscure the real choices available. For example, the Forest Service maintains that its 1980 national program proposal under the Resources Planning Act was the one that maximized present net value. However, the analytical exercise was so tightly constrained that many alternatives that would have had a higher present net value were never considered, while others that were considered were not reported in the public documents.[63] Greater efforts now are needed to make better use of budgeting to improve management.

Personnel

Effective management of personnel is another absolute key to successful government performance, as was shown by Herbert Kaufman's classic 1960 book, *The Forest Ranger.*[64] One instrument is the training and socialization of bureau personnel before or during government service. Forestry and range curricula have changed over the years, as in their increased emphasis on economics, but much more needs to be done. These schools generally still provide too little training in policy, law, business management, recreation, real estate, or public relations, forcing the public land agencies to go elsewhere for these skills or to retrain employees once they are there. Some specialists are best drawn from other curricula, such as experts on operations research, minerals, human resources management, and intergovernmental relations. Some needed specialties are still blocked by obsolete civil service rules. Despite years of effort by the land agencies, the Office of Personnel Management still does not recognize training in parks and recreation as adequate for mid-level park ranger positions.

Directly managing personnel is equally important to successful governmental performance. Although the most visible result of the 1978 Civil Service Reform Act was the creation of the Senior Executive Service, probably more significant were the tools it gave these and lower–level managers to get results from their subordinates, and the new whistle– blowing channels it gave all employees to combat waste and corruption.[65] Another result has been the formalization throughout the government of systems for evaluation of employee performance by supervisors. Much progress still needs to be made in identifying and measuring appropriate criteria for many public sector jobs.

A promising instrument in public agencies is management by objectives, which establishes formal goals for an organization and provides for periodic review of progress made toward them. Secretary Watt has introduced such a system for monitoring bureaus in the Department of the Interior, and a few bureaus, including the Bureau of Land Management, have introduced a similar system for monitoring the field. So far, however, these efforts have been restricted to pursuing specific policy goals rather than used to improve management generally.

Leadership

A final essential in managing the public lands effectively is leadership. Ideology is useful in steeling political leaders to the need for change within government, but to actually carry out this change they need a lighter touch that gains the support or at least the cooperation of bureau personnel. The Reagan Administration gained office partly by deriding the bureaucracy, but unless the objective is simply to destroy a bureau, this approach is not very helpful in management. As the 1981–83 debacle of Anne McGill Burford at the Environmental Protection Agency showed, an administrator who does not trust subordinates is unlikely to be effective no matter what are his or her objectives. A philosophy that assumes the worst about government is not well suited to get the most out of what government there is. Public servants respond best to leaders who show some understanding and respect for their tasks. Effective government depends in part on the morale and sense of purpose instilled by positive leadership from elected officials and political executives.

Leadership is needed, too, from career executives. The temptation for them is not, as for political executives, to be too critical of government, but rather not to be critical enough. In protecting and promoting their bureaus, career executives may resist painful but needed changes that could be brought about most successfully only by their leadership. As I have argued, some major changes are needed to adapt the public land bureaus to new tasks and to prove wrong those who claim that the bureaus in principle cannot be managed successfully. Only through strong, enlightened leadership will these bureaus avoid reacting in a simply defensive way, and will they recognize the public lands debate as an opportunity to improve management for the long haul.

Notes

1. This paper is a revision of remarks presented at a conference on "Selling the Federal Forests" held at the University of Washington, April 22–23, 1983. For expansion on many of the points made here, and for a list of the sixteen symposia referred to, see: Christopher K. Leman, "The Revolution of the Saints: The Ideology of Privatization and Its Consequences for the Public Lands," in *Selling the Federal Forests,* ed. Adrian Gamache (Seattle: University of Washington Press, forthcoming).

2. Michael Walzer, *The Revolution of the Saints: A Study in the Origins of Radical Politics* (Cambridge:Harvard University Press, 1965).

3. Kevin P. Phillips, *Post-Conservative America: People, Politics, and Ideology in a Time of Crisis* (New York: Random House, 1982).

4. Edward Shils, "The Concept and Function of Ideology," *International Encyclopedia of the Social Sciences,* ed. David Sills (New York, N.Y.: MacMillan and Free Press, 1968), vol. vii, p. 73.

5. Consider the following statements: "there is very little, if any, intellectual or philosophical justification for federal ownership and management of commercial timber lands in the U.S." (Barney Dowdle, article in Gamache *Selling the Federal Forests*); "We have yet to encounter a thoroughly convincing argument that federal government ownership and control of vast amounts of land is superior to private ownership." (Richard L. Stroup and John Baden, "Endowment Areas: A Clearing in the Policy Wilderness?" *Cato Journal* 2 [Winter 1982]); "Until we begin to privatize the public lands we will not have accomplished anything of real economic or moral value." (Steve Hanke, *Reason*).

6. Jonathan R. T. Hughes, *The Government Habit: Economic Controls from Colonial Times to the Present* (New York: Basic Books, 1977) p. 7.

7. Ibid., p. 9.

8. For the attack on Schurz, see Barney Dowdle's chapter in Gamache, *Selling the Federal Forests.*

9. Two excellent biographies of Schurz are: Hans L. Trefousse, *Carl Schurz: A Biography* (Knoxville: University of Tennessee Press, 1982); and Claude Moore Fuess, *Carl Schurz, Reformer* (Port Washington, N.Y.: Kennikat Press, 1963; originally published in 1932).

10. Fuess, *Carl Schurz: A Biography,* p. 248; Trefousse, *Carl Schurz, Reformer,* pp. 182–240.

11. I am indebted to Marion Clawson not only for this point, but for calling my attention to the distinction between socialism and rentierism.

12. ABC News/*Washington Post* Poll, April 1983.

13. The evolution of the Asset Management Program can be followed by comparing successive versions of the *Fact Sheet*, Property Review Board, Executive Office of the President.

14. Steve Hanke, "The Privatization Debate: An Insider's View," *Cato Journal* 2 (Winter 1982): 662n.

15. Daniel Poole, contribution to symposium on "Sale of Public Lands," *American Forests* (April 1983).

16. See, e.g., *The God That Failed*, ed. Richard Crossman (New York: Harper, 1950).

17. A classic work on this subject, with extensive bibliography on other treatments, is Jeffrey L. Pressman and Aaron Wildavsky, *Implementation*, 2d ed. (Berkeley, Cal.: University of California Press, 1979).

18. Robert H. Nelson, "Modern Policy Lesson from the History of Past Public Land Policies," U.S. Department of the Interior (July 1979).

19. See generally, Paul W. Gates and Robert Swensen, *History of Public Land Law Development* (Washington, D.C.: Government Printing Office, 1968).

20. Steve Hanke, "Land Policy," in *Agenda '83: A Mandate for Leadership Report*, ed. Richard N. Holwill (Washington, D.C.: Heritage Foundation, 1983), p. 184.

21. In fairness to Hanke, he has cited the privately owned Eiffel Tower in arguing that "from a technical point of view, there is no reason why private ownership cannot be used," even for national shrines. See "The American Debate: Should We Sell Our Public Lands to Private Owners?" sponsored by the Roosevelt Center for American Policy Studies, Washington, D.C., August 29, 1982 (broadcast on Cable Satellite Public Affairs Network).

22. See, e.g., Stroup and Baden, "Endowment Areas."

23. Adolph A. Berle and Gardiner C. Means, *The Modern Corporation and Private Property* (New York: MacMillan, 1935); and Edward S. Herman, *Corporate Control, Corporate Power* (Cambridge, England: Cambridge University Press, 1981).

24. Two such references to the Empire State Building can be found in Richard L. Stroup, "In Defense of Asset Management: The Privatization Component," in Committee on Energy and Natural Resources, U.S. Senate, 97th Congress, 2nd Session, *Workshop on Land Protection and Management* (June 1982), pp. 160, 170.

25. Interview, April 18, 1983.

26. Gene Wunderlich, *The Facts of Agricultural Leasing*, USDA, Economic Research Service, Staff Report (forthcoming); Linda K. Lee, *Linkages Between Landownership and Rural Land*, USDA, ERS, Agricultural Information Bulletin 454, p. 5.

27. Executive Order No. 12411, "Government Work Space Management Reforms," *Federal Register*, March 31, 1983.

28. U.S. Department of Transportation, "Agreement Reached for Redevelopment of Union Station," March 2, 1983.

29. This section is based on the conclusions of my article on "The Revolution of the Saints," in Gamache, *Selling the Federal Forests*.

30. Joseph Schumpeter, *Capitalism, Socialism and Democracy*, 3d ed., (New York: Harper, 1950).

31. This argument is also made in Christopher K. Leman, *The Collapse of Welfare Reform: Political Institutions, Policy and the Poor in Canada and the United States* (Cambridge, Mass.: MIT Press, 1980), chap. 8.

32. See Christopher K. Leman and Robert H. Nelson, "Ten Commandments for Policy Economists," *Journal of Policy Analysis and Management* 1 (Fall 1981).

33. Interview, February 16, 1983.

34. USDI, *A Year of Change: Preparing for the 21st Century* (1982).

35. *Small Tracts Act*, 96 Stat. 2535 (January 12, 1983).

36. From USDI, *Public Land Statistics*, volumes for 1970–82.

37. Marion Clawson, *The Federal Lands Revisited* (Washington, D.C.: Resources for the Future, forthcoming).

38. Ibid.

39. Better Government Association, "Disposal of Federal Surplus Properties," report to the Senate Committee on Governmental Affairs (February 25, 1982).

40. Bruce Babbitt, "The Sagebrush Rebellion," *Catalyst* 7, no., 3, p. 10.

41. See, e.g., Donald J. Bieniewicz and Robert H. Nelson, "Planning a Market for Federal Coal Leasing," *Natural Resources Journal* (July 1983).

42. See Dowdle, in Gamache, *Selling the Federal Forests.*

43. Wunderlich, *The Facts of Agricultural Leasing.*

44. United States and State of Minnesota, "Beltrami Island Lease," August 2, 1940, and various subsequent addenda.

45. Christopher K. Leman, "The Canadian Forest Ranger: Bureaucratic Centralism and Private Power in Three Provincial Natural Resources Agencies," presented to annual meeting of the Canadian Political Science Association, Halifax, Nova Scotia, May 1981.

46. Steven L. Yaffee, "Roles of Nonprofit Organizations in Managing Public Lands," in *Proceedings of the 48th North American Wildlife and Natural Resources Conference,* ed. Kenneth Sabol (Washington, D.C.: Wildlife Management Institute, forthcoming).

47. See Sally K. Fairfax, "Beyond the Sagebrush Rebellion: The BLM as Neighbor and Manager in the Western States," this volume, Ch. 3.

48. Philip Metzger, "Public-Private Partnerships for Land Protection," in Sabol, *Proceedings.*

49. Surface Mining Control and Reclamation Act of 1977.

50. See William E. Schmidt, "U.S. May Lose Control of Western Rangeland," *New York Times* (February 14, 1983).

51. Generally on the importance of delegation by government agencies, see: Lester Salamon, "Rethinking Public Management: Third–Party Government and the Changing Forms of Government Action," *Public Policy* 29 (Summer 1981); see also General Accounting Office, *Civil Servants and Contract Employees: Who Should Do What for the Federal Government?* FPCD–81–43 (June 19, 1981).

52. Public Land Law Review Commission, *One Third of the Nation's Land* (Washington, D.C.: Government Printing Office, 1970), pp. 128–29.

53. General Accounting Office, *Public Rangeland Improvement—Slow, Costly Process in Need of Alternate Funding,* RCED–83–23 (October 14, 1982), p. 36.

54. See Marion Clawson and Jack L. Knetsch, *Economics of Outdoor Recreation* (Baltimore: Johns Hopkins University Press for Resources for the Future, 1966), chap. 14.

55. Office of the Secretary , USDI, "A bill to amend the Land and Water Conservation Fund Act of 1965, as amended, to provide for the administration and utilization of recreation fees, and for other purposes" (February 22, 1982).

56. See, e.g., Barry S. Tindall, "Fees and Charges in Recreation and Parks: A Review of State and Local Government Actions," in Committee on Energy and Natural Resources, *Workshop on Land Protection and Management.*

57. Sally K. Fairfax, " 'I Talk to the Trees . . .': Scholarly Influence on Public Lands Policy, 1951–76," in Randy Simmons, ed., *Politics vs. Policy: The Public Lands Dilemma* (Durham, N.C.: Duke University Press, forthcoming).

58. Fairfax, "Beyond the Sagebrush Rebellion."

59. See generally, Christopher K. Leman and Robert H. Nelson, "The Rise of Managerial Federalism: An Assessment of Benefits and Costs," *Environmental Law,* 12 (Summer 1982); and Western Interstate Energy Board, "Shared Decision Making on Federal Lands and Resources" (1981).

60. Generally on this trend, see Richard B. Stewart, "The Reformation of American Administrative Law," *Harvard Law Review* 88 (June 1975) and *Citizen Participation in America,* ed. Stuart Langton (Lexington, Mass.: Lexington–Heath, 1978). On public participation in natural resources, see "Symposium on Public Participation in Resource Decisionmaking," *Natural Resources Journal* 16 (January 1976).

61. An article on this system by its architects is: Laurence E. Lynn, Jr. and John H. Seidl, "Bottom–line Management for Public Agencies," *Harvard Business Review* 20 (January–February 1977).

62. See Lee L. Gremillion, James C. McKenney, and Philip J. Pyburn, "Program Planning in the National Forest System," *Public Administration Review* (May–June 1980).

63. See Christopher K. Leman, *Resource Assessment and Program Development: An Evaluation of Forest Service Experience with the Resources Planning Act, with Lessons for Other Natural Resources Agencies*, USDI, Office of Policy Analysis, Discussion Paper Series (1980).

64. Herbert Kaufman, *The Forest Ranger: A Study in Administrative Behavior* (Baltimore: Johns Hopkins University Press for Resources for the Future, 1960).

65. Civil Service Reform Act of 1978.

6

Maximizing Public Land Resource Values

RICHARD GANZEL

Resource values associated with public lands have always been recognized as valuable by American society. From the earliest days of independent governance, these resource values have been allocated among prospective claimants as a tool of conscious public policy (Gates 1968). A pool of prospective settlers, especially among the poor, ensured that many of the risks and social costs entailed in relocating into a strange, dangerous land would be borne by individuals and families. Society need bear only the costs of parceling out the frontier heritage, along with occasional outlays for conquest, purchase, defense, and exploratory expeditions.

This mode of public policy implementation worked. Moreover, scholars now conclude that when the government sought to speed and direct the pace of settlement by working through large business enterprises, most notably railroads, these efforts also worked (Mercer 1982). Finally, the notable exercise in state creation out of ceded, purchased, and conquered lands succeeded on a grand scale unimaginable at the beginning of the Republic. These successes are properly emphasized, though each created unfulfilled moral or legal expectations as the purposes behind national policy changed (Ganzel 1982a).

Resources perceived as valuable, by definition always have been scarce. This fact is obscured by the contemporary polemic regarding measurability of natural resource scarcity that conveys more heat than insight (Simon 1981, 1982; Ehrlich 1982), that proceeds in terms of imcompatible paradigms (Dunlap

The author has been engaged in research supported by the State of Nevada and by the Western Office of the Council of State Governments that is reflected in portions of the analysis presented in this chapter. Support is gratefully acknowledged. When appropriate, parallel analyses in funded reports are cited in the text. Time also was made available by the College of Arts and Science of the University of Nevada at Reno to facilitate field research.

1983), and that misses a key aspect of scarcity by insisting upon economic or ecological definitions. If Americans increasingly value wilderness (Nash 1982) or outdoor recreation and an unspoiled environment (Milbrath 1983), scarcity may be manifested in a variety of ways. Higher prices may be one symptom.[1] Another is deflection of competition into the political sphere, to be experienced as a struggle for dominance (Kincaid 1983), an effort to avoid dominance (Deese and Nye 1981), or as increased societal commands through laws and regulations and through increased appropriations to secure their implementation (Nienaber 1983).

Scarcity frequently has been experienced in the form of conflict over public land resources. Guy Benveniste's perspective seems especially pertinent to public lands conflict:

> I believe there is excessive conflict in decision making on behalf of the environment. By this I mean that constant environmental conflict in the legislatures, in the courts, and in the regulatory process itself, is not only time–consuming, inefficient, and expensive, but that, more important, these conflicts generate unforeseen consequences. The whole edifice of regulation emphasizes *procedural* protections which can easily be converted into protective *defenses* by bureaucrats intent on defending petty interests. The increasing role of the courts in arbitrating environmental issues means that long before a conflict reaches the courts, the policy issue has been formulated in a language that fits the way courts consider problems, with more concern with procedural matters than with the outcomes (1981, p. xii).

Though much conflict is inherent in differing values regarding how these resources should be allocated and used, much could be reduced or even eliminated. But a prerequisite is a thorough rethinking of fundamental resource management principles, followed by a carefully crafted program including institutional redesign, administrative devolution, and the selective divestiture of some public lands.

Part I of this chapter outlines four resource management principles and considers the ramifications of two of them by examining how they have been reflected in public land administration and regulation. This analysis culminates in a consideration of changes in bureaucratic roles and agency interaction with Congress. Part II probes the competing logics that have yielded contradictory initiatives and have exacerbated conflicts over the public lands during the Reagan Administration. It emphasizes federalist dimensions of the conflict between the Good Neighbor and Privatization policy strands, and explicates the failure of Asset Management to develop clear, consensual rationales to justify alleged improvements in efficiency. Part III outlines an alternative approach, emphasizing the third and fourth principles of resource management. Their implications are explored through three abstracted illustrations of typical cases. Finally, a brief analysis is offered of the effort by Congress to design an approach to one of these cases.

A critical theme running through the analysis is that contemporary multiple–use management principles have been extended far beyond their appropriate roles; useful where resources uses are compatible and as guides for design of large–scale, multiple–purpose *projects*, they impede efficient societal allocations when applied more generally. Congress has applied an effective remedy of zoning and clear management mandates for goals ranging from preservation to habitat management. Arguably, that remedy might be extended to maximize enjoyment of certain other resource values (Nelson 1982, 1983). However, the three cases examined here seem to require distinctive approaches. Ski resort siting involves and affects local and regional planning, necessitating an anticipatory intergovernmental approach. Concentrated recreational utilization requires attention to tenure and conversion stipulations. Urban fringe land conversion entails decisions on the rate and location of development that traditionally have been state and local functions. The federal role arguably should be confined to transitional zoning and timely transfers of land to be sold by others.

I. Principles of Resource Maximization

For the purposes of the analysis that follows, there are four essential principles of resource allocation which underlie the political economy (Stone and Harpham 1982) of resource maximization. A few preliminary stipulations will clarify and distinguish the approach adopted here. It must be emphasized that the analytic framework does not answer the question of which mix of values from a given resource base will be optimal for a society. And, unlike traditional economics, it does not assume that the societal decision will emerge from the cumulative decisions made by individuals in a market setting. Social decisions are made through societal designs of institutions and are subject to repeated change as results are assessed and opinions change. With an approach which starts "at the bottom" with specific land parcels, the particular values associated with each parcel receive emphasis. The tendency to cumulate countless local value compromises inherent in the existing approach (Cortner and Schweitzer 1982) may thereby be minimized. If the parcel-by-parcel approach fails to yield society's goal for a particular resource, one can simply work backward and adjust decisions for more marginal parcels until the optimum is achieved. Even if the approach is not adopted, it effectively clarifies the issues at stake in multiple value resource decisionmaking.

The first principle of resource maximization is to isolate for each parcel the value to be optimized. The rationale for such isolation, which entails ranking values present for that parcel, lies in the traditional economic principle of comparative advantage explicated in basic texts. The second principle is the assignment of responsibility for achieving maximization to someone: individuals, firms, or government agencies. It is curious that this obvious principle seems to have been rediscovered in the literature on the "commons" (Hardin and Baden 1977) in recent years. American progressives may, after years of neglect, be reembracing the utility of the market place in resource decisionmaking, at the very moment that some property rights theorists are seeking to demonstrate its superiority for all types of values to be optimized (Baden 1980; Baden and Stroup 1981). The manner in which the United States has applied the principles of value ranking and assignment of implementation responsibility is examined in this and the following section.

The third and fourth principles of resource maximization are linked to the second principle. Respectively, the agency assigned responsibility for maximization must have a commitment to achieve that goal, and must have adequate resources to pursue it in practice. Under the right circumstances, which in many cases must be "reinforced" via regulations or constraints on property values, desires to maximize profits may assure optimal commitment to maximization of commercial or commodity values. So may a desire of an individual to provide for family and heirs. And so may a clear statutory mandate for a government agency. In each case, optimization depends upon commitment, however achieved. But an impoverished individual, or a firm which finds itself in marginal circumstances, is no more likely to be capable of maximizing values in accord with a society's preferences than in a bureaucracy lacking budgetary appropriations for investment. Both principles must be fulfilled if we are to assess comparative capabilities rather than rely upon ideological incantation regarding the virtues of federal, state, or private ownership.

Basic to the complex task of applying principles to actual resource planning is the fact that our resource heritage has not been received in neatly segregated increments. Resources overlap and coexist, as do claims already sanctioned by society (Hagenstein 1982). The existing welter of claims, leases, and permits persuades many that significant public land policy reform is extremely unlikely. But the past will not dictate the future if substantial dissatisfaction exists.

The roots of forest reservation and of the Forest Service lie in such perceived or real dissatisfactions: poor husbandry (Pinchot 1947); fear of forest fires (Culhane 1981, p. 45); resource depletion (Olson 1971); and, more generally, American progressive commitment to doctrines of governmentally directed scientific management of natural resources for their maximization (Hayes 1959). Today, leading analysts compare public agency husbandry of forests unfavorably with that of their private counterparts (Clawson 1976); dismiss earlier depletion scares as manifestations of confusion about what investment decisions are socially optimal under conditions of "frontier" resource abundance; and challenge a progressive ideology that has paled into an ill–defined, mythology–repeating embrace of federal retention of all the public lands as somehow in the public interest (e.g., Wyant 1982). But before one can discuss alternatives meaningfully, one must deal with the first principle: isolating and ranking values to be maximized. It is useful to approach this problem by reviewing what has been done in the past.[2]

Beginning in 1872, Congress "reserved" large areas of the public lands for national parks, forests, and monuments, thereby closing them to various forms of entry and claim. Eventually distinctive management mandates were formulated for the different areas set aside. Mandates have undergone steady revision. Instructive for present purposes is the case of the forest reserves.

The authority for forest reservation was approved in 1891, followed in 1897 by the mandate to maximize watershed and timber values via a newly created Forest Service. This Forest Service mandate reflected the consensual judgment of public and private foresters and scientists of that era, who had been clamoring for its creation. They regarded good timber management and good watershed management as opposite sides of the same coin— as reinforcing, as well as compatible, values. The agency mandate therefore clearly followed the maximizing principle of isolating, and then ranking, values according to their compatibility. Other uses, such as grazing or recreation, were subject to agency discretion regarding the extent to which they had negative impacts on the two dominant values. If the impact was deemed negligible, they would be permitted (and eventually even encouraged as good public relations and as a fuller implementation of the utilitarian credo of scientific management) (Ganzel 1983a, pp. 125–27).

Twentieth century evolution of Forest Service policy, and subsequently of its management mandate, is away from clearly ordered rankings. The elevation of other values, especially recreation, to equality with timber and watershed, and the maximization of agency discretion in making specific allocations among and within particular forests comprising the system were main goals. Not only did modified practice precede revision of the mandate, but the Forest Service then practiced the time-honored tradition of sailboat politics, allowing the political winds generated by recreationists to push it into lobbying for a new discretionary mandate. That goal was enacted as the Multiple Use Sustained Yield Act of 1960 (Dana and Fairfax 1980).

Sustained yield concepts retained their original linkage to the scientific roots of forest management (Vaux 1982). However, lacking any obligatory ranking of competing values, or clear notion of how noncommensurable values are to be compared, multiple—use management at best is bereft of principles and adrift in the winds of pressure politics. At worst, a reminder of the warning of Benveniste quoted above, substance has become the captive of process. Under laws of the 1970s, decisions enshrined in a plan adopted in accordance with guiding regulations can be defended in court. Those failing to meet such process requisites are subject to successful challenge in the same courts (Coggins and Evans 1982). Central to the new planning processes and project evaluation processes is the assurance of ample opportunity for well—organized interests to influence outcomes. Assuring such influential roles obviously means that much of the substantive content of multiple–use allocational decisions is devised essentially *ad hoc*. It is determined whenever a plan is formulated or revised, and especially whenever a private project is formulated that would utilize Forest Service lands.

It most decidedly had not been the intention of the trained foresters in the Service to plunge headlong into the thickets of interest group pressure politics. Instead, the agency's twin goals were to strengthen the role of professional expertise, and to ward off predatory "land grabs" that were transferring prime recreational lands to other agencies and threatened permanently to commit its specially protected lands into wilderness. But after having spent decades helping to raise public consciousness regarding nontimber values within its domain, it failed in the effort to defend that terrain. At the same time, it failed to insulate itself from the new forces calling for detailed environmental reviews and for comprehensive public land use planning.

Beset by such pressures in the 1970s, the Service responded by developing an elaborate computerized planning system, which appears to have added little to its ability to exercise decisive control over choices among competing values (Leman 1980, 1982). Indeed, a further irony is that its abstracted but quantified national plans may be more vulnerable to the shifting priorities of determined Secretaries of Agriculture, working via budgetary manipulation and revision of long-term targets for timber output (U.S. Department of Agriculture, Forest Service 1982).

As the Forest Service commenced the juggling exercise that is multiple use management, Congress moved in the opposite direction. At least it did so when it could make up its mind. Otherwise, it sought clarification via further study—notably in its creation of the Public Land Law Review Commission in 1964. The lasting congressional legacy of the 1960s and 1970s did not rest upon the reports commissioned. Instead, it is found in congressional actions that established the primacy of one or another value for particular land areas.[3]

Despite my efforts to be highly selective, my "short list" of actions fitting into the category of innovative dominant use classification and management would include:

1. the Cape Cod National Seashore Act of 1961, creating an urban/recreational park via purchase and condemnation;
2. the Land and Water Conservation Fund, created in 1964 to assist state recreation programs;
3. the Wilderness Act of 1964, creating the process for the assignment of lands to a National Wilderness Preservation System;
4. the National Wildlife Refuge System Administration Act of 1966, consolidating and clarifying administrative principles of protected habitats;
5. the National Historic Preservation Act of 1966, introducing cultural values into the protected category;
6. the Wildlife Protection Act of 1966, adding refuges for endangered species to those of "harvested" species;
7. the Wild and Scenic Rivers Act of 1968, protecting streams with wild, scenic, or recreational values;
8. the National Trail System Act of 1968, creating state and metropolitan trails, national recreation trails, and national scenic trails;
9. the Redwood National Park Act of 1968, reallocating prized commercial redwood stands from timber to recreational/preservation priority;
10. the National Environmental Policy Act of 1969, introducing reviews overseen by a nonland–oriented agency to public land–use planning;
11. the Geothermal Resources Act of 1970, creating a resource classification process and a leasing process to permit commercial development under federal mineral law rather than state water law;
12. the Alaska Native Claims Settlement Act of 1971, making major land reallocations and reserving 80 million acres as a potential park and preserve area if future action finalized that tentative judgment;
13. the Endangered Species Act of 1973, extending protection to critical habitat;
14. the Trans–Alaska Pipeline Authorization Act of 1973, preempting further challenges under NEPA and inspiring the subsequent unsuccessful effort to broaden preemption through an Energy Mobilization Board;
15. the Eastern Wilderness Act of 1974, extending the Wilderness Act of 1964 into the eastern third of the nation;
16. the Federal Coal Leasing Amendments enacted in 1976, effectively terminating the leasing moratorium imposed five years earlier;
17. the Clean Air Act Amendments of 1977, effectively creating a category of pristine air not to be degraded and assigning to it international parks and large national parks and wilderness areas;
18. the Surface Mining Control and Reclamation Act of 1977, setting national standards for strip mining and initiating a program of restoration for abandoned mines; and

19. the Coal Conversion Act of 1978, barring construction of new power plants fueled by oil or gas and promoting conversion of existing plants to coal, thereby increasing pressure upon a key public land resource.

In political terms, the 1970s carried through 1980, during which the 96th Congress completed its actions and the Carter Administration served out its term. That extension would add three exceptionally important enactments to the list:

20. the Energy Security Act of 1980, creating the U.S. Synthetic Fuels Corporation and thereby adding to the pressures upon public land resources;
21. the Santini–Burton Act of 1980, authorizing land sales on an urban fringe, with most proceeds financing added protection of the Lake Tahoe Basin;
22. the Alaska National Interest Lands Conservation Act of 1980, placing 104 million acres into national parks and preserves.

In Chapter 3 of this volume, Sally Fairfax describes the efforts of the Bureau of Land Management to emulate the Forest Service by opportunistically manipulating the limited authority contained in the Classification and Multiple Use Act to circumscribe the overwhelming priority of grazing on the lands it administered. The BLM may have been correct in recognizing the relative significance of the values involved. Certainly the congressional consensus that had sustained the Taylor Grazing Act was no longer apparent. Though the Public Land Law Review Commission report, *One Third of the Nation's Land* (1971), emphasized commodity values, it can be read in contrast with earlier commissioned reports as a compromise document highlighting many other values as well. Looking only at economic values, it seems clear that recreational values of Bureau of Land Management lands are much more important than grazing values (Nelson 1982). Mineral values dwarf both and provide the hidden nationalist motivation for federal retention and management of the bulk of the lands administered by the agency.

The list presented above of significant congressional actions with long–term consequence notably excludes the Federal Land Policy and Management Act of 1976 (FLPMA). If it had been included, the reason would have been that it extended review for wilderness suitability to BLM lands. In contrast, its counterbalancing of multiple use with deference to state and local *governmental* input needs to be understood as an indication of the lack of consensus in Congress on just why these lands were to be retained in perpetuity.

Indeed, not all *were* to be retained. The Recreation and Public Purposes Act (R & PP) of 1926, under which communities, counties, and states can obtain land for a wide variety of public needs, was reaffirmed. Mining companies are still able to perfect patents on mine sites and to obtain land for support communities when that is necessary or convenient. And only four years after enacting FLPMA, Congress resurrected its tradition of using land disposal when it served clear public purposes. The Santini-Burton Act of 1980 in fact finds two public purposes, responsiveness to the desire of communities for expansion and the desire to protect fragile Lake Tahoe ecology, and links them through sales that fund purchases.

Rather than terminating the history of disposal of unreserved lands, Congress merely awaits formulation of acceptable rationales for particular new disposals. That element of its stance toward the turf administered by the BLM can be captured in nomenclature. These millions of acres remain merely "public" lands (U.S. Department of the Interior, Bureau of Land Management, 1982), not rangelands as desired by many of the Bureau's range scientists, and certainly not anything as definitive as the "national interest" lands set aside to serve clearly delineated national purposes in Alaska. Similarly, the Bureau's real management mandate seems to be (1) the performance of studies in response to project proposals, and (2) oversight, financing, and acting as one among many participants in the planning exercises that eventually make resource allocation decisions. As it has come to understand that it is merely the participant with ready cash to fund needed studies, it has found it easier to participate as a partner in cooperative resource management (Wilson and Lundberg 1983).

Over these two decades, the manner of exerting influence on the agencies seems to have changed significantly. The days of a Senator McCarren are gone. Congressman Seiberling can still use his key committee role to induce aggressive Park Service initiatives toward other public turf and toward individuals who own property within the parks (*Wall Street Journal* 1982). But other members of Congress can secure General Accounting Office studies to blunt such initiatives and retain decisionmaking flexibility (1979, 1981). The successors to McCarren can still counter initiatives to raise grazing fees, and even convert them into a rationale for public investments and user investments out of existing fees—the essence of the Public Rangelands Improvement Act of 1978. No one, however, is in a position to dominate an agency or even influence decisively the terms of discourse. Perhaps the real, albeit ironic, virtue of multiple–use concepts is that they strengthen the collective role of Congress at the expense of particular interest groups and their allies in Congress.

II. The Reagan Quandary

The broad–based groundswell of western opposition to the public land policies of the Carter Administration (Ingram and Cortner 1981) was reflected in Ronald Reagan's sweep of the region. However, the nature of the mandate conveyed was far from clear. Although the region in a number of respects may be regarded as conservative, that conservatism has found ample room for a thriving partnership between business and government— especially the federal government (Wiley and Gottlieb 1982; Garreau 1981). Clearly, *that* role of government was not a target of serious regional opposition, not from supporters of Reagan at least. Nor has a strong shift of sentiment been documented that would now endorse the performance by states of roles hitherto performed by the federal government (Wilcox 1983a, 1983b). Indeed the new dominance of state governments in the region by Democratic governors following the election of 1982 can be read as an effort to ensure that policy concerns not be sacrificed to the extraneous ideological commitments of some of the advisors who surround the president. The states had been used to check Carter; they were similarly available to check Reagan.

The Reagan Administration's honeymoon with the West was brief. The widely opposed Krulitz Opinion on federal reserved water rights was withdrawn. The expensive and intrusive public land planning process was corraled through revised regulations and truncated budgets. The manifest tilt toward environmentalist values was replaced with a tilt toward development (Friends of the Earth 1982) that, however, quickly went beyond both national and regional sentiment. Secretary of the Interior James Watt elicited a warm regional response with his promise to follow the path of the Good Neighbor in dealings with the states (Ganzel 1982b).

There was unintended irony in the embrace of a slogan that had been coined in the 1930s. Then, it symbolized a shift in policy toward Latin America from a stance widely perceived as imperialistic to one of cooperation. Try as it might, however, it was hard for the United States to be a mere sovereign equal when it worked with lesser sovereigns. Would it be different in working with the "sovereign" states of the West? Would it defer to the states on coal policy? On oil and gas leasing? Could it abide in all cases by its promise to respect state prerogatives in water law administration when it had a congressional mandate to ensure that there was adequate water for wildlife as well as wild horses and burros? The answer to each of these questions had to be obvious. By law, the executive agencies had to enforce existing law and their diligence would be watched closely. Consequently, these conflicts of national and regional interests would plague any administration under virtually any version of federal law realistically imaginable.

But the administration stumbled into a regional conflict that was as unnecessary as it was unexpected. Nevada had elicited considerable sympathy in the West when it sought to reestablish the obligation of the federal government to retain an active disposal policy (Ganzel 1982a). Secretary Watt therefore saw opportunity in facilitating free or low cost transfers to state and local governments. He made his attitude clear to federal officials who oversaw the planning process. He wanted stalled R & PP transfers consummated quickly. At the same time, he extended an invitation to western state governments to submit lists of new requests. Though many states were unprepared, or saw the request as one that would be available indefinitely, their prompt responses totaled almost a million acres requested (Davis 1982; Ganzel 1982b). At the same time, Secretary Watt added his support to an effort dating back to the Carter Administration that was designed to settle pending state land claims (Ganzel 1983b). Together with increased consultation and selection of administrative appointees who were imbued with the traditional ethic of cooperative resource development, these initiatives suggest the main elements of the Good Neighbor Policy (Koch 1981).

Trouble began with the presentation of the proposed budget for fiscal year 1983, which contained estimated revenues of $1.2 billion from land sales. On February 15, 1982, supportive testimony by David Stockman indicated that much higher revenues were planned for future years (1982). The same day, President Reagan issued Executive Order 12348, creating an interagency Property Review Board with authority over review and selection of properties to be sold.

Apparently because the common task was selection of property to be sold, and because one mutual goal was revenue maximization, it was deemed appropriate to lump public lands together with surplus land and related improvements that had served specific governmental purposes. That fateful step entailed two strategic errors. It created a commonality of interest among opponents to particular sales that otherwise would not have existed. Ordinarily, for example, public land interest groups would have had little interest in a coveted military base on the beach of Waikiki in Hawaii. Now, interest groups coalesced. The second error was that it launched an internal struggle for control of the Property Review Board that pitted economists in the Office of Management and Budget (and their allies elsewhere in the executive branch) against the land agencies (Hanke 1982). By late August, the latter declared victory.[4] Meanwhile, confusion reigned and administration supporters were dismayed at the loss of momentum and opportunity.

But even in the absence of that institutional blunder, the Asset Management Program to be overseen by the Property Review Board was saddled with a congenital defect that was not remedied by reestablishment of land agency control over public land review. It lacked a *politically* persuasive rationale. In an era marked by relative balance among contending interest groups, each wishing to maximize a different value, those who hope to achieve significant innovations have two choices. They must persuade affected parties on the merits of proposed changes or offer them valued rewards as part of a more comprehensive package (Thurow 1980). The proponents of land sales could do neither.

The merits of some land sales or disposals may be substantial. But the administration was advocating a general program of sales and it seemingly lacked an appreciation of the particular instances that possessed distinctive merits. Instead, it relied heavily on alleged commercial efficiencies and explained them in terms of the virtues of private property and a market economy, as Christopher Leman describes more fully in Chapter 5 of this book. A hunter seemingly asked to give up access to a favored hunting site on public lands needs more than abstractions to be persuaded. More was not forthcoming. Indeed, the obvious glee experienced by James Watt in taunting environmentalists transformed their skepticism into angry resistance.

A second illustration underlines the vacuum that substituted for applicable principles in the sales proposals. A rancher with established grazing privileges on the public lands might long for ownership, instead of facing continuing uncertainty over possible shrinkage of grazing allotments, and also for relief from the sometimes sizeable costs of interacting with a federal landlord (Dobra and Uhimchuk 1982). In addition to confronting new local property taxes after achieving ownership, however, the rancher was being asked to bid against an unknown number and variety of prospective owners of the land which now provided relatively secure privileges (Gregg 1982). Among themselves, ranchers lamenting their profit levels commonly joke that they all are fools. Very few, however, were foolish enough to endorse competitive sales of grazing lands. If hunters were unwilling to risk a few weekends of pleasure through sales of hunting lands, ranchers were not going to risk their way of life. Certainly not to reduce the national debt by a few billions while budget deficits were adding hundreds of billions to the burden. And not to affirm a private property ideology regardless of its mode of application at their expense.

Lost in the shuffle of arguing about improving the efficiency of economic endeavors such as ranching, by more firmly establishing property rights, was an important fact. Grazing on public lands entails federal subsidies of hundreds of millions of dollars annually, with no end in sight (Nelson 1979). Ranchers receive a portion of the subsidy, but not the bulk of it. Much fee income goes to state and local governments as shared revenues. However, the bulk of the annual federal cash outlay has gone into the planning and administrative process that now supports multiple uses of these lands. Hence, it is the hunters, fishers, hikers, rockhounds, and so forth who, from the perspective of the federal budget, freely use resources that are not free of administrative costs.

In contrast, state governments ordinarily recover a portion of recreational costs by licensing some of these activities. Logically, if administration of these lands were transferred to the states, more users would certainly be required to pay the real societal costs of enjoying their values. Alternately, if the lands were to be transferred (preferably also at low or no cost) to ranchers, with access for other users guaranteed and other protections assured, a privately administered fee structure would emerge. Realistically, however, many users would continue to enjoy free use under the private alternative because collection costs would be excessive for many private managers. Unable to tax or levy license fees, those managers would be forced to raise fees for some users, with resulting inequities and in many cases management that emphasized only values that could produce collections. For these reasons, state management would be preferable in many cases.

A third logical alternative would be a realistic, enforced federal fee schedule for all users of present grazing lands that serve additional purposes. The alternative is not promising (Nelson 1982), even though the Reagan Administration may finally succeed in raising entrance and user fees for preserves such as parks (U.S. Department of the Interior, National Park Service 1982). Realistic fees assist in rationing especially scarce resources and in allocating various public land resources. The existing substitute for graduated fees is an automated ticketing procedure that is biased toward those able to plan far in advance, including governmental and business conference organizers. Though its superiority on "equity" grounds is dubious, it would have to be dismantled over the objections of its beneficiaries. Even more fundamental, however, is the predictable resistance of states to further federal intrusion into their traditional police powers.

Though lists of BLM and Forest Service lands marked for possible sale have been issued by the Reagan Administration, it now seems probable that only small sales will occur and that their rationale will be the long–accepted but rarely implemented goal of ridding federal holdings of administrative nuisances. Waters have been muddied too much to secure prompt approval of authority for new sales, and a persuasive rationale has yet to be articulated by the Reagan Administration.

Despite the much greater attention accorded land sales and asset management, Secretary Watt's Good Neighbor initiatives remain in somewhat muted effect. They have worked to counteract conflicts over mineral leasing and to quiet fears generated by the threat of land sales. Recently, regulations have been issued which would allow permittees under contract to manage their grazing allotments for ten-year periods without interference (*Public Land News*, May 13, 1983).

No overall assessment of the impact of the Reagan Administration's approach to public land policy can be made on the basis of the few elements discussed here. However, it seems clear that the general effect of the Good Neighbor element of that policy has been to facilitate and initiate a number of positive small changes and to remove particularly provocative thorns that irritated intergovernmental relations. Moreover, at least some users of the public lands in addition to ranchers now see federal policy as more sensitive and more responsive. Taken alone, these small steps bring a measure of common sense to administration. Of course they cannot be taken alone, however. Gains must be weighed against sharp conflicts over mineral leasing, intensified conflict with some users over environmental and wilderness policies, and the failure to justify land sales as an important element of resource value maximization. Part Three addresses such maximization justification.

III. Another Approach to Maximization

The Reagan Administration has failed to articulate suitable goals to justify changed ownership or administrative responsibility for public land resources. Its failure may be traced not only to the desire to maximize federal revenues achieved through sales but also to the very notion of the federal government conducting those sales. But its failure does not imply that no good reasons exist that would justify substantial changes in federal public land policies, including sales and transfers. Examples could be multiplied at length to illustrate the discrepancy in practice between the four principles of effective resource policy enumerated in Part One of this analysis and actual practice. A few will suffice, since the point is to urge responsiveness of reforms to specific complaints, and use of the principles formulated as guides rather than rationales.

Yellowstone National Park

The summer situation, as Secretary Watt has noted, cries out for reform. Existing physical structures and roads are deteriorating despite diligent efforts of the park staff. The scene evokes a "metropolis in the woods" sensation once the awe inspired by the natural splendor has subsided. More investment is needed, but it will merely offset impacts of overuse and increase the resort to artificial structures (Sax 1980) in an effort to cope. The park must be protected through rationing. Current entrance fees do not assist in that process. So low that they are mere nuisance charges incurred while eliciting information from the gatekeepers, they fail to persuade any prospective users to visit instead less–crowded sites nearby. At the same time, these low charges ensure that nonusers of the parks will pay for what maintenance gets funded.

Timber Production

We commonly think of the escalation of energy prices during the 1970s as a severe crisis. Yet the price of softwood lumber used for construction of homes rose 787 percent between 1955 and 1977 (Wolf 1981, p. 460)! The roots of this price escalation lie in (1) population growth, (2) low levels of investment in new growth in public forests, and (3) reallocation of forests to nontimber uses. Meanwhile, federal policy subsidized consumption by aiding those who cut timber, construction firms, and both homeowners and renters. Subsidies must be eliminated while a vigorous timber production program is undertaken by some suitable agent.

These two examples indicate that change may be desirable in the ownership or management (or both) of various public lands. In the remaining pages of this section, three cases are presented that are illustrative of the need to apply the third and fourth principles that were outlined in Part I, to the crafting of new policy. In presenting these cases, simplification will be achieved by an arbitrary statement of the key resource values involved and by an equally arbitrary statement of priorities. Attention will thereby be focused on the nature of the commitment of alternative implementing agents to maximization of prioritized values and on the adequacy of the resource base upon which such agents can draw.[5]

CASE 1: SKI RESORT SITING. It is possible to abstract a prototypical prospective western ski resort. The resort would be on a large scale and would be comprehensively planned to include residential and "destination resort" (restaurants, shops, etc.) elements. The base facilities would be mainly on private lands, with a few facilities and most ski runs on slopes under Forest Service jurisdiction. Access roads would have to be constructed into mountainous terrain or enlarged from existing arteries. Finally, the siting process would be driven by the prospective developer, who would have to acquire crucial private rights to accommodate the development and accumulate adequate capital to launch serious planning and construction. Quality of slopes, though important, ranks behind these requisites in importance. These characteristics in combination yield a policymaking challenge with three main dimensions.

The local and regional stake in a prospective project is substantial. Most mountainous areas are filled with low income communities and governments searching for an augmented tax base. That makes the project nearly irresistible. Second, through road access arteries and the secondary service opportunities associated with commuting users, the impacts of the project spill over into a wider region. Third, there are obvious environmental impacts directly associated with the project, and others spread throughout the affected region. How are these factors handled under existing land management patterns and practices?

Although the Forest Service engages in some anticipatory planning by trying to identify prospective ski sites within its holdings, its relevant decisionmaking is primarily reactive and usually facilitative. Involved personnel acknowledge that their response essentially entails environmental assessment under National Environmental Policy Act guidelines. Moreover, that assessment ordinarily does not ask those proposing a project to suggest two or more alternate sites or activities, as the agency itself would be expected to do in proposing revised land uses. This makes the initial tier of the review *pro forma*. The second tier review presumes that a project will be approved unless assessment uncovers serious ecological impacts that cannot be suitably mitigated or unless an intense public outcry of opposition emerges. Consequently, the review seeks negative impacts and suggests remedies to be attached as conditions to approval of use of Forest Service lands. Review is largely confined to the project itself.

Ordinarily, the Forest Service defers to local and regional governmental entities. They must deal with broader regional impacts and implications. Where such capabilities exist, the large scale of projects and their solid financial underpinnings make it possible for local entities to bargain for modifications and cost sharing, with the result that these projects are analogous to large urban projects (Nelson 1977) and to Western energy projects. However, long lag times between costs and collection of tax revenues, as well limited opportunities for local leverage because sites already have been determined unilaterally, and local entities have only limited planning capabilities, means in most cases that many spillovers will "get through" the interstices. Federal planners lack sufficient interest to cope with regional project impacts; local planners lack resources and sometimes jurisdiction to deal with the same impacts.

There are no obvious unilateral federal answers to prospective regional degradations that predictably will be associated with many ski resort projects. If one looks only for federal remedies, the reality may be excellent projects with reduced quality in the supportive hinterland. But if one instead asks what role would be most helpful from the federal public land agency, useful reforms become more apparent. The Forest Service could increase the rigor and realism of anticipatory planning, allocating an annual budgetary allotment to site review of its holdings. The goal would be designation of zone boundaries within which projects *probably* would gain approval or within which projects would be unacceptable. For probable sites, a model set of conditions could be drawn up, upon acceptance of which included lands (that is, responsibility for their administration) could be transferred to regional administrative bodies. One condition would be more comprehensive review of spillover environmental impacts beyond the project. Another might be regional transportation planning.

One additional difference likely to result from regional administration is in the method of charging rental fees for the public lands utilized. The Forest Service ties fees to resort profitability on a year–by–year basis. A flat, fixed fee, even at rates that reflect current average fees, would effectively preclude some marginally feasible projects that now obtain approval. Another difference is further progress (Healy and Rosenberg 1979; Popper 1981) toward an increased state role in ensuring that planning capabilities are adequate to meet needs and to spread the benefits of projects more equitably across local jurisdictional lines.

CASE 2: CONCENTRATED RECREATION ADMINISTRATION. Many sites with special recreation potential have been developed by federal agencies: the Forest Service, the BLM, the Corps of Engineers, and the Bureau of Reclamation. Some of these are the result of innovative multipurpose planning to take full advantage of opportunities associated with water or power projects. Others reflect federal responsiveness to recreation interest groups or to requests from state and local governments. In some of the latter instances, close intergovernmental cooperation and even joint funding have been involved in the design and creation of facilities.

Most of these recreation sites have only ordinary, rather than extraordinary, aesthetic or geological qualities. Some have relatively high maintenance costs because of their comparative remoteness. Their "gate fee" revenue potential is low, though there are exceptions. Very few are known to more than a handful of people outside the immediate vicinity. In short, almost all have only minimal national interest value. In contrast, many have intense local support and some generate needed income for nearby service providers.

These mundane sites and facilities would seem to be prime targets for transfer to state park systems or to local governments. Acreage limitations on Recreation and Public Purposes Act no or low cost transfers might be adjusted to facilitate orderly transfer of these public lands. Two associated problems can be handled with minimal creativity. One apparent problem seems to be an opportunity on closer examination.

States are reluctant to assume new burdens. They might be more agreeable if (a) transfers included the transfer of any associated mineral rights to the states for administration, and (b) guarantees were attached for the federal government to continue current levels of dollars being spent on the site for five or ten years. Minor local resistance could be satisfied by creation of state payment in lieu of tax programs that would substitute for current federal payments. The dollars involved are not great, but they are important locally.

What about the federal stake in existing facilities, which may be substantial? This seems an excellent analogy to various federal programs that subsidize creation of desired public infrastructure. The goal is to improve local productivity, well-being, or recreational pleasure. An ownership role entails continued subsidy, not a stream of profits. In contrast, states and localities capture indirect benefits in local taxes and fees. For them, operation approaches or even achieves a break-even point. It also can be linked more effectively to tourism promotion efforts, with the promise of even better future prospects.

Recreational homesites on federal lands are an interesting special case, and have been identified by the Reagan Administration as warranting further study and possible disposal. Long unwanted by federal agencies, especially foresters, they utilize only a small number of acres per site yet yield welcome tax revenues for rural counties. Some states with sizeable state lands operate homesite programs; although they are difficult to administer, programs are being implemented in other states. Federal administration reflects a dogged judgment that even though homesites may be the best use of those sites, their administration is inconsistent with agency mission priorities. Not surprisingly, this lack of commitment is reflected in suboptimal implementation.

Sites are assigned via revocable permits, rather than fixed-term, renewable leases. No formula has yet been devised for valuation assessments that captures the reduced value inherent in insecure title adequately to receive agencywide implementation, much less congressional endorsement. Supervisory budgets are low and services minimal. The final indignity is that fees generated go into the general fund, so agency personnel have minimal incentive to innovate or improve the site.

Nevertheless, little is novel about such homesite clusters. State land agencies handle a wide variety of similar fixed-site uses, relying on permits, leases, and very long-term leases or sales to achieve particular goals. The Arizona State Urban Lands Act (Gammage 1982) is a noteworthy example of state innovation in this area. Moreover, the model of codes governing planned unit developments in urban areas already exists if the option selected is to sell the land sites. Such codes contain nonconversion clauses and detailed qualitative standards that must be met before the governing board can reassume ownership powers. A sales strategy also would require agreement on a formula that recognizes the stake of current permittees in the value of the land.

What should be the federal share, if any, in the revenues that might be generated by a state land agency decision to *sell* such sites? No reliable data has become publicly available that would show if the administering agencies have profited or lost money on homesite administration. But much of the value in a sale would reflect local developments on nearby private lands which increase demand for the homesites, rather than positive federal actions. One researcher has suggested that the federal share be limited to the value of the site for its original or preconversion use, with the remaining shares presumably going toward funding state programs and local services (Behan 1983). A large federal return would involve capture of unearned rents, a small problem in this case but a much larger one in the final case examined.

CASE 3: URBAN FRINGE TRANSITIONS. As urban communities grow in the West, many of them begin to press toward federal or state lands. Some, especially small communities, are landlocked. Inevitably, there are pressures to convert some public lands from current low-intensity uses into commercial, industrial, or residential uses. On the other side are pressures to keep public lands in federal hands, to control speculation, and to assist local planners in structuring the path of urban expansion in accord with their plans and with water, sewer, and communication facility extensions.

Only in exceptional cases does the land at issue have significant national interest value; the opposite usually is the case, with permitted users struggling to control trespass, vandalism, and other manifestations of urban encroachment. On the other hand, low–cost land can be an extraordinary developmental tool, so long as it is added slowly enough so that existing market values are not threatened. Neither the national society nor the state has a strong public interest in retained public ownership that will withstand public scrutiny (national regions that are struggling against population loss may resist any policy that will even indirectly add to their problems). Both state and local governments have a vital interest in seeing that the transition from public to private ownership is handled with minimal problems.

Their interests are not identical in practice, at least if one takes early experience with implementation of the Santini–Burton Act as illustrative. Congress sought to ensure full BLM cooperation with local government in land sales in the vicinity of Las Vegas, Nevada. It has succeeded in that goal. But it also sought to sell substantial numbers of acres annually. That goal has been frustrated by federal deference, because it elevates the goals of local planners above those of developers.

Planning departments (and local elected officials) ordinarily face pressures only when a development is being proposed, not when large blocks of land are being selected for prospective sales. Therefore, planning operates in a relative political vacuum. One practical result has been BLM's difficulty in finding buyers for the lands mutually approved by it and local planners for sale. A second has been to misdiagnose the problem as excessively high prices or lack of expertise in marketing. The real problem is misidentification of desired lands.

The Santini–Burton urban fringe land sales initiative is a positive step that should be generalized into a federal program to assist urban fringe transitions in the West. But the approach needs fundamental revision. The federal effort to capture full fair market value from sales should be abandoned. The sales responsibility should be transferred to state land agencies or to local governments. New procedures to facilitate sales should be adopted in state or local implementations of sales programs. A new formula for revenue allocation should be devised. What justifies these suggestions?

The federal effort to capture revenues involves a classic example of unearned rents extracted by an absentee landlord. The current value of fringe land depends on urban growth, not current use. Land yielding grazing fees of a few dollars (or less) per acre leaps in value if conversion of use is permitted. Moreover, federal capture drains funds out of the local economy. Some of the value has been added by long–term developmental subsidies. There also are federal administrative costs entailed in identification of lands for transfer. Finally, an increment might be included to compensate other regions of the country—even if they are being freed of a subsidy burden. The total federal share probably would not exceed one–fourth. The new administering agency might receive another fourth. The final half might be split between recreation funding and general support revenues for state and local planning.

The sales procedure should aim at a simplified version of fair market value that would make room for developer nomination of prospective parcels, negotiated cost–sharing when public facilities are not available but needed, flexibility on the size of parcels to be sold, and provisions requiring sale of at least minimal numbers of acres to assure creation of a real market of willing sellers. There is no justification for complex public–bidding provisions or for multiple bids to establish alleged fair market value. Those provisions are appropriate for known mineral values, not for dusty plots that need private (and additional public) investments before they will become productive lands. Finally, speculation is intrinsic to the process and should be welcomed. Not only will the public thereby capture some value from the lands now. Additional transactions will in many cases be needed before the optimal commercial use of a parcel is determined.

Conclusion

This paper has outlined four principles of effective public land resource value maximization and has used these principles to critique existing approaches. It has argued that land parcel values must be clearly prioritized, that an implementing agent must be designated, that the agent selected must have an ideological, self–interested, or statutory commitment to the priorities established, and that the agent must have adequate resources to pursue value maximization.

The analysis of policy trends and selected case studies suggests that if maximization is to be pursued, substantial policy reform must be undertaken. Included in those reforms will be devolution of some federal administrative responsibility to local, regional, and state governmental entities. Also included will be a broadened land sales program with most revenues going to nonfederal beneficiaries. There are many more possible reforms that can be justified. Like these, they should follow specific discussions of needs or opportunities unmet by current federal public land policy.

Notes

1. A comprehensive economic calculation of real price changes over long periods of time necessarily would have to count the portion of prices paid indirectly through the political efforts and governmental activities discussed here. Though complex and subject to reliance upon estimates, such a calculation would be interesting and useful. Without it, economic commentaries on the question of increased scarcity lack credibility.

2. Except where specific interpretations are cited, this brief synopsis follows standard historical accounts while emphasizing the valuation decisions made and the implementing mechanisms selected.

3. This list includes only major innovative congressional acts that, in my judgment, established important precedents or resolved important conflicts over which resource should receive priority. The synopses attempt to encapsulate that significance and ignore other provisions of the laws cited regardless of their significance.

4. This claim emerged vividly in exchanges regarding trends in public land policy at the 1982 Salt Lake City annual session of the Western Conference of State Legislators. Land agency representatives dismissed emphasis upon and criticism of economic rationales within the Administration as irrelevant to the true intentions of the Property Review Board. Their sensitivity reflected the intensity of the struggle within the inner circles of advisers, the newness of the victory, and (as it appears in hindsight) the incompleteness of the triumph.

5. The first of these cases is based on long–term, self–interested observation, supplemented by interviews and exchanges with relevant responsible officials. The second and third cases are extracted from analyses reported to the Council of State Governments on contract. However, the emphasis here differs somewhat and these presentations contain additional observations that go beyond the report. Detailed citations are available in that study as well.

References

Baden, J. 1980. *Earth Day Reconsidered*. Washington,D.C.: Heritage Foundation.

Baden, J., and Stroup, R.L. 1981. *Bureaucracy vs. Environment*. Ann Arbor: University of Michigan Press.

Behan, R.W. 1983. Private discussion.

Benveniste, G. 1981. *Regulation and Planning: The Case of Environmental Politics*. San Francisco: Boyd & Fraser.

Clawson, M. 1976. *The Economics of National Forest Management*. Baltimore: Johns Hopkins University Press for Resources for the Future.

Coggins, G.C., and Evans, P.B. 1982. Multiple use, sustained yield planning on the public lands. *University of Colorado Law Review* 53: 411–69.

Cortner, H.J., and Schweitzer, D.L. 1982. Local production and national budgets: Gluing together public planning for forestry. Presented to the Western Political Science Association, San Diego, California.

Culhane, P.J. 1981. *Public Land Politics*. Baltimore: Johns Hopkins University Press for Resources for the Future.

Dana, S.T., and Fairfax, S.K. 1980. *Forest and Range Policy*. 2nd ed. New York: McGraw–Hill.

Davis, J.A. 1982. Congress decidedly cool to Reagan land–sale plan. *Congressional Quarterly Weekly Report* 40: 1689–90.

Deese, D.A., and Nye, J.S. 1981. *Energy and Security*. Cambridge: Ballinger.

Dobra, J.L., and Uhimchuk, G.C. 1982. Some economics of sagebrush and sovereignty. In R. Ganzel, et al., *State Sovereignty as Impaired by Federal Ownership of Land*. Carson City: State of Nevada Legislative Counsel Bureau Bulletin No. 82–1, pp. 52–85.

Dunlap, R.E. 1983. Ecology v. exemptionalist: The Ehrlich–Simon Debate. *Social Science Quarterly* 64: 200.

Ehrlich, P. 1982. Simon & Ehrlich. *Social Science Quarterly* 63: 381.

Friends of the Earth, et al. 1982. *Indictment: The Case Against the Reagan Environmental Record*. San Francisco: Friends of the Earth.

Gammage, G. 1982. The state urban lands act. *Arizona Bar Journal*.

Ganzel, R. 1982a. Public lands conflict and the future of federalism. In R. Ganzel, et al., *State Sovereignty as Impaired by the Federal Ownership of Land*. Carson City: State of Nevada Legislative Counsel Bureau Bulletin No. 82–1, pp. 4–25.

———. 1982b. Research perspectives and policy initiatives. Report to the State of Nevada Office of the Attorney General.

————. 1983a. Public land reform alternatives. In R. Ganzel, ed., *Resource Conflicts in the West*. Reno: University of Nevada Public Affairs Institute, pp. 124–41.

————. 1983b. Rationalizing public land use patterns: Experiments and options. Report to the Western Office, Council of State Governments.

Garreau, J. 1981. *The Nine Nations of North America*. New York: Avon Books.

Gates, P.W. 1968. *History of Public Land Law Development*. Washington, D.C.: Government Printing Office.

General Accounting Office. 1979. *The Federal Drive to Acquire Private Lands Should Be Reassessed*. Washington, D.C.: General Accounting Office CED–80–14.

General Accounting Office.1981. *Federal Land Acquisition and Management Practices*. Washington, D.C.: General Accounting Office CED–81–135.

Gregg, F. 1982. The federal lands today: Uses and limits. Presented to the Rethinking the Public Lands Workshop sponsored by Resources for the Future, Portland, Oregon.

Hagenstein, P.R. 1982. The federal lands today: Uses and limits. Presented to the Rethinking the Public Lands Workshop sponsored by Resources for the Future, Portland, Oregon.

Hanke, S.H. 1982. The privatization debate: An insider's view. *Cato Journal* 2: 653–62.

Hardin, G., and Baden, J. 1977. *Managing the Commons*. San Francisco: W. H. Freeman and Company.

Hayes, S.P.1959. *Conservation and the Gospel of Efficiency*. Cambridge: Harvard University Press.

Healy, R.G., and Rosenberg, J.S. 1979 *Land Use and the States*. 2nd ed. Baltimore: Johns Hopkins University Press for Resources for the Future.

Ingram, H.M., and Cortner, H.J. 1981. Uncommon interests in policy initiation: The case of the Sagebrush Rebellion. Presented to the Western Political Science Association, Denver, Colorado.

Kincaid, J. 1983. The content of body and soul: Resource scarcity in western political theory. In S. Welch and R. Miewald, eds., *Scarce Natural Resources*. Beverly Hills: Sage Publications, pp. 25–46.

Koch, K. 1981. Reagan shifts U.S. policies on public land management. *Congressional Quarterly Weekly Report* 39: 1899–1904.

Leman, C.K. 1980 Resource assessment and program development: An evaluation of forest service experience under the resources planning act, with lessons for other natural resource agencies. Washington, D.C.: United States Department of the Interior, Office of Policy Analysis.

————. 1982. Formal vs. *de facto* systems of multiple use planning in the bureau of land management: Integrating comprehensive and focused approaches. In National Research Council, *Developing Strategies for Rangeland Management*. Boulder: Westview Press.

Mercer, L.J. 1982. *Railroads and Land Grant Policy: A Study in Government Intervention*. New York: Academic Press.

Milbrath, L.W. 1983. Images of scarcity in four nations. In S. Welch and R. Miewald, eds., *Scarce Natural Resources*. Beverly Hills: Sage Publications, pp. 105–23.

Nash, R. 1982. *Wilderness and the American Mind*. 3rd ed. New Haven: Yale University Press.

Nelson, R.H. 1977. *Zoning and Property Rights: An Analysis of the American System of Land Use Regulation*. Cambridge: MIT Press.

————. 1979. An analysis of the revenues and costs of public land management by the interior department in 13 western states. Washington, D.C.: United States Department of the Interior, Office of Policy Analysis.

————. 1982. The public lands. In P.R. Portney, ed. *Current Issues in Natural Resource Policy*. Baltimore: Johns Hopkins University Press for Resources for the Future, pp. 14–73.

————. 1983. A long term strategy for the public lands. In R. Ganzel, ed. *Resource Conflicts in the West*. Reno: University of Nevada Public Affairs Institute, pp. 107–23.

Nienaber, J. 1983. Two faces of scarcity: Bureaucratic creativity and constraints. In S. Welch and R. Miewald, eds., *Scarce Natural Resources*. Beverly Hills: Sage Publications, pp. 151–77.

Olson, S. 1971. *The Depletion Myth*. Cambridge: Harvard University Press.

Pinchot, G. 1947. *Breaking New Ground*. New York: Harcourt Brace.

Popper, F.J. 1981. *The Politics of Land Use Reform.* Madison: The University of Wisconsin Press.
Public Land Law Review Commission. 1970. *One Third of the Nation's Land.* Washington, D.C.: Government Printing Office.
Public Land News. May 13, 1983.
Sax, J.L. 1980. *Mountains Without Handrails.* Ann Arbor: University of Michigan Press.
Simon, J. 1981. *The Ultimate Resource.* Princeton: Princeton University Press.
―――. 1982. Simon & Ehrlich. *Social Science Quarterly* 63: 381.
Stockman, D.A. 1982. Statement to the Senate Committee on Governmental Affairs. February 25 (xerox).
Stone, A., and Harpham, E.J. 1982. *The Political Economy of Public Policy.* Beverly Hills: Sage Publications.
Thurow, L.C. 1980. *The Zero Sum Society.* New York: Basic Books.
United States Department of Agriculture, Forest Service. 1982). *Report of the Forest Service, Fiscal Year 1981.* Washington, D.C.: Government Printing Office.
United States Department of the Interior, Bureau of Land Management 1982. *Managing the Nation's Lands.* Washington, D.C.: Government Printing Office.
United States Department of the Interior, National Park Service. 1982. *Federal Recreation Fee Report 1982.* Washington, D.C.: National Park Service.
Vaux, H.J. 1982. The application of sustained yield to areas of mixed ownership. Presented to the Sustained Yield Symposium, Spokane, Washington.
Wall Street Journal (1982). February 24, May 25, and June 14, 1982.
Wilcox, A.R. 1983a. Conflict and utilization of the public lands: Legitimacy and consensus in Nevada. In R. Ganzel, ed., *Resource Conflicts in the West.* Reno: University of Nevada Public Affairs Institute, pp. 45–49.
―――. 1983b. The Sagebrush Rebellion: Utilization of public lands. In S. Welch and R. Miewald, eds., *Scarce Natural Resources.* Beverly Hills: Sage Publications, pp. 227–41.
Wiley, P., and Gottlieb, R. 1982. *Empires in the Sun.* New York: G.P. Putnam's Sons.
Wilson, L.U., and Lundberg, F. 1983. Cooperative management on the public rangelands. In R. Ganzel, ed., *Resource Conflicts in the West.* Reno: University of Nevada Public Affairs Institute, pp. 94–106.
Wyant, W. 1982. *Westward in Eden.* Berkeley: University of California Press.

7

Realizing Public Purposes without Public Ownership: A Strategy for Reducing Intergovernmental Conflict in Public Land Regulation

JOHN G. FRANCIS

This chapter offers a proposal for the selective disposition of uses to the public lands. It should be emphasized from the outset that the form of use disposition proposed is neither the outright sale of the public lands suggested during the Reagan Administration nor a recommendation for extensive organizational improvements in the existing structure of federal land management. The proposal for selective land use disposition is assessed first in terms of difficulties confronting existing federal land management policies and second in the context of its capability to advance or retard the realization of four main objectives for the federally held lands—objectives that have over the past eighty years gathered substantial public and congressional support. They are: (1) the management of renewable natural resources according to the principles of sustained yield; (2) the prudential development of nonrenewable mineral resources; (3) the maintenance of public access and recreational use; and (4), the preservation of the largely undeveloped and largely western landscape afforded by the public lands. Because it promotes these four purposes and avoids difficulties confronting existing policies, the selective disposition of uses to the public lands should be seriously considered by policymakers.

The land policy discussed below is restricted to the "public lands." The "public lands" are defined in the Federal Land Policy and Management Act[1] to include any land or interest in land owned by the United States within the several states and managed by the Bureau of Land Management (BLM), without regard to how the federal government originally acquired ownership.[2] The proposal described below could be extended to federal lands administered by other federal agencies, particularly the Forest Service.[3] The initial restriction to the public lands is justified on the grounds that these lands have been the focal point of recurring controversy. The recent Sagebrush Rebellion, for example, was directed almost entirely at the federal lands administered by the BLM.[4]

I. Objectives for the Public Lands

From our beginnings as a nation, when the public lands stretched from the Blue Ridge to the nation's boundaries, the public purposes that should govern federal land policy have been intensely debated. At different periods in our history, different purposes have predominated. Some, such as raising revenue from the lands, have been recurrently important, while others, such as promoting settlement, have entirely faded. Moreover, purposes have been thought varyingly relevant, depending on the lands at issue. Areas of great natural beauty were singled out early for aesthetic preservation, whereas economic development has been the predominant purpose for vast stretches of the public lands. However, four purposes that prevail today—sustained yield, prudential development, public recreational access, and preservation—are each widely supported for application to all of the public lands.

Described in the introduction to this volume are nineteenth century efforts to raise revenue, promote economic development, and encourage settlement of the public lands. An additional purpose that gained support in the nineteenth century was to set aside areas of great natural beauty as well as recreational interest. The concern in this analysis is to establish the changing agenda of purposes for the public domain that took place beginning with the turn of this century.

Settlement began to recede as a major purpose for the public lands at the turn of the century. In 1891, Congress authorized a new system of forest reserves.[5] This withdrawal of land from settlement was a turning point in changing the agenda for the public lands away from the promotion of settlement towards conservation of natural resources.

Theodore Roosevelt's administration, 1901–09, initiated a program of active federal management of national forests and the supervision of subsurface fossil fuel development.[6] The emergence of the Forest Service under Pinchot during this era of conservation exemplified the new confidence that governmental institutions could undertake active management of renewable resources.[7] For Pinchot, conservation meant a set of compatible uses to be managed on a sustained yield basis. Timber production using scientific management techniques, for example, could be sustained at a high yield for generations to come, together with the preservation of watersheds. During Pinchot's tenure, the concept of compatible multiple uses gained increasing acceptance.

This shift toward retention of the lands and active management was controversial. The imposition of Forest Service grazing fees, for example, was attacked vigorously by grazers.[8] The substantial expansion of the "national forests," as the reserves were now called, was denounced as socialism and a violation of the rights of the states in which the forests were located.[9]

Conservationist concerns were also increasingly evident in the treatment of fossil fuel resources. The ratification of existing mining practices in the 1872

Mining Act had never extended to coal. While it was transforming forestry into a major federal responsibility, Theodore Roosevelt's administration in 1906 withdrew some 6 million acres of coal land from homesteading. This withdrawal was justified not only on the ground that individuals and groups had sought to profiteer from the coalrich homesteading lands but on the need for rational administration and orderly development of these nationally valuable but exhaustible resources.[10] The U.S. Geological Survey argued at the time that the nation needed to determine the best uses for the remaining public lands, because the best lands had already been taken up by private individuals and groups.[11]

This interest in the prudent management of exhaustible resources and the withdrawal of coal lands from entry prompted debates in Congress over conflicting purposes for the public lands. Proponents of homesteading argued that coal seams were often located under valuable agricultural lands—now entirely withdrawn from entry. Congress resolved the conflict between retaining coal in federal control and allowing for private agricultural development by severing the surface and subsurface estates. In a series of acts between 1909 and 1916, the surface lands were reopened for homesteading while fossil fuel rights were retained in federal ownership.[12] The retention of subsurface rights further complicated the mosaic of western landownership and foreshadowed the shape of controversies to come, as conflicts in public purposes for the same lands intensified.

With the 1920 Mineral Leasing Act came the formal recognition of federal leasing as a means for the orderly development of fossil fuel and some other mineral resources.[13] The act achieved some delineation of how leases were to be awarded, and an acceptable division of royalties between states and the federal government. Controversies on both counts would recur, but not on the scale that marked grazing policies, an irony given the much greater but exhaustible subsurface wealth. The passage of the 1920 act culminated a shift in the management of fossil fuels and fertilizers: it abridged the right to explore and develop and emphasized systematic utilization with a fair return to the public. Only hardrock mining, still today governed by the Mining Act of 1872, continued to be guided by the nineteenth century public purposes of settlement and full development.

In the decades after Theodore Roosevelt's administration, support increased for the objectives of management of renewable resources on the principle of sustained yield, and the prudent management of exhaustible resources. The strategy devised by the Roosevelt Administration for realizing these two purposes was increasingly accepted, even by its early critics. The commitment to federal ownership of the subsurface estate at least in fossil fuels became so well established that by the late 1920s and early 1930s the Hoover Administration's proposal to turn the public lands over to the states in which they were located did not include the subsurface mineral rights.[14] The omission, in part, rested on the assumption that conservation of these important minerals required continued federal ownership.

In the 1930s, prompted by widespread concern over the deterioration of the public rangelands, Franklin Roosevelt issued an executive order withdrawing the grazing lands from entry. The Taylor Grazing Act, passed in 1934, can be seen as a further elaboration of Theodore Roosevelt's strategy of reclaiming the

federal lands and expanding active federal management of renewable resources.[15] The act, followed by the formation of the Grazing Service, was met
with reluctant acceptance by the western stockmen, who on the one hand feared
federal interference but on the other hand supported federal aid to restore the   .
rangelands.[16]

At the end of the 1930s, the Grazing Service encountered increasing political
resistance to raising the grazing fees. In 1946, the Service and the old General
Land Office, which had been principally responsible for land disposition, were
replaced and their duties combined in the new Bureau of Land Management.[17]
The BLM, however, did not escape the influence of the stockmen. In its first
years, the Bureau became a textbook illustration of "agency capture," serving the
interests it had been designed to regulate.[18]

During the 1940s, the concepts of multiple use and sustained yield continued
to enjoy growing support among public land policymakers. The administrative
concept of multiple use, however, masked the fact that since Pinchot's day
demands for uses of the public lands had become increasingly varied and
perhaps incompatible. Indeed, multiple use was still largely assumed to mean
multiple productive uses of the land's natural resources; aesthetic enjoyment
and recreational use were not regarded as falling under the multiple–use aegis
until at least a decade after World War II.[19]

Yet in the period after World War II, new public purposes became increasingly evident, reflecting the growing importance of leisure time and recreation,
the aesthetics of open spaces, and the increasing importance of the environmentalist movement. The last, in contrast to the conservation movement of the
Progressive era, emphasized the natural landscape as a value in itself, rather
than as a resource base to be prudently managed.[20] Uses of the public lands, of
course, have never been confined to natural resource development alone.
Hunting and fishing as well as enjoyment of the landscape are uses that are
consistent themes in American literature and painting. But the extent of
recreational use and its burgeoning importance to western states' economies
became much more apparent after World War II.[21]

The recurring dissatisfaction of ranchers and other resource users again
found expression in congressional investigation in 1946. Hearings conducted by
the House Interior Committee's subcommittee on the public lands revealed
increasing recreational concerns, to the surprise of many. Peffer concludes from
the testimony "that in the West recreation has become second to water in
importance as a public domain function." Recreation had policy implications:
"The large number of local sportsmen's and wildlife organizations at the
hearings indicates that a new pressure group has come into being, a force which
promises to be more influential in retaining the public domain in continued
public ownership."[22] Peffer's prediction was borne out by the rapid growth in
recreational demands, which has accompanied the remarkable population expansion in most of the western states over the past three decades. Parks and
other recreational facilities grew in step with the population. The diversity of
recreational uses burgeoned as well, from relatively "soft" uses such as hiking,
backpacking, or fishing, to "harder" uses such as downhill skiing, snowmobiling,
and off–road vehicle motoring.[23] Some observers estimate that as urban centers
grow and mineral resources are exhausted, recreation will become the dominant
use of the public lands.[24]

The steady growth of the environmentalist movement in the 1960s, along with heightened interest in the preservation of the aesthetic values of the western landscape, represent the most controversial of the four purposes articulated at the outset of this paper: preservation of the public lands as they are or, in some cases, as they were. The imagery of nature as a quasi–religious experience expressed in the 1964 Wilderness Act is a legislative expression of preservationist values.[25] The National Environmental Policy Act of 1969 (NEPA) mandates consideration of aesthetic and other unquantifiable values in assessing the environmental impact of major federal actions.[26] Both the Public Rangelands Improvement Act[27] and the Surface Mining Control and Reclamation Act[28] also pursue the value of restoring the lands to their natural state. It is reasonable to conclude that with sustained yield, prudent development, and recreational use, preservation stands today as a major objective that policymakers will continue to consider.

With the Federal Land Policy and Management Act of 1976, Congress formally recognized the enlarged number of public purposes governing management of the public lands.[29] The report on the Senate version of what became FLPMA concluded that the BLM's efforts

> to fulfill its myriad responsibilities, particularly its basic management responsibilities for the national resource lands, have been impeded by its dependence on a vast number of outmoded public land laws which have developed over the earlier period in American history when disposal and largely uncontrolled development of the public domain were the dominant themes.[30]

Congress sought, therefore, to give the BLM a statutory basis for ordering conflicting demands for the public lands.[31] Indeed, the BLM regards FLPMA as its statutory charter, placing it on an equal footing with the Forest Service.[32]

In FLPMA, Congress followed the model that had evolved for the Forest Service since Pinchot's time. It charged the Bureau to administer the lands according to the principles of multiple use and sustained yield. Multiple use was an appealing concept for Congress in large part because of its plasticity; it can embrace an expanding number of uses as acceptable for a specific set of lands.[33] "Multiple use," however, is not in itself a public purpose for the lands; it is an umbrella under which other purposes can be drawn. The concept, moreover, fails to capture the often contradictory nature of the growing number of uses to which the lands can be subjected.[34] At the time it passed FLPMA, there is little evidence that Congress saw the difficulties that would continue to plague the Bureau. There is some indication that the Bureau itself has yet to come to terms with the conflicts underlying FLPMA:

> "The Act" reflected changing public priorities that had evolved over a decade, the increasing recognition that the public lands are a national heritage and the belief that decisions about how these lands are used should encompass the broadest possible public participation and debate. FLPMA established the national policy that the public lands would be retained in federal ownership and managed for all Americans under the principles of multiple use and sustained yield.[35]

The evolution of these four public purposes—sustained yield, prudential development, recreational access, and preservation—suggests the increasing complexity of the debate over the public lands. Yet the underlying assumption that continued federal retention and active federal management of the lands are strategies needed to achieve these purposes persists. The four objectives are intensely supported by groups interested in the future of the public lands. However, the likelihood of contradiction among the four purposes raises the important question of the ability of governmental agencies to defend and realize these purposes with any success.

Over the past four decades, as interest in the public lands has become more diverse and more intense, the federal government has assumed responsibility for these four purposes. It has, for example, explicitly managed the lands for their preservation and at the same time their resource development. On the one hand, the federal government has regulated against environmental deterioration of the lands while on the other hand it has actively promoted energy development. The federal government in seeking full responsibility for the actual realization of these four objectives has, as Charles Schultze observed about governmental intervention in output decisions, been taxed beyond its limits.[36] And it has stretched thin the delicate fabric of political consensus.

II. Some Difficulties with Federal Ownership and Management of the Public Lands

A. *Introduction*

Controversy over public land management policies has increased steadily over the past decade. Traditional user groups such as ranchers and miners have seen some federal environmental and land management acts and court decisions as serious threats to their livelihood. The settlement in 1974 of *National Resources Defense Council* v. *Morton*,[37] which resulted in the preparation of livestock grazing environmental impact statements under NEPA, many ultimately calling for reduced grazing to improve range conditions, met with great resistance from cattlemen. FLPMA was regarded by some grazers as reducing their importance on the federal lands and as weakening the influence of ranchers in their dealings with the federal agencies by formally recognizing permanent federal retention of the lands.[38] The miners saw both FLPMA and federal environmental legislation as obstacles that not only delayed mineral exploration and development but seemed to threaten any mineral development at all in some locations.[39] This growing dissatisfaction among traditional users, in conjunction with a resurgent interest in states' rights, fed the Sagebrush Rebellion—the movement in western states to claim the public lands as state property.[40]

Environmentalist and recreationist groups also have been critical of federal land management. Their criticism is not so much that federal land agencies have been lax in implementing environmental standards as that the agencies have not performed according to the mandates of law, in part because they often seem to be pursuing conflicting goals.[41]

In large measure, the federal management dilemma is to be held accountable for all that is perceived wrong with the lands but rarely to be praised for any accomplishments. At least four sources of the present dissatisfaction with federal land management can be singled out. First, as a result of the vast size and increasingly diverse uses of the federal holdings, federal land agencies confront intensified challenges to any active land management policy. Second, conflicts among the four purposes discussed in the first part of this paper have not been resolved into a clear mandate for federal land management. Third, the perceived insecurity of surface land tenancy such as leasehold or grazing rights is a nagging source of concern to the land users. Fourth, the anomaly that federal landownership is disproportionately located in the West has produced a special and often difficult relationship between the western states and the federal presence.

B. *The Size and Diversity of the Federal Domain*

There is a Brobdingnagian scale to the responsibilites of the BLM. The BLM manages nearly 400 million acres of land, approximately 60 percent of which are located in Alaska. The overwhelming remainder of the BLM lands are located in the eleven contiguous western states.

The BLM's traditional responsibility is the management of the federal grazing lands; nearly 155 million acres of these are grazing districts that fall under the Taylor Grazing Act. In 1980, 12,793 operators held 14,676 licenses or permits to graze 17,533,745 head of livestock.[42] Mineral lease revenues, however, are by far the most important income source for the federal government from the public lands. In 1980, there were 105,913 mineral leases in effect for the public lands, embracing 10,198,276 acres.[43] More than 95 percent of these are oil and gas leases. In addition, despite the Forest Service, the BLM manages some 24 million acres of commercial forest land, including some of the nation's most productive forests.[44] For both recreational and preservationist reasons, the BLM maintains a wide variety of wildlife, in over 324 million acres of big game habitat populated by nearly 70 million big game animals.[45] Finally, in 1980 the Bureau had authority over 406 recreation sites and responsibility to protect areas of natural beauty and cultural significance.[46] An ongoing plan of the Bureau's is to provide recreation for residents in sixteen major western cities.[47]

C. *Conflicting Directions in Public Land Policy*

By the 1950s, it had become evident that the BLM faced increased and increasing demands for the public lands that extended beyond its initial focus on managing the rangelands. Clawson and Held warned of difficulties for the Bureau as more groups sought increased uses "incompatible with all that other groups seek," land proposed to replace the BLM as land manager with a government corporation designed for greater freedom and flexibility in resolving conflicts than a government bureau.[48] In the late 1960s, the Public Land Law Review Commission offered its own recommendation for dealing with the enlarged dimensions of "multiple use."[49] The commission believed that although existing statutes authorized land management agencies to consider and allow diverse uses of the public lands, the statutes failed to provide clear guidance for allocation of the lands among competing uses. The commission recommended recognition of the "highest use and best use" of particular areas of land as dominant over other authorized uses.[50] Other uses were permitted only to the extent compatible with the best use.[51] This proposal, criticized as a means to allow greater resource development, was not incorporated into FLPMA, nor did FLPMA otherwise resolve the difficulties of establishing any preference ordering among the vast array of acceptable uses for the public lands.

In FLPMA, Congress declared thirteen policies to govern management of the public lands.[52] One of the thirteen stipulates that management is to be based on multiple use and sustained yield.[53] Another requires nonmarket values to be taken into consideration in public land management:

> The public lands be managed in a manner that will protect the quality of scientific, scenic, historical, ecological, environmental, air and atmospheric, water resource, archaeological values that, where appropriate, will preserve and protect certain public lands in their natural condition; that will provide food and habitat for fish and wildlife and domestic animals; and that will provide for outdoor recreation and human occupancy and use . . . .[54]

The BLM, however, is also charged to manage the lands in a manner that recognizes the public's need for domestic natural resource production.[55] This is a remarkable catalog of values for a single agency that is also required to receive fair market value for the use of the public lands.[56]

The Bureau's ability to impose its own preference ordering by relying on expert opinion and rational planning to guide land use, moreover, was diminished by the requirement in FLPMA for public participation in the process of decisionmaking for the public lands.[57] Land use planning, for example, is a central component of FLPMA.[58] The Bureau is charged with preparing a continuous inventory of the public lands and their resources, including recreational and scenic values. Land use plans are to be prepared for the various tracts of the public lands.[59] Plans are to be developed with public participation, by notice and comment rulemaking. Federal landuse planning, moreover, is to be coordinated where possible with state and local land use plans.[60] Tension is thus built into the planning process. On the one hand, FLPMA calls for rational, scientific landuse planning, by experts; on the other hand, it assigns an important role to public participation, thus adding an inherently political dimension to the planning process.[61]

Nor does the prospect of judicial involvement seem likely to present a solution to the heavily charged political atmosphere that surrounds the BLM. Although the last two decades have been a period of increased litigation for public land managers, the courts have not heretofore offered much guidance to the Bureau in resolving multiple–use conflicts. Indeed, the courts have shown little interest in analyzing the multiple–use/ sustained–yield statutes.[62]

While FLPMA is the central statutory authority for the BLM, its managerial task remains complicated by a number of other federal laws that govern the public domain. In particular, the Wild Free–Roaming Horses and Burros Act of 1971,[63] the Public Rangelands Improvement Act of 1978,[64] the Endangered Species Act,[65] NEPA of 1969,[66] the Antiquities Act of 1906,[67] and the National Wilderness Preservation Systems Act[68] all continue to govern the BLM. Such laws reflect the conflicting concerns of specific interest groups about livestock, wildlife, and the wilderness itself. For example, stockmen regard the Endangered Species Act as an act for the preservation of predators.

In the years following the enactment of FLPMA, the Bureau has found itself in a climate of political volatility. The emergence in 1979 of the Sagebrush

Rebellion suggested the Bureau's exposure to political attack was as great as it had been in the 1940s. Some environmentalists argue that the Rebellion was not primarily motivated by states' rights but by the shift in managerial policies promised in FLPMA.[69] Regardless of the explanation, six years after its enactment FLPMA appears not to have strengthened the Bureau's autonomy. The Bureau's traditional major client, the stockmen, have demonstrated resurgent political strength.[70] The change in administrations in 1980 raised the specter of fresh discontinuities in public land management policies and called into question the assumption that federal management is the appropriate mechanism to realize public purposes on the public lands.

D. *Insecurity of Land Tenure*

The most long–standing and undoubtedly most politically active user group on the public lands is the grazing community. Their relationship with the BLM has been characterized by resentment of and dependence on federal authorities, who in turn have sought to escape their dominance.[71]

The grazing system assigns allotments, pasture areas on which the stockman is allowed to graze a specific number of animals for a given number of months of the year. Over the past three decades, decisionmaking about the number of cattle to be grazed and the time of the year grazing is to be allowed has shifted from the nearly exclusive purview of the stockmen to the active participation of the federal land managers. Permit distribution requires the stockman to control private grazing land to be used in conjunction with the public grazing land. The economic value of these base lands is partially contingent on continued access to the public lands. Grazing permits are alienable and inheritable and in practice they have been relatively permanent. They are, however, of limited duration and could be ended without compensation on expiration. Stockmen, therefore, are perennially concerned that calls for reduction in grazing allotments or vigorous protection of the public domain could curtail or even extinguish their claims to the public lands.[72]

This concern has been the major impetus for the notable recurring political activity of the stockmen. Since World War II, public land ranchers have repeatedly led the attack on federal retention of the public lands. Sometimes they have proposed turning the lands over to the states and sometimes they have supported sale of the lands directly to the grazers themselves. They have repeatedly drawn attention to their sense of the precariousness of their public land tenure and since the 1970s have protested the burdens of complying with federal environmental regulations. They have held the federal government accountable for the steady decline in the number of head of cattle grazing on the public lands.[73] Ranchers have turned regularly to Congress for aid in resisting grazing fee increases and often have been successful in obtaining congressional intercession on their behalf.[74] In recent years, moreover, stockmen have turned with some success to the courts in seeking support for their interests in the lands.[75]

The last two decades have also seen a sharp increase in the influence of recreationists and environmentalists on public land policy.[76] The ranchers often see these groups' interests as antithetical to their own concerns for continued economic viability of the public rangelands.[77] The conflicts between ranchers and recreationists, however, should not be overstated. Their respective interest group representatives have been allies in certain important legislative battles such as the Public Rangelands Improvement Act and the Surface Mining Control and Reclamation Act.[78] Nonetheless, the rise of powerful recreational and environmental lobbies on public lands issues has contributed to the insecurity that many public lands ranchers have concerning their place on the public range.

Public lands stockmen pose something of a paradox. Western stockmen are a relatively minor part of the western states' economies but they continue to remain formidable political forces within the western states, in Congress, and most recently in the executive branch. On the other hand, political skill and success have not generated the type of land tenure they seek.

Moreover, although the grazing land tenure system has been quite productive in politicizing ranchers, it has been of little economic benefit to the stockmen, let alone the nation. In addition to the hazards of interest group politics, problematic management of the range itself may stem from the insecurity of grazing rights. Many critics of the leasing policies applied to the public lands argue that the range is overgrazed and that some areas have shown little improvement after four decades of federal management.[79] These critics explain the condition of the range as a tragedy of the commons: the public lands, they argue, are overused or underimproved because of the absence of clearly defined, enforced, individually assigned property rights.[80]

The tragedy of the commons analysis, however, has not been straightforwardly applicable to the public grazing lands since the passage of the Taylor Grazing Act in 1934. The act established grazing lease arrangements for the appropriate public lands, which set specific animal grazing units per month. Nonetheless, it may be argued that there is sufficient ambiguity in these leasing arrangements to permit application of the "commons effect" to the public range. Public land grazers, given the restrictions on their land use, are not likely to make certain improvements on the grazing lands or to reduce their respective herds in order to permit long–term improvement in the condition of the range. They would not do so for the following reasons. First, their leases are defined in terms of an agreed amount of livestock forage consumption; therefore, to reduce the number of livestock is to diminish the value of the lease. Second, certain range improvements, particularly fencing of pasture lands, may be resisted by federal land managers as an effort to deepen private claims to public land. Third, efforts to seed the range with grasses that appeal exclusively to livestock may be rejected by federal land managers as harmful to wildlife. Fourth, ranchers may not wish to improve forage capacity in general, because it may only advantage other range users who cannot be excluded from the range, namely protected wildlife, free riders on the open range enjoying the contributions of others.

What has developed on the public grazing lands over the years is a distinctive land tenure system that is neither based on the individual property rights model preferred by public choice theorists,[81] nor a form of commonly held property governed by the informal collective rulemaking favored by institutionalists, nor the dark visions of the tragedy of the commons. The existing grazing system is a curious form of land tenure that is alternately diluted or strengthened by shifting interpretations of federal policy and changing levels of influence on federal policy making by surface resource user groups. It is a system of remarkable ambiguity that has captured many of the less desirable aspects of nonmarket and market systems. The ambiguity and absence of delineated responsibilities pose ongoing problems in the realization of public land goals.

E. *The Disproportionality of Land Ownership*

Nearly every published account of public land politics recites the litany of federal landownership in the western states. One–third of the nation's lands are in federal ownership. Approximately half of the federal holdings are in Alaska, and 90 percent of the remainder are in the eleven contiguous western states.

This disproportionate location of federal land in the western states continues to be a source of friction in intergovernmental relations. The scale of federal landownership has long been regarded by its critics as a premier states' rights issue. In the nineteenth century, the question of federal land disposition was often debated in terms of a policy choice between direct sale or transfer to individual citizens and the transfer of the lands to the respective states of location.[82] Opponents of a strong federal government, notably Calhoun, argued for the transfer of the land to the states,[83] contending that any policy of direct land disposition to individuals would only strengthen the federal presence. The shift in the twentieth century to a policy of retention of the public lands intensified the fear of an overwhelming federal presence. There is an echo of Calhoun in Hoover's plan to turn the public lands over to the states to promote states' rights.[84]

The putative justification for the Sagebrush Rebellion was that substantial federal landownership in the western states impaired state sovereignty, especially in the following areas.[85] First, states cannot regulate wildlife within their borders because Supreme Court decisions have affirmed the supremacy of federal authority on federal lands at the clear expense of state wildlife management.[86] Second, serious questions have been raised about the legal ability of the states to extend environmental regulation to the federal lands except on terms found consistent with federal regulations. Federal requirements may preempt local regulations with the potential to interfere with federal purposes. In this fashion, a state or community that seeks to limit or to restrain its development may find its concerns brushed aside.[87] Third, the intricate mosaic of land tenure in the West may frustrate the realization of state objectives for state–owned parcels—whether those purposes be resource development or preservation—because the use to which state lands can be put unavoidably is shaped by the regulations governing the intermeshing federal lands.[88] Fourth, the federal lands extend from densely settled areas to nearly uninhabited areas.[89] Federal ownership of land in urban areas is a nagging concern for local authorities in devising comprehensive urban planning, especially if the land has not been put to any apparent national purpose. Fifth, the federal government does not speak with a single voice on questions of policy or management. Many policies and many agencies are involved in the management of the public lands. Both shifting policies and the shifting fortunes of the agencies involved can result in shifting practices. The states are obliged to come to terms with proposals for radically different uses for the same lands, from preserving wildlife to basing missiles to dumping chemical and nuclear wastes.[90] Under existing practices, such shifts in use are frequently constrained only by the preparation of an environmental impact statement.

These difficulties with federal landownership and management in the West reflect the complexity of pursuing objectives in an atmosphere of heightened political activity by groups expressing conflicting interests and values. Although not unusual for governments, this condition is made immensely more complicated by declining confidence in governmental institutions. The promise of the New Deal that federal intervention would improve social problem–solving has been met in recent years with cynicism and frustration.[91] Increasing dissatisfaction has resulted from the federal government's efforts to serve as policy planner, regulator and land manager. It is not that public values for the lands have lost their luster but that governmental realization of them has been tarnished.

III. Selective Disposition of the Public Lands:
A Proposed Strategy

This proposal for selective disposition of uses to the public lands should be understood in a framework that accepts the public purposes discussed above as controlling activity on the public lands.[92] The proposal fully recognizes that the federal government must play an important and continuing role in formulating strategies that can best accommodate these often competing public purposes for the federal lands. What is sought is a strategy to avoid both the difficulties of increasingly disappointing direct federal land management and outright surrender of the lands to the market as the sole means for protecting long–held public land values.[93] The proposed strategy utilizes a mix of public planning and private incentives to realize important public ends.

Quite simply, it has become progressively more difficult for the federal government to retain the confidence of concerned publics in achieving the accommodations necessary to realize public purposes on the public lands. What may be necessary for public purposes to guide policies for the public lands is to shift the challenge of accountability from the federal government to user groups themselves. The disposition of property rights to selective uses of the lands can create entitlements that hold owners of the uses accountable for actions on the lands with adverse consequences for other users. In this way, accountability for the condition of the public lands can be shifted from the federal government to user groups, without derogating the federal government's ultimate responsibility for land use planning.

At the outset, the proposal for selective disposition of uses to the public lands would require classification of the lands managed by the BLM. A four–part classificatory scheme is suggested here:

A. lands near or in populated areas that are isolated parcels or are unlikely to be used for grazing or resource development because of shifting population patterns;

B. grazing lands, the single largest category of land use administered by the BLM;

C. timber lands administered by the BLM; and

D. the remaining public lands.

A. *Urban Public Lands*

Federal ownership of public lands in urban areas has been a particular irritant to state and local governments in the West. None of the four public purposes described above, moreover, provides strong justification for retention of these lands in federal ownership. Isolated urban parcels are unlikely to be utilized for their renewable or nonrenewable resources, or to be retained for wilderness values.

Outright private sale of all of these lands, however, would not be justified. As the 1982 debate over privatization of the public land revealed, an important subset of these lands has come to play a major role in community expectations such as potential park, shcool or public building sites. Even if no current public use is apparent for these lands, long–term planning may anticipate the need for the lands as a resource to meet changing community needs. Yet surely there are some parcels for which no such public need is likely to arise. Such lands should be sold by the federal government once the land has been cleared under the criteria set out in FLPMA. The revenues generated should be dispersed according to existing federal statute.

Orderly disposition of the urban public lands, should proceed as follows. States wishing to assume control to particular parcels of urban lands for public purposes should be allowed to submit land use plans for the parcels. Like state implementation plans under the Clean Air Act,[94] these plans should be submitted only on the basis of full opportunity for public hearing and participation in their development. The federal government, after insuring adequate development of a state plan, should turn the land over to the state. A reasonable constraint on state requests is that in exchange for the gift of urban public land to the state, the state would acknowledge formally an obligation to the federal government, to provide equivalent parcels or the cost of obtaining them in the case of determined national need.

Absent submission of a state plan, however, the remaining urban public lands should remain under federal management. These lands are potentially valuable community resources, and their disposition such as by immediate private sale would not be justified except on the initiation of a plan developed with an opportunity for full community participation.

B. *Grazing Lands*

The existing pattern of grazing permits should be replaced by a system in which the right to graze on specified acreages is sold outright. The proposal does not recommend outright sale of the some 155 million acres of grazing land; only the grazing use of the land should be sold. Title to the grazing use, however, should pass completely to the new owner, and be both alienable and inheritable. Ownership of the use should not be contingent on continued utilization, nor should the fact that the land has become unsuitable for grazing extinguish the property right to the use. Were the federal government to wish to cut off the grazing use, it would have to resort to the power of eminent domain, as when it seeks to terminate other property rights. With the sale of the grazing use must come the recognition that other users or potential users must recognize constraints on their own activities on the lands.

Changing conditions may reduce a rancher's interest in livestock, or alternative uses of the land may become more attractive. Under this proposal the only way to terminate an existing land use would be to reconvey the land use to the designated federal agency. The owner of the use would be compensated on fair market value rates for the grazing use lands, figured on the basis of their leased lands in conjunction with the private base ranch. The federal agency, together with the designated state and local planning agencies, would, if appropriate, change the land use classification and place the new use on the market. Thus, the grazing use owner could conceivably keep his or her base ranch lands while receiving fair compensation for the grazing use of public lands, without having captured the economic benefits of development of the lands. It would be a publicly determined policy change to extinguish the grazing use and substitute a new land use—from a mobile home park to a new national park.

Sale of a grazing use, however, does not entail transfer of the right to use the lands exclusively. Sale is not to preclude recreational use nor wildlife use of the public lands used for grazing. The grazing use should be sold subject to recreational and wildlife easements that are to run with the lands to which grazing rights are to be sold.

Existing holders of grazing permits should be offered the first opportunity to purchase the grazing uses. In order to give further recognition to the four fundamental values that should guide public land policy, especially public recreational access, the grazing uses could be sold at subappraised valuations in exchange for the rancher's grant to the public of scenic and recreational easements on the base ranch. Otherwise, grazing uses should be sold at appraised market values.

Under the proposal, both the rights of purchasers of the grazing uses and the rights of the public are to be legally enforceable. Stockmen will have the enforceable property right in their grazing uses. Individual members of the public, in turn, should be granted the right to sue a rancher for violating recreational or wildlife easements.[95]

Under this proposal, an enlarged role may also be envisioned for state agencies in monitoring and protecting recreational activity on the grazing lands. In conjunction with federal agencies, state agencies, with knowledge of local conditions, may help to work out accommodations between recreationists and ranchers in the use of the grazing lands.

Accommodation of the sale of grazing uses with the protection of habitats for wildlife poses by far the most complicated problems for the proposal. Grazing uses are to be sold subject to wildlife easements, but a method must be set up for determining to what extent the two uses are to constrain each other. Moreover, when wild animals compete with livestock for grazing, the intervention of state and federal agencies may be required to defend the wildlife.

To realize a strategy for the protection of wildlife that is consonant with this proposal to use private means to further public purposes, a program of inducements and penalties is suggested. To develop the program, the existing grazing districts should continue to be used as units of analysis in assessing land use activities. For each district, a committee should be established of representatives of the various interested groups, such as wildlife groups or stockmen. The committees then should assess the existing ratio in their district of livestock and wildlife animal unit months, that is, the amount of forage necessary for the complete sustenance of one animal unit for the period of one month.[96] If the existing ratio found in the district is judged to be a satisfactory balance between wildlife and livestock, then if the number of wildlife AUMs declines while the number of livestock AUMs increases the designated state agency may levy a fine on the stockmen within the district. If on the other hand the committee determines that the wildlife AUMs should be increased, an incentive payment will be offered to stockmen within the district. The penalty payments as well as the incentives could be paid into and drawn from licensing fees for hunting. The individual rancher's contribution to the district's fine, or share in the district's incentive payment, would be correlated with the rate of increase or decrease in his or her livestock use of the range. This policy may allow for shifts in grazing patterns. Some grazing districts may see an intensification in livestock production, whereas in other districts increases in wildlife grazing may be obtained. It would be the task of designated state and federal agencies to monitor wildlife patterns and for the wildlife committees to review periodically the relationship between wildlife and livestock use on the public lands. This policy of incentives and penalties to encourage desirable ratios between livestock and wildlife does not eliminate problems of accommodation, but it does begin to devise local and private responses to the national concerns for wildlife and recreational activity on the public domain.

C. *Timber Lands*

The timber lands owned by the BLM are chiefly the O and C forest lands found in western Oregon. The surface use for timber should be sold, in the same manner as the grazing uses. Sale of the timber uses should be subject to recreational and wildlife easements.

An additional constraint is necessary, however, in any disposition of surface timber use. Where the forest lands serve as watersheds for local communities, the states, and regional areas, any use of the timber must be consistent with preservation of the watershed. Timber uses must be limited by a watershed easement in addition to the other limitations imposed on sale of grazing uses.

D. *Limited Natural Resource Development Lands*

Significant portions of the BLM–managed lands are mountainous or marginally suited for either grazing or logging. These lands should be designated as primarily lands for recreational use and preservation in their natural state. Characterizing their primary use as recreational and preservational reflects two concerns: the growing interest in the western landscape as a value in and of itself, and the growing value of the recreational use of the public lands.

Retention of these lands in federal ownership is recommended, although their administration could be designated to state agencies. It is unlikely, however, that states would be willing to take on the responsibility for lands with limited surface productivity.

E. *A Note About Minerals and Fossil Fuels*

The subsurface wealth should be retained in federal ownership. The policy initiated during Theodore Roosevelt's administrations and expanded thereafter is the recognition that extractive minerals are an important source of national wealth and that their ownership has major political ramifications in every nation. The policy of federal ownership with shared royalties to the states and important taxing powers left to the states has so far allowed a reasonable working accommodation of the potent political problems associated with the uneven areal distribution of nonrenewable underground wealth.

Conclusion

The public lands, largely by virtue of historical accident, present a great opportunity to sustain public objectives. The issues of public values and public ownership of the lands should not be confused. Retention of the lands under federal ownership and management should not be an end in itself if the land tenure system cannot realize an accommodation of conflicting public values for the lands. Nor can large–scale privatization as a means to extend the power of the marketplace be the principal means of realizing the public choice for the public lands. The public lands are an unusual blend of local and national perspectives and private and public uses. What the proposed strategy has sought is to retain the public heritage of the lands while shifting some of the responsibility for the lands to local and private users.

Notes

1. 43 U.S.C. §§ 1701–82 (1976).

2. Id. § 1702(e). This definition of the public lands conforms to the definitions most generally employed in the public lands literature.

3. After the BLM, the Forest Service is the second largest federal land management agency, with approximately 187 million acres under its control. Although the Service is primarily responsible for managing the national forests, it also manages some grazing lands.

4. Between 1979 and 1981, 16 western states considered bills claiming or supporting state ownership of the public lands administered by the BLM. Five states enacted legislation claiming the federal lands as state property. Of the five, only Wyoming's statute included lands administered by the Forest Service.

5. General Revision Act of 1891, 26 Stat. 1103 (currently codified at 16 U.S.C. § 471 (1976)). Section 24 authorized the President to proclaim land covered by timber or undergrowth, wholly or in part, as a public reservation.

6. S.T. Dana and S.K. Fairfax, *Forest and Range Policy: Its Development in the United States*, 2d ed. (1980). Chapter 3 gives an account of the Roosevelt–Pinchot era.

7. See S.P. Hays, *Conservation and the Gospel of Efficiency: The Progressive Conservation Movement, 1980--1920* (1959); T. Roosevelt, *An Autobiography* (1925). Chapter 9 of the autobiography gives T.R.'s views on conservation.

8. Grazing fees constituted a major source of revenue for the Forest Service. T.R.'s plan sought to create an independently financed service drawing its income from the forest reserve receipts. In 1906, Congress turned 10 percent of the gross receipts over to the states in which the forests were located, weakening the Service's financial autonomy. See S.T. Dana and S.K. Fairfax, *Forest and Range Policy*, at 90.

9. See The Proceedings of the Public Lands Convention, Denver, Colorado, June 18–20, 1907, compiled by Fred Johnson, Permanent Secretary. Resolutions 2, 7, 16, and 17 expressed concern over the policy of federal land retention and its consequences for states' rights and existing grazing practices.

10. See B. Hibbard, *A History of Public Land Policies* (1965), Ch. 25. See also S.P. Hays, *Conservation and the Gospel of Efficiency* at 86, n.62, for the observation that retention was not designed to preserve coal for future generations but to provide rational contemporary development.

11. United States Geological Survey, The Classification of the Public Lands 7 (1918).

12. An Act to Protect Surface Rights, 35 Stat. 844 (1909); An Act to Provide for Agricultural Entries on Coal Lands, 26 Stat. 583 (1910); An Act to Provide for Stock Raising Homesteaders, 39 Stat. 862 (1916).

13. Mineral Leasing Act of 1920, 41 Stat. 437 (codified as amended at 30 U.S.C. §§ 181 et seq.). See Office of Technology Assessment, Management of Fuel and Nonfuel Minerals in Federal Lands: Current Status and Issues 80–89 (1979), for a discussion of successive mineral acts.

14. See R.L. Wilbur & A. Hyde, *The Hoover Policies* 229–40 (1937), for a succinct account of Hoover's public land and conservation proposals. In part the Hoover proposals were rejected by Western governors because of the reservation of the subsurface rights. See Hearings on H.R. 5840 Before the House Committee on the Public Lands, 72d Cong., 1st Sess. 22 (Statement of Hon. George H. Dern, Governor of Utah).

15. Taylor Grazing Act, 48 Stat. 1269 (1934).

16. See P. Foss, *Politics and Grass* (1960).

17. See W. Voight, *Public Grazing Lands: Use and Misuse by Industry and Government* (1967).

18. E. Hanks, A.D. Tarlock, & J. Hanks, *Cases and Materials on Environmental Law and Policy 501* (1974). The example given relies on evaluations of the BLM during the 1940s and 1950s.

19. Coggins, *Of Succotash Syndromes and Vacuous Platitudes: The Meaning of Multiple Use, Sustained Yield for Public Land Management*, 53 U. Colo. L. Rev. 229, 240 (1981).

20. See, e.g., B. Commoner, *The Closing Circle* (1972); R. Dubos, *A God Within* (1972).

21. E.L. Peffer, *The Closing of the Public Domain: Disposal and Preservation Policies*, 1900–50, 337 (1951).

22. Id. at 339.

23. See Bureau of Land Management, Public Land Statistics 1980, 70–71 (1980) for the distribution of recreational uses on the public lands.

24. E.g., Kirschten, "The Federal Landlord in the Middle," 10 *National Journal* 1928–31 (1979).

25. 16 U.S.C. § 1131 (1976).

26. National Environmental Policy Act of 1969, Pub. L. No. 91–190, 83 Stat. 852 (codified at 42 U.S.C. §§ 4321–47).

27. 43 U.S.C. §§ 1901–08 (Supp. IV 1980).

28. 30 U.S.C. §§ 1201–1328 (Supp. IV 1980).

29. See 43 U.S.C. § 1701(a)(1)–(13) (1976).

30. S. Rep. No. 94–583, 94th Cong., 1st Sess. 34 (1975), reprinted in *FLPMA Legislative History* at 99.

31. Id. at 24, reprinted in FLPMA Legislative History at 89; H.R. Rep. No. 1163, 94th Cong., 2d Sess. 1–2 (1976), reprinted in FLPMA Legislative History at 431–32; H.R. Rep. No. 1724 (Conference Report), 94th Cong., 2d Sess. 57 (1976), reprinted in FLPMA Legislative History at 927.

32. Bureau of Land Management, Four Year Authorization Report to the Congress for Fiscal Years 1982–1985 [hereinafter *BLM Four Year Authorization*] at 13 (1980).

33. Coggins, *Of Succotash Syndromes and Vacuous Platitudes*, at 241. Nearly two decades earlier, the Forest Service had fought successfully for the enactment of the Multiple Use Sustained Yield Act of 1960, 16 U.S.C. §§ 528–31 (1976), that recognized the expanding number of uses for the forest lands but in the traditional format of multiple use. According to S.T. Dana and S.K. Fairfax, *Forest and Range Policy*, at 204, this format allowed the Service discretion in determining the pattern of uses. The Sierra Club objected to the Act on the grounds that it set insufficient legislative standards to mandate the Service to include noneconomic values.

34. Id.

35. *BLM Four Year Authorization* at 15.

36. See C. L. Schultze, *The Public Uses of Private Interests* (1977), for a discussion of the difficulties with past governmental regulatory efforts.

37. *Natural Resources Defense Council* v. *Morton*, 388 F. Supp. 829, 840 (D.D.C. 1974).

38. See Special Hearings on Rangeland Management Policy and Wood Energy Development Before the Senate Committee on Appropriations, 96th Cong., 2d Sess. (1980); Hearings on Livestock Grazing on the Public Lands Before the Subcommittee on Parks, Recreation and Renewable Resources of the Senate Committee on Energy and Natural Resources, 96th Cong., 1st Sess. (1979).

39. Critics of federal land management policies have argued that federal policies have so delayed mineral exploration and development that the nation's industries are increasingly dependent on importing needed minerals from abroad. See Office of Technology Assessment of the United States Congress, Mineral Accessibility on Federal Lands: Interim Report (1976). See also Bennethun & Lee, "Is Our Account Overdrawn?," Mining Congress Journal (Sept. 1975); Everett & Associates, Withdrawal of Public Lands from Access to Minerals (report prepared for the Public Lands Study Group 1980).

40. Francis, *Environmental Values, Intergovernmental Politics and the Sagebrush Rebellion*, ch. 1, this volume.

41. Since the inception of active federal responsibility for the condition of the range in the 1930s, and especially since the environmental legislation of the past 15 years, environmentalists have offered two grounds for criticizing the Bureau's management of the range. First, the Bureau has been insufficiently diligent in stopping overgrazing. See Sheridan, *Western Rangelands, Overgrazed and Undermanaged*, 23 Environment 15 (1981). Second, the Bureau has failed to charge fair market value grazing fees to the stockmen for use of the public range. See. e. g., Environmental Policy Center et al., Alternative Budget Proposals for the Environment, Fiscal Year 1983, at 40 (1982) ("Charging market values for livestock grazing leases on federal lands would bring in substantial additional revenues and would also help to protect the badly overgrazed public lands from further damage.")

42. Bureau of Land Management, Public Land Statistics 1980, at 76.

43. Id., at 94.

44. Id., at 55.

45. Id., at 74.

46. Id., at 68.

47. Id.

48. M. Clawson & B.R. Held, The Federal Lands: Their Use and Management 346–74 (1957).

49. Public Land Law Review Commission, *One Third of the Nation's Land* 19 (1970).

50. Id., at 48.

51. Id., at 51.

52. 43 U.S.C. § 1701(a)(1)–(13) (1976).

53. Id., § 1701(a)(7).

54. Id., § 1701(a)(8).

55. Id., § 1701(a)(12)

56. Id., § 1701(a)(9).

57. Id., § 1701(a)(5).

58. See S. Rep. No. 583, 94th Cong., 1st Sess. 45–46 (1975), reprinted in FLPMA Legislative History at 110–11; H.R. Rep. No. 1163, 94th Cong. 2d Sess. 5–6, reprinted in FLPMA Legislative History at 435–36.

59. 43 U.S.C. § 1711(a) (1976).

60. Id., § 1712(c)(9).

61. The Bureau recognizes the complications of determining the level and extent of public participation:

> Not surprisingly, therefore, the increasing value and the demands upon the resources of the public lands have become a topic of escalating debate among Westerners concerning who should make public land use decisions, how much decisions should be made, and whether the interests of the West can be effectively addressed at the national level.

BLM Four Year Authorization at 18. See B. Ackerman & W. T. Hassler, *Clean Coal, Dirty Air* (1981) for a discussion of the shift in regulatory models from the New-Deal reliance on agency expertise insulated from central political control to the contemporary model of regulation in which Congress takes the role of setting standards. The contemporary model places regulation very much in the political process.

Environmentalists have been among the strongest supporters of public participation in land use planning, in the belief that it legitimates consideration of their interests in the lands. The politicized nature of public participation is reflected in the Interior Department's effort to encourage participation under the Carter Administration. The Department contracted with the League of Women Voters to devise a strategy to increase public participation. In many Western states, the League is identified with environmentalist positions. Its contract was canceled in the first year of the Reagan Administration.

62. Coggins, Of Succotash Syndromes and Vacuous Platitudes, at 231.

63. The Wild Free–Roaming Horses and Burros Act of 1976, Pub. L. No. 92–195, 85 Stat. 649 (1971), *codified as amended* at 16 U.S.C. §§ 1331–40 (1976 and Supp. IV 1980).

64. The Public Rangelands Improvement Act of 1978, Pub. L. No. 95-514, 92 Stat. 1803, *codified at* 43 U.S.C. §§ 1901–08 (Supp. IV 1980).

65. The Endangered Species Act, Pub. L. No. 93-205, 87 Stat. 884 (1973), *codified as amended* at 16 U.S.C. §§ 1531–43 (1976 and Supp. IV 1980).

66. NEPA of 1969, Pub. L. No. 91–109, 83 Stat. 852, *codified at* 42 U.S.C. §§ 4321–61 (1976).

67. The Antiquities Act of 1906, 34 Stat. 225, *currently codified at* 16 U.S.C. § 431 (1976).

68. The National Wilderness Preservation Systems Act, Pub. L. No. 88–577, 78 Stat. 890 (1964), *codified as amended* at 16 U.S.C. §§ 1131–36 (1976 and Supp. IV 1980).

69. R.M. Cawley, *The Sagebrush Rebellion* (unpublished doctoral dissertation, Colorado State University, 1981); Francis, *Environmental Values, Intergovernmental Politics, and the Sagebrush Rebellion.*

70. Fairfax, *Beyond the Sagebrush Rebellion: The BLM as Neighbor and Manager in the Western States,* Ch. 3, two volumes.

71. P. Culhane, *Public Land Politics: Interest Group Influence on the Forest Service and the Bureau of Land Management* (1981). Culhane suggests the BLM was successful in escaping the dominance of the stockmen. The events since 1979, however, may reopen the question.

72. See G.D. Libecap, *Locking up the Range,* chs. 5 and 6 (1981).

73. See sources cited in note 38. Much of the rancher concern was expressed after the BLM's release of the draft environmental statement for a grazing district in New Mexico, July 28, 1977.

74. See W. Voight, *Public Grazing Lands.*

75. See Fairfax, *Beyond the Sagebrush Rebellion.*

76. The range of uses on the lands is reflected to some extent in the composition of the National Public Lands Advisory Council and its predecessors. The Council was created to provide a cross–sectional representation of the public served by the Bureau. Between 1940 and 1975, stockmen's representation fell from 100 percent to 28 percent as the number of recreationists, environmentalists, and local governmental representatives were increased. Some measure of the shifting political atmosphere facing the BLM is found in the Council reorganization under Secretary Watt. The user group criterion was abandoned in membership selection, perhaps in order to facilitate the selection of members who share a common outlook on land use decisions. See Cawley, *The Sagebrush Rebellion: Advising the United States Department of Interior in Western Land Management* (Western Social Science Association, 1982).

77. The controversy over the condition of the rangelands has generated a complex and costly triadic relationship among the ranchers, the environmentalists and the Bureau over the appropriate set of forage and grazing policies. Stockmen see the Bureau as insensitive to market conditions. Some critics of the Bureau's forage policies believe they are unscientific. Regardless of the Bureau's competence, rangeland improvement programs have been a major factor in the Bureau's budgetary expenditures for what may seem as having produced little in the way of improvement. See USDA, USDI, and CEQ, *Rangeland Policies for the Future* (1979), for a discussion of the scientific issues of rangeland improvement. See also Kremp, "A Perspective On BLM Grazing Policy,"in J. Baden and R. L. Stroup, eds., *Bureaucracy vs. the Environment* (1981). For a discussion of the budgetary and

administrative problems of range improvement, see R. H. Nelson, *The New Range Wars: Environmentalist Versus Cattlemen for the Public Rangelands* (1980).

78. Surface Mining Control and Reclamation Act of 1977, 30 U.S.C. §§ 1201–1308 (Supp. IV 1980). Section 1304 strengthened the position of the surface owners in the lands where the surface and subsurface estates have been severed in the 1909–1916 statutes. Strip mining is permitted only with the consent of the surface owners.

79. See note 77.

80. Economists who are persuaded by the tragedy of the commons argument stress the importance of market choice and the dangers of bureaucratic decision making for economic performance and the preservation of political liberty. Garrett Hardin's *The Tragedy of the Commons*, 162 Science 1243–48 (1968), has served as the principal exposition of the argument as an explanation of rangeland conditions. See generally G. Tullock, *The Politics of Bureaucracy* (1965); Baden, *The Environmental Impact of Government Policies*, 4 Policy Report 1 (1982); Hanke, *Grazing for Dollars*, 1982 Reason 43.

In sharp contrast to the free market/public choice school are the institutionalists who stress the limited applicability of the tragedy of the commons argument. See, e.g., Bromley, *Property Rules, Liability Rules, and Environmental Economics*, 12 Journal of Economic Issues 43 (1978); Randall, *Property, Institutions and Economic Behavior*, 12 Journal of Economic Issues 1 (1978). See Runge, *Common Property Externalities: Isolation, Assurance, and Resource Depletion in a Traditional Grazing Context*, 63 American Journal of Agricultural Economics 595 (1981), for a specific critique of the tragedy of the commons argument.

81. G.D. Libecap, *Locking up the Range*, at 18.

82. P. W. Gates, *History of Public Land Law Development* (1968); V. Carstensen, ed., *The Public Lands: Studies in the History of the Public Domain* (1968); B. Hibbard, *A History of Public Land Policies*.

83. See P.W. Gates, *History of Public Land Law Development*, at 525.

84. See R.L. Wilbur & A.M. Hyde, *The Hoover Policies*, at 230.

85. Although there appears to be little in the way of persuasive constitutional interpretation that the federal government is in any way obligated to transfer the public lands to the respective states or to acknowledge state law as controlling over federal property. See Leshy, *Unraveling the Sagebrush Rebellion: Law, Politics and the Federal Lands*, 14 U.C. Davis L. Rev. 317 (1980); Note, *The Property Power, Federalism and the Equal Footing Doctrine*, 80 Colum. L. Rev. 817 (1980).

86. *Kleppe* v. *New Mexico*, 426 U.S. 529 (1976). See also Davis, *Impacts of Federal Land Ownership and Management on Nevada's Ability to Manage Wildlife*, in Report of the Nevada Legislature's Select Committee on Public Lands, State Sovereignty as Impaired by Federal Ownership of Land (1982).

87. See Percival, *State and Local Control of Energy Development on Federal Lands*, 32 Stan. L. Rev. 373 (1980); York, Fairfax and Twiss, *Federalism and the Public Lands: The State Role in Managing BLM Lands in California* (prepared under contract with the California State Lands Commission, 1981).

88. Western Governors' Policy Office, *Issues and Considerations in the Management of the Public Lands* (1980).

89. Francis, "Environmental Values, Intergovernmental Politics, and the Sagebrush Rebellion," chapter 1, this volume.

90. Testimony on SB 240, AB 413, Nevada Department of Conservation of Natural Resources, Division of State Lands (1979).

91. See Advisory Commission on Intergovernmental Relations, The Federal Role in the Federal System: the Agencies of Growth, the Condition of Contemporary Theories and Collapsing Constraints (1981).

92. I am indebted to Robert Nelson for his useful comments on this proposal. He is, of course, not responsible for the recommendations that follow.

93. Steve H. Hanke had an important role in getting privatization on the Reagan Administration agenda. He, also, is a major contributor to the debate on privatization that has followed. Hanke supports the selling of the public grazing lands to the stockmen who currently hold the grazing permits to the lands. See S. H. Hanke, "Privatize Those Lands,"

Reason, March 1982. "Grazing for Dollars," *Reason*, July 1982. "The Privatization Debate: An Insider's View," *The Cato Journal*, 2, Winter 1983.

94. 42 U.S.C. § 7510 (1976 and Supp. IV 1980).

95. This would be analogous to the right to sue provision of the Clean Air Act, 42 U.S.C. § 7709 (1976).

96. AUMs are currently used by the BLM in granting grazing permits.

PART TWO
SELECTED NATURAL RESOURCE ISSUES

Overview:
Selected Issues in Resource Policy

American perspectives on resource use have undergone substantial revision over the past two decades. The new perspective couples heightened consciousness of ecological and human impacts of resource use with an appreciation of the finiteness of many resources. The result is an effort to reshape state and national laws and institutions. Although goals are widely shared, there is considerable uncertainty about the best methods of pursuing them and about the effects of implementation upon the malleable American version of federalism.

In many resource use areas, Congress has concluded that national standards should be set for both existing and new activities. It also has concluded that a significant share of the costs of compliance should be borne by the general public, either through grants or tax write-offs. At the same time, it has sought to involve state and local governments in the administration and enforcement phases of the effort. The results have been mixed: business has criticized high compliance costs; environmentalists have criticized the slow pace of implementation and the frequency of concessions to polluters; and economists have criticized the failure to build compliance incentives into the approach. Relatively little attention, however, has been devoted to the basic problem of the impacts of resource policy on federal–state relationships.

Many analysts in the 1970s concluded that the environmental movement and successive energy crises would inevitably lead to greater centralization of government. However, the contributions to Part Two document the surprising resilience by state governments that traditionally have exercised major policy and regulatory powers. Clearly, states are unwilling to be mere vehicles for implementation of policies formulated in Washington, D.C. Moreover, they demonstrate considerable capacity to shape policy, take independent action, and undertake innovations that include regional cooperation. Much remains to be sorted out, and differences among states may be as troublesome as the need to develop financial capacity and new regulatory institutions. Nevertheless, these analyses imply a pattern of federalism marked by increasingly assertive, competent, and innovative state governments.

In Chapter 8 Walter A. Rosenbaum begins Part Two with an analysis of what may be the most complex problem in natural resource management—nuclear waste policy. Unlike more traditional resource uses, in which regulation came long after use was established and a national role was superimposed upon state initiatives, the national government preempted nuclear policy and its regulation from the outset. Federal policy has gone through several phases, from outright promotion to more restrictive oversight by the Nuclear Regulatory Commission. Waste disposal policy has become a central focus for further reform efforts, generating a complex politics that Rosenbaum explicates and analyzes critically. His treatment emphasizes that the failure at the beginning of the nuclear age to devise institutional arrangements between the federal government and the states for nuclear waste disposal has made federal–state cooperation increasingly difficult.

National environmental law sets air and water quality standards and eases implementation by assuming a large share of the costs of compliance. Within that context, much administrative responsibility has been delegated to the states. What would happen if federal standards were relaxed and federal funds no longer available? James Lester, with Patrick Keptner, provides an answer in Chapter 9 in the form of a comprehensive review of current statutory authority, institutional capacity, and budgetary priorities of the states. The conclusion is that all of these elements would need substantial upgrading to maintain the effort at air and water quality that now depends on both components of the national governmental role.

Two contributions examine complementary aspects of water resource policy. The traditional national role has been to underwrite and engineer major projects approved through a style of politics dubbed "distributive" by analysts. Does this tradition, marked by conflict avoidance and vitally essential to continued urbanization in the arid West, remain viable? In Chapter 10 Henry P. Caulfield, Jr. analyzes water resource policy its historic roots, the nature of challenges that have surfaced in the last two administrations, and then suggests that western states will need to bear more of the financial and political costs of future projects. In Chapter 11 Terry D. Edgmon and Timothy De Young survey both riparian and prior appropriation approaches to ground water management by states and find considerable innovation as well as variation among states. They examine the structure of decisionmaking within a state over water and conclude that the extent of centralization is not related to the aridity of an area, contra to Wittfogel's classic theses. If there is a pattern emerging among the diversity, it is an increased reliance on agency–based decisionmaking.

The next two contributions focus upon an emerging regionalism in public utility regulation, no longer a quiescent subject dominated by accountants and engineers. In Chapter 12 Lauren McKinsey describes the early experience of the multistate Northwest Power Planning Council, created by federal law enacted in 1980. He notes the reform orientation of the legislation and concludes that this national mandate has been essential to the effort to obtain serious consideration for conservation and for decentralized alternatives to central power generation to be assessed on their merits. In a region accustomed to both cheap energy and integrated planning implicit in the role of the Bonneville Power Authority, old lessons die slowly. In Chapter 13 Richard Ganzel uses a case study of the now–shelved Allen Warner Valley Energy System to probe the motives and strategies of actors in a de facto energy supply region. The involvement of coal resources from both state and federal lands, the proximity of national parks and wilderness preserves to proposed facility sites, and the involvement of development–oriented states and localities as well as an aggressively reformist California

Energy Commission indicate the difficulty of formulating regional or national policies responsive to issues that emerge from the study.

The nature of the future pattern of federalism depends significantly on the capacity of states to raise adequate revenues. That capacity, in turn, rests upon the acceptability of various levies within the dominant political culture. In the West, resource extractive industries have become suitable targets for such levies because many of the resources leave the region and corporate headquarters are located elsewhere. These rationales are complemented by consciousness of the impacts of extraction. Do taxes on resource severance violate constitutional prohibitions against interference with interstate commerce? James J. Lopach provides an analysis in Chapter 14 of the tests relied upon by the Supreme Court in upholding Montana's levy on coal production. The state's victory suggests that all states retain considerable latitude in shaping levies that tap available resources so long as they are applied in a nondiscriminatory fashion and bear some relationship to related governmental services.

These seven studies suggest that the role of states in the federal system is being strengthened in many aspects of natural resource management, as well as in their capacity to shape policies. One notable result is that state leaders are aggressively asserting their interests in ways that challenge federal preeminence and even preemption. The complexity of the issues precludes simple ideological cleavages. State assertiveness is neither liberal nor conservative, although it is practiced by leaders who embrace both labels. In any event, resource federalism promises to continue as a lively arena in which reform notions get tested in practice.

8

Nuclear Wastes and Federalism: The Institutional Impacts of Technology Development

WALTER A. ROSENBAUM

In December, 1982, Congress passed the Nuclear Waste Policy Act of 1982 (Pub.L. 97–425) intended to resolve more than a decade of conflict between Washington and the states over the disposal of the nation's rapidly growing military and civilian nuclear wastes. The act, however, created a truce rather than a settlement. The legislation temporarily suspended the controversy within the federal system by providing procedures intended to resolve technical and institutional issues associated with waste management. The legislated resolution of the institutional issues will, in effect, create a new infrastructure within the federal system, a statutory and governmental arrangement dividing powers between Washington and the states and defining their respective roles in the future management of nuclear wastes. Although the legislation appears to give the states sufficient power to protect their interests within the institutional planning process—an interest most states have felt keenly at jeopardy in the past—the prospect that the states will emerge from the process with a strong or secure role in future waste management remains at best very uncertain. Moreover, the process of institutional planning may, unless properly pursued, protract the waste management conflict to the point where the federal government may be compelled to assume greater authority for its resolution.

The broad impact of this legislation upon U.S. federalism can be best understood by examining the major problems raised for the federal structure by the waste management issue, the response of Pub.L. 97–425 to these issues and, finally, the importance of procedures for resolving institutional questions in the future development of federal arrangements for nuclear waste siting.[1]

Nuclear Waste and the Federal System

The institutional implications of nuclear power development were seldom given careful or explicit attention in the early years of governmental planning for nuclear technology after 1945. The "generally negligent and uncommunicative attitude" of the Atomic Energy Commission (AEC) concerning institutional questions, in the words of a recent study by the National Academy of Science, hardly permitted any attention to the implications for federalism.[2] After World War II, the federal government largely preempted the management and regulation of radioactive wastes generated by military activities. The Atomic Energy Act of 1954, authorizing the development of private nuclear utilities, vested in the AEC most of the responsibility for regulating the siting and management of civilian nuclear facilities and for controlling the reprocessing of spent nuclear fuels from commercial reactors. While the states could, and often did, create their own licensing procedures for civilian nuclear facilities, these were usually done under authorization of the AEC, or under terms that did not challenge the federal government's preemption of fundamental regulatory powers over site operations. Federal preemption of nuclear power regulation continued when the AEC's regulatory responsibilities were assumed by the Nuclear Regulatory Commission (NRC) in 1976.

Controversy over the safety of reactor sites, the adequacy of utility regulations and the disposal of nuclear wastes did not emerge as significant public issues until the late 1960s. The initial controversies, focused primarily on siting decisions and reactor safety, culminated in the AEC's publication of the so-called "Rasmussen Report" of 1976 intended to demonstrate that a reactor core "meltdown" or other catastrophic accident was highly improbable. Nonetheless, the controversy over reactor safety continued, fed by growing problems with the disposal of nuclear wastes first becoming evident in the early 1970s. As the safety issues magnified, attention began to turn to problems of institutional design for the regulation of nuclear power including, especially, issues about the relative roles of state and federal governments in the regulatory process.

Long before it was recognized, however, nuclear technology had, in fact, been creating over the last four decades significant and persisting transformations in American governmental and legal structures. Nuclear technology's continuing pressure upon government for institutional and statutory accommodation to itself—a sort of political *lebensraum*—is a reminder that, in Edward Wenk's terms, major technologies create "social–institutional couplets" that often amount to new structures joining technology to government.[3]

The Failure of Institutional Planning

The unprecedented and largely unanticipated problems of nuclear waste management, first evident in the late 1960s, are gradually forcing state and federal officials to define for the first time their respective rights and responsibilities for the management of these long-lived and hazardous wastes. The impact of nuclear waste management on the federal system is evidence of a continuing failure by federal technology planners to give sufficient attention to the long–term institutional impacts of technology development even while they examine other implications through various techniques of benefit/cost or risk/benefit assessment.[4] This social myopia is partially the result of the AEC's early preoccupation with what it believed to be the far more important, and more politically attractive, task of proving the technology's commercial viability between 1955 and 1970.[5] Indeed, institutional problems were largely nonissues until the 1970s because the technology was widely presumed by its governmental and scientific planners to be free of wastes that would create highly visible and controversial institutional questions.

With the breakdown of the nuclear fuel cycle's "back end" in the early 1970s, however, the institutional problems suddenly became salient and their solution imperative. The cycle's "back end," as originally conceived by optimistic federal and industry planners, would involve (1) the reprocessing of spent fuel from nuclear utilities to recover uranium and plutonium that could subsequently by enriched and recycled through reactors again; (2) the solidification of high–level wastes from fuel rods for permanent disposal; and (3) burial of low–level wastes. This "back end" cycle itself produces significant qualities of radioactive wastes, particularly highly radioactive residues from the fuel reprocessing.

A Growing Problem

Discussions between Washington and the states over waste management are conducted in an environment where consciousness of the waste's rapidly growing volume produces relatively short planning horizons. The enormous publicity accorded these often highly toxic wastes, together with the great public interest and apprehension about them, adds further to the sense of urgency about their management. Such an environment is not congenial to protracted intergovernmental negotiations, to the careful and elaborate examination of technical issues and other procedures that extend the time for waste management planning. The states, environmentalists, and other interests favoring a strong state and local government voice in nuclear waste siting fear that this sense of urgency may become a political level to advance the authority of the federal government at the expense of the states in nuclear waste siting if state and local interests do not defer rather readily to federal planning.

The nature of nuclear wastes, as we shall shortly observe, affects the institutional planning for their management. Five types of waste are commonly identified as management problems but there are, in fact, six: a great variety of nuclear facilities must also be decommissioned within the coming decades. The five most frequently identified nuclear wastes include:[6]

SPENT FUEL. Used commercial reactor fuel assemblies presently stored in racks in cooling ponds near the reactor sites. Although not technically classified as "waste" under current federal law, these assemblies will have to be so treated in the absence of commercial reprocessing facilities.

HIGH-LEVEL WASTES. The radioactive liquids created by the reprocessing of reactor fuels. These highly dangerous substances are presently stored in temporary holding tanks above and below ground in the States of Washington, Idaho, South Carolina, and New York.

TRANSURANIC WASTES. Highly radioactive byproducts of reactor fuel and military waste reprocessing. Among the extremely long-lived elements are plutonium

239, with a half–life of 24,000 years and Americum with a 7300 year half–life. These materials are presently stored with other high–level wastes in Washington state, Idaho, South Carolina, and New York.

LOW-LEVEL WASTES. Any material contaminated by radiation and emitting low levels of radiation itself. This category includes worker's clothing, tools, equipment, and other items associated with electric power generation, medical treatment or research materials, and other items used in nuclear activities. Low–level wastes are currently sent to repositories in South Carolina, New York, and Nevada.

URANIUM MILL TAILINGS. The low–level wastes from uranium mills and mines, resembling sand. Substantial amounts of abandoned tailings from mast mining and milling operations exist in 11 predominately western states.

The substantial amount of this waste presently awaiting final disposal is expected to increase greatly within the next few decades, as Table 8.1 indicates:

Disposal of some wastes is already approaching a critical issue. The American Planning Association has recently warned that current facilities for low–level waste disposal at Beatty (Nev.), Barnwell (S.C.) and Richland (Wash.) are approaching their limit. The Beatty site is likely to be closed within the next

Table 8.1   Estimated Amount of Nuclear Wastes Generated in the U.S., 1981–2000

Fuel cycle step	Type of waste	Current annual amounts	Current cumulative amounts	Projected cumulative amounts for year 2000
		(tons)		
Uranium mining and milling	Uranium tailings	10,000,000	35,510,260	400,000,000
Conversion	Low level	2,530	6,945	99,208
Enrichment	Low level	413	1,200	16,535
Fabrication	Low level	7,771	30,093	375,668
Nuclear reactor operation	Spent fuel Low level	1,618 71,099	8,318 464,624	89,820 3,566,528
Nuclear fuel reprocessing	High level Transuranic	— —	3,296 73,546	(a) (a)

a Under current U.S. policy, a moratorium on reprocessing spent reactor fuel is in effect.

Source: U.S. General Accounting Office, "Coal and Nuclear Wastes: Both Potential Contributors To Environmental and Health Problem," *Report* No. EMD-81-132 (September 21, 1981), p. 14.

several years and the Department of Energy (DOE) estimates that 5 to 7 new sites will be needed by 1990.[7]

Additionally, many facilities associated with nuclear technology will have to be decommissioned over the next century. Specifically, at least 75 presently licensed nuclear utilities, a substantial portion of the 90 additional facilities under construction or planned, 75 research reactors, nuclear ships and 44 fuel cycle facilities including fuel fabrication and conversion plants and uranium mills must be decommissioned. Neither the (NRC) nor DOE presently require comprehensive planning for decommissioning during the design and building of nuclear facilities; neither agency has promulgated standards for such decommissioning. In effect, as a recent report by the U.S. General Accounting Office noted, the U.S. still lacks a national decommissioning policy. And the essential discussions between federal and state officials leading to a cooperative strategy for such activity have yet to be undertaken.[8]

What Role for the States?

No problem has preoccupied state governments involved with nuclear waste planning more than the issue of their legal and political role in the development and implementation of a waste management strategy. This is a generic issue, compounded of several considerations which together will shape the relative powers of state and federal governments in the federalized arrangements to be created under the Nuclear Waste Policy Act.[9]

Constitutional Issues

The states have virtually no ability to erect strong statutory barriers against the disposition of nuclear wastes within their jurisdictions by the federal government, although they may be able to share some regulatory authority over details of management. Since 1945, Congress has shown a determination to preempt the management of nuclear wastes. While permitting the states some delegated authority over low–level wastes, "mill tailings" and a few other waste matters, the Congress showed little inclination to follow the federal pattern in other areas of energy management involving generous allotments of regulatory authority to the states. The Atomic Energy Act of 1954, as amended by the Energy Reorganization Act (1974), the Department of Energy Reorganization Act (1977), and the Uranium Tailings Radiation Control Act (1978) are the fundamental statutory sources of federal regulations affecting nuclear waste. This legislation gives the federal government responsibility for the regulation of high–level and low–level wastes while permitting agreements with the states for the management of low–level wastes.[10] The federal government has entered into agreements with 26 states, allowing them under the Atomic Energy Act to license low–level waste disposal sites and uranium–milling facilities and to decommission nuclear facilities within their jurisdictions. However, Washington retains responsibility for both the interim storage and final disposal of all military and federal R & D wastes, private sector high–level wastes, and abandoned uranium mining materials. Thus, waste management—particularly the responsibility for the most dangerous high–level and transuranic substances—remains in law almost exclusively a federal function.

Although more than thirty states had attempted by 1980 to legislate some controls over nuclear power management within their jurisdictions, any attempt to prohibit categorically the siting of waste repositories within their boundaries, or otherwise to interfere substantially with federal waste practices, would be constitutionally untenable. Generally, the federal government would appear to be protected from state encroachment on most of its nuclear waste management activities by the constitutional doctrines of intergovernmental immunity, the Supremacy Clause, and the Interstate Commerce Clause.[11] In addition to sharing regulatory authority with the states for low–level waste management, as previously noted, the federal government has permitted the states to designate highway routes for transportation of nuclear wastes within their borders (but, significantly, the Department of Transportation has also promulgated final uniform standards for nuclear waste transportation that largely preempts state rules). Even a promise, affirmed by both presidential candidates in 1980, that Louisiana would have a veto over any proposed waste repository in Gulf Coast salt domes is vulnerable at any time to congressional repudiation.

Federal Inconsistency and Insensitivity

A second major state preoccupation is the volatility of past federal policies concerning the state role in nuclear waste siting and, especially, Washington's frequent insensitivity to state viewpoints. Since 1971, the federal government has attempted at least five times to solve its nuclear waste problems by strategies involving the states. The experiences are not reassuring about the future of cooperative federalism in the nuclear arena:

1. THE LYONS, KANSAS SALT VAULTS. In 1970 the AEC decided, over objections from Kansas scientists still studying the site, to deposit plutonium contaminated debris from Idaho in salt mines near Lyons. The decision was strongly protested by local, state and congressional representatives from the affected area. Evidence was discovered that the sites might not be secure from water infiltration and from other geologic problems threatening the integrity of the containment area. After two years of futile effort to persuade Kansas officials otherwise, the AEC abandoned the project at a cost of $100 million and more than a decade's preliminary work. The Lyons incidents were among the earliest waste controversies to illuminate the conflicting federal and state viewpoints concerning final responsibility for site selection. It also alerted state officials and many concerned environmental organizations to the substantial unsolved scientific issues in proving site safety.

2. THE SAVANNAH RIVER PLANT WASTES. In 1974 the federal government moved approximately 23 million gallons of high–level military wastes to its Savannah River facility for experimental disposal in vaults located in bedrock beneath the plant. The Savannah River site, a sprawling 300–acre facility used by the AEC to provide plutonium and tritium for nuclear weapons, was also a storage facility for high–level and transuranic military wastes. Objections by the governor and other political leaders, in addition to technical problems with the geological stability of the bedrock, persuaded the AEC to abandon its effort to site the wastes permanently underground at Savannah. However, public officials, envi-

ronmentalists, and scientists continue to object to the federal government's failure to remove the high–level wastes initially scheduled only for "temporary" siting at Savannah in the absence of a permanent repository. The relative powerlessness of the state government to remove these wastes, and growing apprehension about the security of temporary storage arrangements, have been major issues in all subsequent discussions between Georgia and federal officials.

3. THE WEST VALLEY CASE. In 1976, the private corporation managing the West Valley reprocessing facility, Nuclear Fuel Services, announced that it was vacating the plant, idle since 1971, and transferring authority for the site management to a wholly unprepared New York State. Left at the site were about 600,000 gallons of high–level liquid wastes in two temporary storage drums, a spent reactor fuel storage facility, two low–level waste burial grounds, and a contaminated plant. For the succeeding six years, New York and the federal government engaged in protracted negotiations over responsibility for disposal of the nuclear wastes on site and the plant decommissioning. Although the federal government was not prepared legally, technically, or institutionally to assume management of the wastes, it finally entered into an agreement in 1982 to accept 90 percent of the cost for facility management and most of the technical burden for waste management.[12]

4. THE ERDA INITIATIVE. In 1976, the Energy Research and Development Administration (ERDA), created by Congress to assume the AEC's research and promotional tasks, proposed a plan for investigating waste disposal sites in 36 states. Most states objected to the proposal. Several attempts to prohibit by law various activities anticipated for the ERDA project. "In the face of such political opposition and impediments to progress at the state level, ERDA soon drew back into a narrower approach of searching for a single salt site in a state that would not object to a repository within its borders."[13]

By the mid–1970s the issue of final waste repository selection, and especially the respective authority of federal and state governments in the process, had emerged as a major technical and political impediment to any future planning for civilian nuclear power development. The federal government's inconsistency and clumsiness in dealing with state sensibilities on the waste–siting issue continued:

5. THE WIPP PROJECT. A Waste Isolation Pilot Project (WIPP) was originally authorized by Washington for location at the Los Medanos site, near Carlsbad, New Mexico. It was to be a permanent repository for defense transuranic wastes temporarily stored in Idaho; President Nixon had promised the Governor of Idaho in 1970 the wastes would be removed. The Carlsbad site was essentially an experimental repository. To quiet official apprehension in New Mexico, the Secretary of Energy had assured the state it would have a veto right over any waste disposal site in its borders. In February, 1980, President Carter announced that work on the proposed site would be cancelled while alternative sites were investigated and, if built, the New Mexico site would be licensed by the NRC. However, Congress reversed these commitments in 1981 by requiring WIPP to be an unlicensed facility solely for military waste. In January, 1981,

DOE announced it would proceed without an agreement with New Mexico to develop the site. Although a lawsuit by New Mexico compelled DOE to allow the state technical review of site plans and assured the state its other concerns would be noted, DOE retained ultimate authority to make final decisions on site selections and planning.

The tenuous legal position of all the states in the waste management process adds further to apprehension about Washington's policy irresolution. Little has happened recently to fortify state confidence in federal planners. In West Valley, for instance, the "final" agreement between New York State and the federal government proved not to be final at all. In late 1981, Congress and DOE successfully compelled New York to renegotiate the contract originally signed in 1980 because Washington believed the state had received too generous compensation for use of the West Valley facility; New York's income was reduced from $35 million to $12 million. It was distrust born from experience that prompted South Carolina Governor Richard Riley to complain: "There is a basic law of nuclear waste often overlooked. All waste remains where it is put"—a reference to millions of gallons of high–level wastes still residing at Barnwell long after its "temporary" siting was supposed to end.[14]

"Consultation and Concurrence"

State and federal governments have increasingly come to rely upon the concept of "consultation and concurrence" as a procedural arrangement to facilitate institutional planning for waste management without any consensus upon the substance of the term. In mid–1978 President Carter created an interagency task force, subsequently known as the Interagency Review Group (IRG), representing policy officials in 13 federal administrative agencies and departments and charged it with recommending to him a long–term strategy for nuclear waste management in the U.S. The IRG's report has been the most explicit statement in favor of the elusive "consultation and concurrence" whose ambiguities have not deterred the states from espousing it.

According to the IRG, "consultation and concurrence" is a federalized strategy that assumes:

> the development of cooperative relationships between states and relevant federal agencies during program planning and site identification and characterization . . . through the identification of specific sites, the joint decision on a facility, any subsequent licensing process and through the entire period of operation and decommissioning.[15]

In brief, "consultation and concurrence" is a prescription for de facto federal–state collaboration in all technical and institutional aspects of nuclear waste siting. Included in this generous conception of the state role was the notion that the states "as a matter of policy" should be given "the right to suspend federal activities" even though a state veto was rejected. Here, as noted, constitutional realities preclude any such preemptive state veto. What "consultation and concurrance" does seem to imply to the states is an arrangement based on an implicit or explicit understanding that state objections to a waste repository can be overridden only by extraordinary, and presumably difficult, federal action.

But no state is assured this federal deference. Thus, the concept remains ambiguous and its meaning will depend, in good part, upon Washington's willingness to concede to the states a major role in the planning process.

President Carter's creation of the IRG suggests how deeply rooted and chronic had become the conflict between Washington and the states over the waste siting and how little confidence the states were prepared to place in Washington's unilateral regard for state concerns on the matter. Further, the IRG was an effort to overcome what had become a persistent congressional failure to create a permanent statutory formula for the respective role of federal and state agencies in siting decisions. The evolution of the siting conflict stands in sharp contrast to the patterns of "cooperative federalism" and mutual accommodation which are assumed to prevail in current federal–state relationships. It also resurrects in modern technological terms the same sort of issue that lead John Calhoun to propose more than a century ago the use of "concurrent majorities" in cases where it appeared a national majority wished to impose its will upon a dissenting state. There is little doubt that if such concurrent majorities were required for the siting of permanent nuclear waste repositories, virtually no state would today provide the necessary affirmative vote to permit such a site within its jurisdiction.

The Temptations of Irresponsibility

State leaders, anxious to discover a politically attractive solution to the contradictory demands they frequently experience in dealing with the waste issue, may be strongly motivated to adopt institutional and technical arrangements that appear to remove from them the responsibility for deciding whether waste sites will be located within their jurisdictions. The cross–pressures arise from the anticipation in many localities that new waste sites will become a cornucopia of federal dollars pouring into the local economy and, at the same time, antipathy from many other interests, including a substantial portion of the public, to becoming the nation's "waste dump."

This urge to avoid responsibility for a final decision on the location of waste sites was apparent in the report prepared by the State Planning Council on Radioactive Waste Management, a commission appointed by President Carter to represent the states' viewpoint on the waste issue. One reason the commission recommended that states and Indian tribes should not have a veto over the location of high–level waste sites was, according to one informed observer, a "recognition that most state elected officials would find acquiescence in a proposal to site a repository in their state very difficult politically and, therefore . . . they would rather be able to pass the ultimate responsibility and blame on to others."[16] The difficulty of dealing with the waste issue otherwise is suggested by the experience of New Mexico's governor who was constrained from opposing the WIPP project with the vigor he preferred by the enthusiasm for the project among political leaders in Carlsbad.

Under such circumstances, state leaders might be attracted to several strategies that substantially surrender responsibility for a siting decision without appearing to do so: (1) blocking a decision by forcing a legal confrontation with Washington leading to a judicial determination that the federal government may preempt the designation of a waste site, thereby placing the state in the position of the protesting and aggrieved party; (2) avoiding responsibility by permitting the decision to be made on the basis of technical studies that provide criteria for site selection that appear to give each state a reasonable chance of avoiding selection—for instance, by allowing a decision to be reached through comparison of several, or many sites; or (3) by shared responsibility through a regional compact permitting a siting decision to be made by a consortium of state governments. Thus, one problem in interpreting institutional solutions to the waste management issue is understanding the extent to which the covert purpose is to abet the desire of many state leaders to avoid the responsibility for waste site determinations they profess to want.

Such suspicions may also be unfounded. The states of Nevada, Washington, and South Carolina have accepted responsibility for the management of low–level waste sites under agreements with the NRC. Relatively few technical problems have arisen under these arrangements; further, local interests prefer that the primary regulatory responsibility rest with public authorities accessible and responsible to the local constituency. The states routinely accept the responsibility for the regulation of low–level wastes at hospitals and research laboratories. The willingness of states to expand their responsibilities to include high– level and transuranic wastes may ultimately depend upon local perceptions of how well the state is compensated for its enlarged responsibilities and, presumably, for increased risks.

Nonsolutions: The Nuclear Waste Management Act

The Nuclear Waste Management Act considerably advances the search for a solution to the problem of nuclear waste siting by specifying, for the first time, an institutional and technical process for site selection that has been accepted by the White House, Congress, and the states. An explicit procedural arrangement eliminates one of the most productive sources of conflict between the states and Washington by eliminating, among other things, the unpredictability of past federal approaches to the siting issue. However, the act resolves few technical or institutional issues about the eventual management of the sites and, consequently, leaves many important implications for federalism unclear. The act accomplishes the following:

1. Requires the Secretary of Energy to study five potential sites for high–level waste repositories and to recommend three of these to the President by January 1, 1985. From these three, the President may select one site as the nation's first permanent high–level waste repository.
2. Requires the Secretary of Energy to study five additional sites, three of which were not in the original five studied earlier, and to recommend to the President by July 1, 1989, three of these sites for designation as the nation's second high–level waste repository.
3. Provides for the siting and construction of "monitored, retrievable storage" (MRS) of nuclear wastes during a period of 50 to 100 years after which the wastes must be removed to permanent sites.
4. Permits a state veto for the location of a high–level waste repository that can be overridden only by a concurrent majority vote in both Congressional chambers.

5. Requires public hearings at each site under consideration for permanent high–level or MRS waste storage.
6. Requires the President to designate the first and second permanent high–level waste sites by March 31, 1987 and March 31, 1990, respectively.
7. Permits the designated waste repositories to accept spent fuel from commercial reactors, high–level waste from fuel reprocessing and military high–level and transuranic wastes.
8. Promises Indian tribes and states the rights to "consultation and concurrence" in the siting process.

The inclusion of military wastes in the materials acceptable at designated permanent repositories may prove an especially volatile issue. Generally, the federal government in the past has sought to keep the civilian nuclear waste issue apart from the military aspects of nuclear research in order not to invite opposition to civilian waste management procedures from those opposed to nuclear weapons. Further, DOE management of military wastes has generally been regarded by the states as less responsive to state viewpoints than the NRC's program for civilian wastes.

From the viewpoint of the states, however, the act provides at least four satisfactory institutional arrangements for choosing waste sites. It requires the federal government to study a substantial number of alternative sites and thereby reduces somewhat the probability that any one state will be selected. It also provides for the creation of MRS sites, which may ultimately diminish the attraction of underground storage, particularly if attractive underground sites become intensely controversial upon their nomination. The act also gives the states a substantial bargaining position through the requirement of concurrent congressional majorities to override a state objection to the designation of a site within its jurisdiction. And the "consultation and concurrence" provisions of the act provide a politically and legally enforceable right of consultation during site investigations. These procedural advantages do not necessarily imply that the eventual institutional design for waste site management will be satisfactory to the states, however. The long– term institutional impacts of the act will depend upon several factors, particularly upon technical considerations and possibly upon provisions for public participation in siting decisions, that may not work to the advantage of the states at all.

The Questionable Primacy of Institutional Issues

There has been a growing tendency among state and federal officials involved in the waste management issue to believe the institutional issues associated with the waste sites ought to be given priority over technical issues. This would mean, for instance, that such matters as the respective regulatory roles of the different governments, the specific responsibilities for site management, procedures for altering agreements about site management and a multitude of other matters should be given at least equal priority with such issues as the suitability of various geologic sites for specific types of waste containment, the vulnerability of sites to various forms of natural disaster, and other technical matters. In spite of the protracted and often inconclusive controversies over technical issues in waste siting, this conviction about institutional priorities suggests a belief that when the institutional design for waste management is sufficiently acceptable to state and federal governments, the technical problems will be solved far more expeditiously. However, this logic seems in several respects to ignore the logic of "social–technical couplets" and particularly that technologies often constrain the forms of institutional adaptation to themselves. More specifically, the manner in which several technical questions will be resolved would seem to imply sufficiently different institutional consequences as to require the resolution of the technical issues first.

Some Technical Considerations

At least two generic technical issues ought to be, in fact must be, solved before the institutional implications to state and federal governments can be known at a site: (1) whether a geological site will be used for storage of spent reactor fuel, high–level reprocessing wastes, or transuranic wastes; and (2) whether surface repositories will be utilized at any site.[17]

The nature of the nuclear wastes to be located at a site will depend, in good measure, upon technical studies suggesting the geological appropriateness of a site for long–term or short–term waste storage. Spent reactor fuel rods, the wastes presently stored in cooling ponds at commercial reactor sites, contain the most long–lived radionuclides, such as plutonium and uranium, that will require containment and supervision for many thousands, or hundreds of thousands, of years. High–level wastes from reprocessing, in contrast, remain toxic for much shorter periods, decaying to safe radioactive levels in 600 to 1000 years. Moreover, the technology to create a barrier system to isolate these high–level wastes with a high degree of reliability seems reasonably assured and, thus, the risks and uncertainties to storage of such wastes may seem less to a potential host than would be the case for spent fuel itself.

Placed in an institutional context, the problems of designing a technically and politically satisfactory arrangement for assuring the security of wastes over tens of thousands of years would seems far more formidable, and the solutions far more difficult to advance persuasively, than would be the case with arrangements for the management of wastes over much smaller periods. There are, however, strong political and institutional criticisms of "short-term" storage. Geological sites to be maintained for as long as a thousand years may be vulnerable to terrorists and sabotage; assuring continuity of technical and political management raises issues of institutional longevity never before encountered in technology management. Thus, while the problem of technical containment of wastes may seem easier in short–term storage, the political and institutional problems may not. Perhaps very different institutional designs would have to be considered for dealing with different time periods of waste containment. In short, it appears unlikely that any potential host state can know what institutional responsibilities and risks it must accept until technical studies can determine for what nuclear wastes it may be considered a host.

Put in the language of political calculation, the federalized arrangements for nuclear waste management will have to include some system for providing equitable compensation to those states bearing the greater risks for waste management. Such an allocation is difficult until state and federal officials can characterize the risks and their distribution. Most importantly, since these broad institutional implications will certainly be apparent to public officials and interest group spokesmen of all persuasions involved in the siting process, technical issues are likely to be highly politicized and technical controversies will become, in effect, struggles over different institutional alternatives to waste management. Indeed, it would seem impossible to prevent technical arguments from becoming polarized among proponents of different institutional alternatives for site management, each using the controversies to advance their own institutional interests.

The use of surface repositories for waste storage—the MRS arrangement—also creates a different set of institutional implications than underground siting. MRS can be used in any state; its selection as a storage strategy for high–level waste, spent fuel rods, or transuranics would considerably relieve the pressure to make accommodations for waste disposal by states with geologically appropriate underground sites. Not surprisingly, the strongest advocates of MRS are states like Utah and Louisiana, who are now experiencing this pressure. Because MRS sites are temporary, problems in creating institutional arrangements for their management would seem to be considerably less difficult, expensive, or risky than would be the case with institutions expected to assure the integrity of containment for thousands of years or more. The experience of West Valley is a caution against any supposition that even these relatively short–lived institutional arrangements can be easily created. However, the problems are sufficiently less intimidating than those associated with more permanent forms of waste to suggest that federal officials may be strongly tempted to resort to MRS if they encounter great difficulty in designating underground sites by the deadlines mandated in the Nuclear Waste Act.

As these illustrations suggest, addressing the institutional issues in advance of the technological ones will probably result in extending the time needed for institutional design of waste management systems because the institutional arrangements are likely to be altered as the technical issues are further clarified. This applies to almost all aspects of the nuclear waste issue. In the case of decommissioning, for instance, it is not possible to determine accurately the costs, institutional problems, or duration of management responsibilities until standards are available prescribing the acceptable radioactive levels at decommissioned facilities. These standards do not exist and are not expected to materialize before the late 1980s because the EPA, responsible for setting the standards, has given the task a low priority. Additionally, creating institutional solutions to waste management too far in advance of the technical determinations can create tenuous and highly misleading solutions that further undermine the credibility of the process when, as is likely, substantial changes must subsequently be made.

The Impact of Public Involvement

The Nuclear Waste Act's explicit requirement for public hearings at every site to be considered for a nuclear waste repository is evidence of the belief among many involved officials that generous latitude for public involvement in the process will greatly facilitate the planning process. Given the importance of technical planning to the eventual institutional arrangements for nuclear waste management, the impact of this public involvement on technical issues merits particular concern.

Those promoting public involvement have, unfortunately, burdened the process with expectations that planning officials might understandably value but are likely to find unfulfilled. In particular, it is expected that (1) public involvement will reduce the scope and intensity of conflict over planning issues, technical and institutional; and (2) public participation will enhance the legitimacy of technical and institutional solutions to problems and thereby encourage greater program support among involved interests.

Evidence is scarce, however, to suggest that broad public involvement in the resolution of administrative issues dependably reduces the duration or intensity of issue conflict. Especially in the case of technical controversies, considerable research suggests that public involvement may encourage the polarizing of technical arguments and discourage moderation in the use or interpretation of scientific data. Often, public involvement in technical controversies leads to greater public apprehension and lack of consensus about such issues if, in fact, the public understands them. Dorothy Nelkin's conclusion after studying public involvement in technical controversies throughout the U.S. and Europe seems well founded: "There is little evidence that efforts to improve public knowledge about uncertain technical issues have actually reduced conflict. Indeed . . . access to information may indeed increase confusion and conflict, for many people are not ready to accept and evaluate the uncertainties inherent in many technical issues."[18]

The author's more recent study of public involvement in the EPA's toxic waste siting programs indicated that public involvement in technical determinations associated with siting toxic waste dumps often resulted in delay, obstruction, and opposition to decisions by regulatory authorities about the location of specific waste sites. Groups opposed to the close proximity of sites within their own geographic area often used technical issues as the first line of defense. It was often difficult for regulatory officials to discriminate between technical controversies rooted in genuine differences of opinion or information about technical issues and generalized opposition to the waste sites expressed in the language of technical objections.[19] Significantly, the predictable opposition of local residents and public officials to the proposed waste sites often caused the eventual determinations of site locations and permit conditions to be made by administrative agencies or judicial bodies acting on appeals from regulatory authorities originally responsible for the siting decisions. In light of the vigorous public opposition to nuclear waste repositories usually found among those geographically near the proposed sites, one might well expect the public participation process under the Nuclear Waste Act to produce analogous conditions, and results, to those experienced in toxic waste site determinations.

Generally, the evidence suggests that public attitudes about technical issues, and the stance of organized interests toward such issues, is often determined by political attitudes, or other values, that resist modification through persuasion by technical data. In the case of nuclear waste siting, it would seem most reasonable to expect, at the least, that public involvement will produce a high level of controversy and protract the technical planning process significantly. Early indications suggest that public involvement procedures will be used by opponents of proposed nuclear waste to extend as long as possible the technical and institutional planning in hopes of maximizing opportunities for opposition. In Washington state, for instance, environmental groups and public officials have both objected that the DOE has allowed insufficient time for public review of proposed guidelines for the siting of high–level repositories. Since the DOE presently hopes to recommend the first three sites for highlevel repositories by 1983, instead of the 1985 deadline under the Nuclear Waste Act, opponents within the state have understandably chosen the public involvement process as an early means of forcing DOE to prolong the determination period. Moreover, opening the planning process to public participation is likely to broaden the scope of conflict by publicizing issues and alerting previously uninvolved interests to issues concerning them.

There is also little evidence to suggest that public participation in administrative planning, whatever its substantive nature, will predictably increase public acceptance of programs or encourage greater public trust of the institutions responsible for planning. At best, what can be said about the impact of public involvement on citizen trust is that the studies are inconclusive; at worst, they provide little reason for confidence that the procedures work as anticipated.[20]

In light of past experience with public involvement, then, there are significant risks associated with its use in technical or institutional planning for nuclear waste management. First, such involvement may prolong the planning process and exacerbate controversies to such a degree that it may encourage a much more aggressive and authoritative federal role in the process as the statutory deadlines for site designation approach. Second, it may promote the resolution of issues through various judicial and administrative appeals processes that remove such determinations from the direct influence of local or state interests. Third, it may intensify and broaden local public and official opposition to proposed nuclear waste sites to the extent that local officials may feel compelled to oppose the sites categorically. Finally, public participation may indirectly but significantly encourage the adoption of the MRS as the most feasible solution to the waste siting issue for the next few years because it involves fewer institutional or technical risks than the alternatives and may be more politically palatable to state and local interests.

Conclusion: The States and Nuclear Waste Planning

The future role of the states in the management of the nation's nuclear wastes will largely be determined by the technical and institutional procedures only now being implemented under the Nuclear Waste Act of 1982. The states would seem assured a strong voice in this planning process through the act's mandate for federal "consultation and concurrence" with state governments and through the right of state negation of a proposed site; however, the states" long–term status in waste management planning seems less secure. If the planning process for waste siting becomes as prolonged and controversial as it is likely to be based on past experience and the generous provisions for public involvement in the present procedures, pressures to meet statutory deadlines for site selection may well confront tive federal role in the planning process and exacerbate controver preferences. State leaders may also feel so crosspressured that they may intentionally or unintentionally force the federal government into making a siting decision thereby removing responsibility from the local leadership.

Currently, no state has expressed a willingness to accept a high– level underground waste repository and only a few have suggested they might favorably consider low–level waste burial grounds (these latter states are primarily those already accommodating low–level wastes). It is conceivable that the planning procedures created by the Nuclear Waste Act may enable the federal and state governments to reach an accommodation leading to permanent underground burial sites in several states. However, the past history of federal–state relations in nuclear waste siting, the relatively weak constituencies for such siting in most states, and the institutional uncertainties involved for the states strongly suggest the probability of continued vigorous state resistance to such sites. The planning process created by the act is likely to exacerbate the conflicts already evident in site location. Since the Reagan Administration is aggressively moving the planning schedule ahead of its statutory pace, many states are likely to believe (or, at least, to assert) that the process has given insufficient time to technical and institutional problems. This increases the likelihood that one, or several, states will vote against a site in their boundaries, thus creating the possibility of a congressional vote to override the states.

An alternative, and perhaps more probable, resolution to the planning impasse will be the federal government's selection of the far more palatable MRS storage option for the present wastes. This option has the virtue of appearing to be a temporary solution under which no state has a permanent waste depository but the federal government is permitted to distribute wastes among a much larger group of states, leaving no state in the position of having to become the principal burial ground for the rest of the nation's nuclear wastes. Indeed, under the MRS arrangement, it may be possible for many states to negotiate agreements to accept in MRS only those wastes generated within their own jurisdictions.

In any event, the long, confused, expensive, and still unresolved problem of creating a satisfactory institutional arrangement for managing the nation's nuclear wastes illuminates the substantial social costs for earlier failures to anticipate the institutional problems and to begin planning for their resolution. The lesson ought to be evident and persuasive: "technological couplets" should be recognized as an inevitability in the development of any new technologies through federal R & D and the planning for the institutional implications of this development ought to be as routine as other forms of risk analysis associated with governmental technology development. Had such anticipatory institutional analysis been conducted concurrently with the technological innovations in nuclear power development, the institutional planning for nuclear waste management could have proceeded through a longer period and with greater capacity by public agencies to generate the technical and institutional research necessary to resolve many of the issues that must now be resolved under extreme time pressures in an atmosphere of crisis and conflict.

Notes

1. A useful summary of the legislative history of proposals leading to the Nuclear Waste Policy Act of 1982 may be found in U.S. Congress, House Committee on Interior and Insular Affairs, Subcommittee on Oversight and Investigations, *Nuclear Fuel Cycle Policy and the Future of Nuclear Power*, October 23, 1981; and ibid., Subcommittee on Energy and the Environment, *Radioactive Waste Legislation* Hearings on June 23, June 25 and July 9, 1981. See also James P. Murray, Joseph J. Harrington and Richard Wilson, "Chemical and Nuclear Waste Disposal," *Cato Journal*, Vol. 2, No. 2 (Fall, 1982), pp. 565–606.

2. National Academy of Science's, National Research Council, Committee on Nuclear

and Alternative Energy Systems, *Energy In Transition: 1985--2010* (San Francisco: W.H. Freeman and Co., 1979), p. 314.

3. Edward Wenk, Jr., *Margins For Survival* (New York: Pergamon Press, 1979), p. 32.

4. On the neglect of institutional analysis generally, see Walter A. Rosenbaum, "The Hidden Risks of Risk Assessment," in Susan G. Hadden, ed., *Institutions and Technological Risk Control* (New York: Kennikat Press, 1983).

5. The AEC's concern for the promotional aspects of commercial nuclear power development is explored in Irvin Bupp and Jean–Claude Derian, *The Failed Promise of Nuclear* (New York: Harper and Row, 1981).

6. Technical data on nuclear wastes is obtained from U.S. Comptroller General, "Coal and Nuclear Wastes: Both Potential Contributors to Environmental and Health Problems," Publication No. EMD–81–132 (September 21, 1981).

7. *Environmental Reporter*, February 13, 1983, pp. 1843–44.

8. U.S. General Accounting Office, "Cleaning Up Nuclear Facilities: An Aggressive and Unified Federal Program Is Needed," Publication No. GAO/EMD–82–40 (May 25, 1982).

9. A useful summary of the issues may be found in E. William Colglazier, Jr., ed., *The Politics of Nuclear Waste* (New York: Pergamon Press, 1982).

10. Harold P. Green and L. Marc Zell, "Federal–State Conflict in Nuclear Waste Management: The Legal Basis," in E. William Colglazier, op. cit., pp. 110–37.

11. Ibid.

12. On the problems of West Valley, see U.S. Comptroller General, "Status of Efforts to Clean Up the Shut–Down Western New York Nuclear Service Center," Report No. EMD–80–69 (June 6, 1980); U.S. Comptroller General, "Further Analysis of Issues At Western New York Nuclear Service Center," Report No. EMD–81–5 (October 23, 1980); and U.S. Congress, House Committee on Insular Affairs, Subcommittee on Energy and the Environment, *Remedial Action at West Valley, N.Y.*, 96th Congress, 1st Session, Serial No. 96–12.

13. Ted Greenwood, "Nuclear Waste Management in the United States," in E. William Colglazier, op. cit., p. 7.

14. Quoted in E. William Colglazier, Jr. and Paul Doty, "Preface," in E. William Colglazier, op. cit., p. xv.

15. Ted Greenwood, op. cit., pp. 35–37.

16. Ibid., p. 47.

17. The importance of these two issues is suggested by the repeated focus upon them during the majority of public hearings on the nuclear waste issues. In addition to Congressional sources already cited in fn. 1, see also: U.S. Congress, House Committee on Interstate and Foreign Commerce, *Nuclear Waste Policy Act*, Hearings on September 24, 1980. H. Rpt. 96–1382, pt. 1; and U.S. Congress, Committee on Interstate and Foreign Commerce, Subcommittee on Energy and Power, *Spent Fuel Storage and Disposal*, Hearings on June 26, 1979.

18. Dorothy Nelkin, *Technological Decisions and Democracy* (Beverly Hills, Calif.: Sage Publications, 1977), p. 96. See also the findings summarized in Mary Grisewz Kweit and Robert W. Kweit, *Implementing Citizen Participation in a Bureaucratic Society* (New York: Praeger Special Studies, 1981).

19. Walter A. Rosenbaum, "The First to Go: Public Involvement in Toxic Waste Management," in James P. Lester and Ann O.M. Bowman, eds., *The Politics of Hazardous Waste Regulation* (Durham, N.C.: Duke University Press, 1983).

20. The literature suggesting this rather pessimistic view of public involvement is reviewed in Walter A. Rosenbaum, "Public Participation As Ritual and Reform," in Stuart Langton, ed., *Public Participation* (Lexington, Mass.: Lexington Books, 1978), pp. 145–171.

9

State Budgetary Commitments to Environmental Quality under Austerity

JAMES P. LESTER with
PATRICK M. KEPTNER

Introduction

Federal environmental policy actions during the 1970s have had a significant impact on the states as a result of specific requirements, money incentives, and the examples set by federal legislation. Despite this fact, the states are not spending significant amounts of money for pollution control even though intergovernmental grants to the states and cities have risen dramatically. Although state fiscal increases for environmental quality control are impressive and clearly represent a "launching" of state programs, they remain miniscule when put into the perspective of total state expenditures for more traditional state functions such as education, public welfare, highways, etc.[1] As Charles O. Jones noted several years ago, "we are not talking about a significant policy commitment for most state governments. The continuum will be from 'little or no commitment' to a 'limited, low priority commitment.' "[2]

It should be noted that Jones's assessment of state commitment to environmental quality was made at a time when the issue of environmental protection was associated with rather strong federal support of state activities.[3] More recently, evidence continues to suggest that we may be entering a phase change or a reinterpretation of environmental imperatives as a new administration challenges the necessity of environmental regulation and shifts its priorities in directions other than environmental.[4] Thus, new federal priorities and diminished fiscal resources now threaten to undermine what little progress has been achieved to date. Whether the states would continue the same level of support for pollution control if the federal government were largely to discontinue its contribution is a proposition most environmentalists would not like to test. Citizen groups, on the other hand, continue to express strong support for the maintenance of existing regulations for environmental protection. Indeed, a recent survey of public opinion by Louis Harris found that "not a single major segment of the (American) public wants the environmental laws made less strict."[5]

This study cannot resolve the debates accompanying environmental protection since they are endemic to a pluralistic system in which no single person or group has a monopoly on the definition of the problem and the proper response to it. What it can do, however, is to identify systematically the broad determinants of state policy responses to environmental pollution and, on the basis of this analysis, to suggest some implications for future state environmental protection activities.

Therefore, the purpose of this paper is to examine the states' commitments to environmental protection activities and to evaluate the utility of several indicators (both new and conventional) hypothesized to be sources of public policy responses. For both theoretical and practical reasons, it is particularly important to understand the determinants of state commitment to environmental quality, especially given the impending decline in fiscal resources through both federal and, inevitably, state budget cuts. First, there are too few empirical studies of environmental protection activities using the fifty states as the units of analysis.[6] Such studies, from a macropolitical perspective, would help us to understand policy differences across the states and the primary determinants of these differences. Second, among those environmental studies available, even fewer examine the relative effects upon policy of socioeconomic, technical and political process factors in this functional area. Both political scientists and policymakers alike are interested in the extent to which institutional reforms—undertaken to improve system performance—make a significant difference in the content of environmental policy outputs.[7]

Three related research questions are addressed in this paper. First, how have the fifty American states responded to the environmental pollution issue in terms of their budgetary priorities for environmental protection during the period 1975–1979? Second, what particular variables account for state policy variations within each year during this period? Finally, does the importance of particular factors (e.g, economic or structural) depend upon such contextual conditions as severity of the policy problem (i.e., saliency or policy demand)? We review the relevant literature before we assess the influence of technical, resource, and especially political (institutional) factors upon the fifty American states' commitments to environmental quality during the period under study.[8]

Explaining Environmental Policy: Four Propositions

Within the literature on environmental policy, there are at least four basic explanations for state policy responses to the problems posed by environmental pollution. These four explanations may be identified as: (1) technological pressures or "needs" for policy, (2) economic resources, (3) partisanship and/or vigorous interparty competition, and (4) administrative–organizational capabilities. Each of these alternative explanations are discussed below.

Technological Pressures: Severity of the Problem

This explanation suggests that rapid and concentrated population growth, extensive industrialization (especially a greater reliance upon the petrochemical and metallurgical industries), and steadily increasing rates of public consumption of goods and services create severe pollution problems which, in turn, bring about strong pressures for environmental protection policies. Thus, an obvious source of regulatory policy differences among the states is the severity of the pollution problem itself.[9]

Recent studies suggest that state environmental control policies are either *not* strongly correlated with the actual level of pollution or that more refined indicators of pollution severity are needed before a final assessment of their effect on state policy is known.[10] However, it may be argued that the relationship between severity of the pollution problem and state policy outputs depends not only on our ability to measure "pollution potential," but also on the pollution threshold itself. For example, some state legislators have indicated a reluctance to adopt stringent air quality standards because the air pollution problem was "not bad enough."[11] Although the evidence is mixed, we expect indicators of technological pressures, particularly the level of industrialization and "pollution potential," to affect a state's commitment to environmental quality. Thus:

> *Hypothesis 1:* The greater the technological need (or severity of the pollution problem), the greater the commitment to environmental quality. Conversely, the less the technological need, the less the commitment.

Economic Resources

A second model posits a direct relationship between the socioeconomic resource base of a political system (national, state, or local) and levels of public policy outputs. For example, it is often suggested that the level of a state's economic wealth sets limits on, or provides opportunities for, the provision of public goods and services by a government for its constituents. This consideration is often overlooked by persons who assume that the failure of government to act in the environmental sphere is caused by states' backwardness or nonresponsiveness to the public policy problem of environmental pollution.[12] Thomas R. Dye has presented the strongest argument for the "resources" perspective. He finds that a state's degree of economic development explains policy activity across a variety of state functions (i.e., state aid to education, welfare, transportation, natural resources, etc.)[13]

An early environmental policy study tested this perspective and concluded that political variables were largely insignificant predictors of expenditures for air pollution control while industrialization, urbanization, and (to a lesser extent) median income accounted for the greater proportion of the variance.[14] More recent studies of state environmental protection activities, as well as a substantial body of comparative state literature, continue to find that measures of state wealth are strongly related to policy outputs.[15] Therefore:

> *Hypothesis 2:* The greater the wealth of a state, the greater the commitment to environmental quality. Conversely, the greater the poverty within a state, the less the commitment to environmental quality.

Political Demands: Partisanship and Competition

A third model is based on possibly the most common generalization in the environmental policy literature: environmental policy making is, to a large extent, explained by political party differences. For example, Dunlap and Gale argue that there are important reasons for expecting significant partisan differences to emerge on environmental issues:

> On the one hand, pro–environmental measures generally are opposed by business and industry, entail an extension of governmental regulation and intervention, and imply the need for "radical" rather than "incremental" policies. On the other hand, traditionally the Republican party, relative to the Democratic, has maintained a more pro-business orientation, a greater opposition to the extension of governmental power, and a less innovative posture toward the use of governmental action to solve societal problems.[16]

On the basis of these presumed views about the influence of party on environmental policy, they hypothesized that Republicans would give significantly less support to measures designed to protect the quality of the environment than would Democrats. In testing this relationship, it has been shown, for the most part, that Democratic partisanship is strongly related to pro-environmental *voting* within some state legislatures and the Congress.[17] Additional studies (as well as reviews of the literature), have examined these earlier findings with regard to nonelite's support for (and state policy enactments of) environmental policies and found that environmental concern continues to be a partisan issue.[18]

It is important to consider the fact that only one of these environmental policy studies is a comparative state project in the sense that all fifty states are included in the analysis.[19] That study, along with fifty state comparisons in other policy domains, suggests a much different aggregate relationship between Democratic partisanship and environmental policy activity. We would still expect partisanship and policy outputs to be significantly related. However, we would *not* expect Democratic partisanship in state legislatures to be associated with high levels of policy activity. Comparative state studies have consistently found Democratic party strength to be negatively related to policy outputs. This finding has, of course, been due to the anomalous character of the southern states. Legislatures in these states have traditionally been dominated by Democrats who ideologically have little in common with their counterparts elsewhere. At the same time, as a group the southern states have normally been outliers at the low end of a variety of policy output measures. Their outlying nature in comparative analyses in terms of both the independent and dependent variables has substantially affected, or perhaps distorted, the results obtained. Thus, we have no basis upon which to expect the relationship to be any different in the environmental policy area. That is, with the inclusion of all fifty states, we expect the opposite relationship between Democratic partisanship and environmental policy activity to that found for single states, a subset of the fifty states, or the U.S. Congress. Thus:

> *Hypothesis 3a:* The greater the Democratic party strength in the state, the less the commitment to environmental quality. Or, the less the Democratic party strength, the greater the commitment.

However, according to more generalized theories of public policymaking, it may also be hypothesized that vigorous two–party competition promotes greater responsiveness to voter anxiety about environmental pollution. This suggests that electoral marginality rather than partisan– ideological orientation increases the likelihood that policymakers, in order to secure electoral support, will take positive steps to deal with various political issues.[20] Empirically, this observed relationship is the result of low levels of policy activity in the noncompetitive southern states. Thus:

> *Hypothesis 3b:* The greater the inter–party competition, the greater the commitment to environmental quality. Conversely, the less the competition, the less the commitment.

Administrative-Organizational Reforms

A fourth explanation focuses upon administrative and legislative reforms as potential explanations of public policy outputs. For example, the movement for administrative reform advocated consolidation of administrative agencies into a small number of departments, organized by function, whose heads are appointed and controlled by the governor.[21] This reorganization would, it was hoped, help to eliminate jurisdictional overlap, jealousies, and conflicts between multiple agencies in a specific functional area. Moreover, consolidation of the bureaucracy would serve to increase the governor's span of control and—if he were able to appoint the head of a particular functional agency—his resulting ability to mobilize the bureaucracy in order to carry out his immediate policy objectives.

However, the available evidence suggests that state executive branch reorganization has little effect on public policy outputs after economic development factors are taken into consideration.[22] Nevertheless, we expect executive branch reorganization in this specific functional area to exhibit an impact upon the establishment of *budgetary priorities* for environmental protection, even though more generalized measures of executive reform may not be linked to *absolute* levels of spending in this area (or other areas). Thus:

> *Hypothesis 4a:* The greater the executive branch reorganization (i.e., the greater the consolidation of the environmental bureaucracy and/or gubernatorial appointive power in the environmental area), the greater the commitment to environmental quality. Conversely, the less the executive branch reorganization, the less the commitment.

A second body of reform literature focuses upon the state legislature. For example, the ability of a "professional" (versus an "unprofessional") state legislature to be more responsive to the needs and concerns of its citizens has been the focus of several empirical studies.[23] Reformers generally argue that "professionalism" will result in legislatures that are innovative in many different areas of public policy, generous in spending and services, and "interventionist" in the sense of having powers and responsibilities of broad scope.[24]

Legislative reformers have had high hopes that reforms would lead to policies better designed to cope with pressing state problems. However, comparative state policy research generally has shown that reforms have not had much of a policy impact.[25] Nevertheless, environmental policy analyses that do not consider the potential influence of state legislative capabilities assume that political party differences affect state policies independently of organizational influences. Presumably, this view of state politics omits a very important linkage in determining state environmental outputs. Thus:

Hypothesis 4b: The more professional the state legislature, the greater the commitment to environmental quality. Conversely, the more parochial the state legislature, the less the commitment.

While each of these models provides some insights into policy formulation, no single model captures the complexity of the public policymaking process or provides a sufficiently comprehensive explanation of the forces influencing the commitment to environmental quality. Thus, instead of pitting these four models against one another, it seems more constructive to ascertain their strengths and limitations in the context of a more complex theoretical framework. In this regard, it will prove helpful to know under what conditions the importance of certain resource, political, and administrative/organizational variables in shaping environmental policy is greater or lesser.

The Measures

Dependent Variable

Our measure of state commitment to environmental quality is based on a ratio of total state government expenditures for environmental quality control compared to total state governmental expenditures. It is calculated separately for each state for each of the years, 1975–1979.

Although state expenditures are a commonly used measure of commitment, there are several limitations of this data which should be noted at the outset. First, current pollution control expenditures of the states rank well below those of the national and local governments. States typically spend the least of all three levels of government.[26] In addition, the private sector within each state provides expenditures for pollution control. Thus, one might rightly question the extent to which judgments might be made about state effort from expenditure data. Perhaps some states are successful in ensuring environmental protection through other means, such as gaining cooperation from polluting industries or through intergovernmental (federal) grants for specific environmental problems where they exist.[27] Second, a portion of state expenditures for pollution control is motivated by economic considerations, such as a desire to qualify for federal matching contributions, rather than a purely environmental commitment.[28] Third, it could be argued that compilations of state laws and enforcement activities in the environmental sphere might provide a better measure of state commitment to environmental protection. These limitations notwithstanding, we argue that this ratio provides an approximate indication of the relative *budgetary priority* attached to environmental protection by each state relative to other competing state functional activities. Moreover, it is expected that our use of a *ratio* expressing a state's budgetary priority for environmental quality control will reflect the concept of *political choice* to a greater extent than would gross expenditure measures (or total state expenditures for environmental protection). That is, we sought to employ a more political measure of state environmental policy in order to offset the bias inherent in relating economic input factors to gross expenditure indicators of policy output. Such procedures are believed to bias findings against political determinants of state policy outputs.[29]

Independent Variables

Table 9.1 describes the conceptual model that guides our study and the variables used in our subsequent analysis.

Table 9.1 The Conceptual Model

Components of the model	Description-rationale	Measure[a]
Problem severity	Pressures resulting from the effects of population and technological growth generate demands for policy action	a. Total state population b. Population density c. State industrialization d. Jones index of pollution potential
Economic	The economic environment provides constraints on (or opportunities for) collective action by state governments	a. Personal income per capita b. Median family income, 1975 c. Percentage of poor families, 1975 d. Environmental aid per capita
Political demands	Environmental pressures give rise to demands which find expression in terms of Democratic partisanship or interparty competition	a. Democratic Party strength b. Interparty competition
Administrative-organizational structure of authority	These demands are then acted upon (or not acted upon) by state institutions which have the technical and/or institutional capabilities to translate policy demands into policy actions	a. Governor's appointive powers for environmental protection b. Bureaucratic consolidation for environmental protection c. Reformism d. Legislative professionalism, 1974
Commitment to environmental quality	States engage in protection of the environment as a consequence of problem severity, economic resources, political demands, and institutional capabilities	Ratio of environmental quality control expenditures ——————— Total state governmental expenditures

[a] All data are concurrent unless otherwise noted, are established for each state, and are aggregated annually for each of the years 1975-1979.

In order to evaluate the technological pressure or "needs" model, four separate measures are utilized: (1) total state population; (2) population per square mile in each state; (3) percent of the labor force in manufacturing (i.e., industrialization); and (4) Jones's index of "pollution potential." Our hypothesis would predict a positive relationship between each type of pressure or "need" for environmental protection and a state's commitment to environmental quality control.

Our second model, based on economic resources, is evaluated by employing three conventionally used measures of internal state economic wealth and a measure of state aid: (1) personal income per capita; (2) median family income; (3) percent of the state population below the poverty line; and (4) per capita federal aid (from the Environmental Protection Agency) for waste treatment facilities construction.[30] The latter measure reflects state resources relevant to this functional area that are provided by sources external to the state, while the three former measures indicate existing (internal) state resources. The resource model hypothesizes that state commitment to environmental quality will be greater in those states that can "afford" it (i.e., greater state resources) and less in those states that cannot (i.e., more poverty).

Measures of the political demands model are Ranney's index of Democratic party strength and a folded Ranney index of the interplay competition. These two measures are based on four components: (1) the average percentage of the popular vote won by the Democratic gubernatorial candidate; (2) the average percentage of the seats in the state senate held by the Democrats; (3) the average percentage of the seats in the state house of representatives held by the Democrats; and (4) the percentage of all terms for governor, senate, and house in which the Democrats had control. Each state's percentages were averaged to develop an index of Democratic party strength.[31] As we discussed above, previous empirical findings in other policy areas involving all fifty states would lead us to expect that Democratic states would give significantly less support to measures designed to protect the quality of the environment, at least when all fifty states are included. This expectation is due to the anomalous character of the southern states.

However, as stated earlier, it may also be argued that vigorous interparty competition stimulates aggressive governmental action. Only in the competition between such parties can the popular will be translated into government action. Thus, our hypothesis assumes that where two–party competition exists, pro–environmental election promises will be translated into state commitments for environmental quality for fear of retribution at the polls.

Finally, our last model (administrative–organizational reform) is measured by an updated version of Grumm's legislative professionalism, and by three measures of administrative reorganization (or reform) for state environmental policymaking.[32] The measure of legislative professionalism is based upon a number of variables believed to capture the degree to which a state's legislature is "professionalized" or "nonprofessionalized." Each state thus received a factor score which was then used as a measure for this particular dimension of legislative capability.

Our administrative reform measures are based on three separate indicators: (1) the extent of gubernatorial appointive power in the environmental area, with higher scores reflecting greater appointive power; (2) state administrative reorganization (consolidation) of the environmental bureaucracy, with higher scores reflecting greater consolidation; and (3) a combined index of these two former measures, with higher scores reflecting states which have *both* consolidated environmental bureaucracies and in which governors are able to appoint the head of this new agency or department. The latter combined index reflects the essential recommendations of the Reform Movement.[33] Thus, our hypotheses would lead us to expect that reforms in the state legislatures or within the state executive would be associated with greater commitment to environmental quality.

The Findings

The basic objective of our analysis is to determine why some states express a greater commitment to environmental quality than others. The four models introduced above provide alternative explanations of state environmental commitment. Initially, these explanations will be separately considered by examining the predictive ability of multiple indicators of each phenomenon introduced above. Theoretically, as well as methodologically, isolated comparisons across the four models are not informative in terms of understanding the overall processes leading to policy activity. However, this preliminary exercise will be used to gain insights which will facilitate the eventual selection of single indicators of each model for an informed assessment of the policy process in a multivariate framework.

The first step in analyzing the research results is to examine the individual state's level of commitment to environmental quality. Individual state scores are presented in Table 9.2.

From an examination of the scores, it is readily apparent that the states are spending very little on environmental protection relative to other state functions. The percentage of total state government expenditures allotted to environmental protection ranges from a low of .1% (e.g., Louisiana) to a high of 4.5 percent (e.g., New Hampshire in 1978). The average percent of expenditures allotted to activities in this area by all fifty states is approximately .7 percent (or less than 1 percent).

Regional variations in Table 9.2 indicate that the twelve northeastern states (relative to all others) give this functional area the highest budgetary priority, whereas the eleven "Deep South" states exhibit the lowest priority for environmental protection (their mean scores are .011 and .003, respectively).[34] The finding that the northeastern states are committing a greater proportion of their total budget to this area is not surprising given the higher degree of industrialization in this region relative to the South or other regions.

The Bivariate Analyses

Table 9.3 examines the four explanations of commitment to environmental quality in terms of the indicators of each model. The best bivariate predictors of commitment are the multiple indicators of severity and resources. Those states experiencing a greater "need" for policy are reacting by committing a larger percentage of state funds to this area. In addition, those states with greater economic resources are more likely to exhibit a greater commitment ($r = .18$ to

Table 9.2 Individual State Scores for the Ratio of State Environmental Quality Control Expenditures/Total State Budgetary Expenditures, by Year

State	1975	1976	1977	1978	1979
1. Alabama	.00214	.00216	.00209	.00201	.00210
2. Alaska	.00551	.00844	.01036	.00933	.00951
3. Arizona	.00347	.00392	.00468	.00456	.00446
4. Arkansas	.00162	.00229	.00317	.00288	.00261
5. California	.00525	.00629	.00627	.00585	.00624
6. Colorado	.00712	.01107	.00526	.00649	.00663
7. Connecticut	.01007	.00810	.01773	.00937	.00676
8. Delaware	.00608	.00842	.00685	.00541	.00967
9. Florida	.00952	.00283	.00301	.00313	.00283
10. Georgia	.00399	.00350	.00322	.00302	.00288
11. Hawaii	.01313	.01425	.01078	.00378	.00283
12. Idaho	.00944	.00797	.00545	.00696	.00620
13. Illinois	.00687	.00858	.01212	.01019	.01036
14. Indiana	.00382	.00434	.00581	.00454	.00553
15. Iowa	.00344	.00477	.00381	.00313	.00245
16. Kansas	.00205	.00222	.00328	.00315	.00263
17. Kentucky	.00375	.00321	.00315	.00317	.00343
18. Louisiana	.00127	.00121	.00140	.00130	.00174
19. Maine	.01192	.00832	.01064	.00968	.00875
20. Maryland	.01328	.01708	.01145	.01119	.00893
21. Massachusetts	.00766	.01157	.00989	.00979	.01093
22. Michigan	.00474	.00394	.00357	.00303	.00452
23. Minnesota	.00476	.00679	.00659	.00607	.01096
24. Mississippi	.00448	.00162	.00177	.00208	.00366
25. Missouri	.00402	.00716	.00738	.00669	.00564
26. Montana	.00604	.00613	.00544	.00629	.00598
27. Nebraska	.00396	.00822	.00659	.00526	.00455
28. Nevada	.00308	.00552	.00747	.01369	.00517
29. New Hampshire	.01760	.01956	.02653	.04453	.03321
30. New Jersey	.00869	.01119	.00814	.00493	.01690
31. New Mexico	.00442	.00611	.00270	.00480	.00315
32. New York	.00836	.01279	.00815	.00734	.00688
33. North Carolina	.00522	.00579	.00437	.00443	.00626
34. North Dakota	.00114	.00166	.00164	.00237	.00252
35. Ohio	.03123	.03411	.03557	.03740	.02845
36. Oklahoma	.00221	.00226	.00245	.00291	.00282
37. Oregon	.00488	.00450	.00748	.00767	.00858
38. Pennsylvania	.00657	.00671	.00640	.00580	.00556
39. Rhode Island	.00828	.00800	.00724	.01007	.00908
40. South Carolina	.00293	.00305	.00304	.00394	.00341
41. South Dakota	.00319	.00303	.00352	.00299	.00307
42. Tennessee	.00420	.00385	.00421	.00417	.00416
43. Texas	.00352	.00317	.00339	.00314	.00309
44. Utah	.00223	.00211	.00248	.00225	.00220
45. Vermont	.00959	.00732	.01820	.01855	.01090
46. Virginia	.00656	.00679	.00565	.00388	.00364
47. Washington	.00748	.01236	.00380	.00720	.00576
48. West Virginia	.00308	.00406	.00374	.00334	.00327
49. Wisconsin	.00799	.00661	.00679	.00431	.00443
50. Wyoming	.00367	.00351	.00451	.00458	.00426

Ratio Calculated as: Total expenditures for EQC
────────────────────────────────
Total state budget expenditures

Source: State and Local Government Special Studies, U.S. Dept. of Commerce, Bureau of the Census, *Environmental Quality Control*, Washington, D.C.: Government Printing Office, (1975-1979), and *Statistical Abstract of the United States*, (1975-1979).

.50) than those with fewer state resources (r = −.30 to −.44). Moreover, the greater the amount of environmental aid provided by the federal government, the greater the priority (r = .12 to .26). Although the findings for the indicators of political demands are less compelling, both Democratic party strength and interparty competition are significantly related to state commitments. However, our indicators of administrative-organizational reform are not, for the most part, significantly related to state commitment. While our combined index of "reformism" and our index of legislative professionalism exhibit a moderately significant relationship to state commitment during the years 1976 and 1979, those findings are not consistent within other time periods. Thus, these few significant relationships must be viewed with caution.

The performance of the economic and political demand models is not surprising. Most comparative state policy research has consistently demonstrated two things. On the one hand, economic resources have been shown to be strongly related to policy outputs in the bivariate case and even after controlling for political process variables in the multivariate situation. On the other hand, Democratic party strength has usually been negatively related to policy outputs

Table 9.3 Pearson's Product-Moment Correlation Coefficients (r) Showing the Relationship between Selected Independent Variables and States' Commitments to Environmental Quality, 1975–1979 (All States)

Independent variables	1975	1976	1977	1978	1979
Model I (Needs)					
State population	.13	.14	.02	−.08	.04
State density	.31[c]	.34[c]	.27[b]	.13	.33[c]
Industrialization	.21[a]	.12	.22[a]	.21[a]	.31[c]
Pollution potential	.24[b]	.24[b]	.19[a]	.11	.19[a]
Model II (Resources)					
Personal income	.18[a]	.35[c]	.29[b]	.14	.24[a]
Median income	.35[c]	.50[c]	.37[c]	.22[a]	.33[c]
Percent poor	−.32[c]	−.44[c]	−.38[c]	−.30[b]	−.36[c]
Environmental aid	.25[b]	.26[b]	.19[a]	.12	.20[a]
Model III (Political demands)					
Democratic Party strength	−.16	−.19[a]	−.30[b]	−.32[c]	−.15
Interparty competition	.18[a]	.23[b]	.23[b]	.17	.31[c]
Model IV (Admin-organ factors)					
Governor's appointive power (env.)	.06	.06	.06	−.05	.02
Bureaucratic consolidation (env.)	.12	.15	.12	.01	.10
Reformism	.12	.20[a]	.15	.06	.18[a]
Legislative professionalism	.14	.19[a]	.10	.01	.10

[a] p = .10
[b] p = .05
[c] p = .01

while interparty competition has been a positive predictor. These earlier findings have been due to the unusual character of the southern states. While the political and economic complexion of the South has obviously not changed markedly during the past decade, our data suggest that neither has their response to this policy problem. As we noted earlier, the southern states are lagging behind all other regions in terms of their commitment to environmental quality. In addition, these southern states are heavily Democratic and lack competitive party systems. Thus, it could be that our bivariate findings for the aggregate relationships between Democratic party strength or interparty competition and state commitment are artifactual in the sense that the eleven southern states are distorting the relationships found in Table 9.3. As a test of this possibility, we excluded the eleven southern states from the analysis and again examined the bivariate relationships. Table 9.4 presents these findings.

As expected, the initial relationships found in Table 9.3 between indicators of political demands and state commitment do not persist in Table 9.4 once we

Table 9.4   Pearson's Product-Moment Correlation Coefficients (r) Showing the Relationship between Selected Independent Variables and States' Commitments to Environmental Quality, 1975-1979   (non-south states)

Independent variables	1975	1976	1977	1978	1979
Model I (Needs)					
State population	.19	.26 c	.12	−.01	.13
State density	.28 c	.31 c	.24 b	.09	.31 c
Industrialization a	.35 d	.24 b	.37 d	.32 c	.42 d
Pollution potential	.26 c	.26 c	.21 b	.12	.22
Model II (Resources)					
Personal income	.06	.23 b	.16	.01	.14
Median income a	.27 c	.39 d	.25 b	.09	.23 b
Percent poor	−.22 b	−.29 c	−.24	−.18	−.26 c
Environmental aid	.23 b	.27 c	.17	.11	.18
Model III (Political demands)					
Democratic Party strength a	−.01	.03	−.17	−.21 b	.04
Interparty competition	.01	−.01	.05	.00	.19
Model IV (Admin-organ factors)					
Governor's appointive power (env.)	.04	.11	.11	−.07	.00
Bureaucratic consolidation (env.) a	.05	.06	.09	−.04	.06
Reformism	.04	.12	.13	.03	.17
Legislative professionalism a	.12	.19	.10	.00	.09

a Variable selected for inclusion in the subsequent analyses.
b p = .10
c p = .05
d p = .01

Note: Does not include the eleven states of the Old Confederacy.

exclude the southern states. The relationships between other indicators and state commitment, however, do persist. Hence, the finding in Table 9.4 are more fully representative of the factors influencing states' commitments to environmental quality. Once again, the bivariate findings in Table 9.4 demonstrate that a state's commitment to environmental quality is a function of the severity of the pollution problem and the economic resources available to the state. It does not, on the other hand, appear to be a function of partisan political demands, the level of interparty competition, or the degree of institutional reform adopted by individual states. Thus, in terms of the hypotheses advanced above, two are supported at the bivariate level and two are not. The severity of the pollution problem provides some degree of pressure for policy action; in addition, states that can "afford" a commitment to environmental quality are likely to do so. However, state commitment to environmental quality is not likely to be supported by Democratic legislatures, Republican assemblies, or even by competitive legislative structures. Similarly, reform efforts to "professionalize" the legislature and/or reorganize (i.e., consolidate) the executive branch for dealing with state environmental protection activities, are also apparently unrelated to state commitments. Moreover, our findings with regard to the impact of legislative and executive reforms are quite consistent with the thrust of most comparative state policy research.[35]

While the foregoing data are somewhat supportive of hypotheses one and two, it is also possible that our observed bivariate correlation may not persist once controls are introduced. That is, an independent variable (in any model) may have a catalytic or depressor effect on another's relationship to state commitment. Thus, the next step in the analysis is to incorporate these four simplistic models into a more comprehensive explanation of the processes leading to state environmental protection activities. The conventional approach followed in the comparative state policy literature suggests the inclusion of five independent predictors into a single, multiple regression model.[36] Such an approach is not directly applicable to these data because, among the independent variables, there is an unconventional indicator. Unlike most comparative state analyses that assess policy outputs only in terms of economic resources and political and/or administrative phenomena, we also have measures of the severity of the problem itself. The inclusion of this indicator requires a modification of the methodology consistent with the theoretical role played by the other independent variables. While there is a moderately strong relationship between the magnitude of the pollution problem and state commitment, this tells us nothing about the policymaking process. Thus, pollution severity provides an ecological (or conditioning) variable that allows us to compare the policy process at work in states where the problem is, and is not, severe. The important question, especially in those states where the problem is more critical, is what aspects of the preexisting policy process are conducive to states' commitments?

The Multivariate Analyses

In order to test the basic proposition advanced above, and to compare the relative importance of economic, political (i.e., partisanship), and institutional (i.e., reform) factors in explaining state commitment, three separate regression analyses were performed for each of the years, 1975–1979. First, Table 9.5

Table 9.5 Multiple Regression Coefficients (Beta Weights) Showing the Relationship between Independent Variables and States' Commitments to Environmental Quality, 1975-1979 (All States)

Independent variables	1975 (N = 50)		1976 (N = 50)		1977 (N = 50)		1978 (N = 50)		1979 (N = 50)	
	B	Beta	B	Beta	B	Beta	B	Beta	B	Beta
Median income	.0001	.31[a]	.0001	.47[a]	.0001	.31	.0000	.17	.0001	.31
Democratic Party strength	-.0035	-.11	-.0039	-.11	-.0087	-.23	-.0126	-.27	-.0027	-.09
Bureaucratic consolidation	.0001	.03	.0001	.03	.0003	.06	-.0000	-.01	.0001	.03
Legislative professionalism	.0001	.02	.0001	.02	-.0000	-.01	-.0003	-.05	-.0000	-.01
Summary:										
Mean (Dependent variable)	.0063		.0070		.0070		.0071		.0066	
Multiple R	.37		.51		.43		.35		.34	
R^2	.13		.26		.19		.12		.12	
R^2	.06		.19		.12		.04		.04	
F-ratio	1.74		3.93[b]		2.62[a]		1.57		1.48	

[a] p = .05
[b] p = .01

presents the regression results for the effects of our economic and political variables upon state commitment for each year among all fifty states. This first set of regressions thus allows us to examine relative effects of our four independent variables without regard for the severity of the problem.

For the fifty state regressions, only state wealth (median income) is consistently related to policy output (betas = .31 to .47). Our political and/or institutional variables are either insignificant predictors (as in the case of our indicators of bureaucratic and legislative capabilities) or weakly associated with states' commitments (as in the case of Democratic party strength). In addition, the nonsignificance of the overall regression equations (except for the year 1976) reflects the fact that the other three independent variables contribute virtually nothing to an explanation of state activity.

However, it is quite possible that political (institutional) variables will exhibit a greater influence in situations where the policy problem is more severe; hence, under such critical ecological conditions, Democratic assemblies, highly organized environmental bureaucracies, and/or professional legislatures will exert their impacts on complex issues that are salient to them.[37] Table 9.6 considers this possibility.

The findings in Table 9.6 show that state resources are again positively related to state environmental commitment (betas = .13 to .49), while Democratic party strength is inversely related to state responsiveness to environmental protection (betas = −.18 to −.52). Our reform measures, on the other hand, are once again insignificant predictors of state policy responsiveness.

It may be tempting to conclude, therefore, that Democratic assemblies are less supportive than Republican ones of environmental quality in highly industrialized states; however, this finding is partially explained by the fact that seven of the highly industrialized states are southern states (Alabama, Arkansas, Georgia, Mississippi, North Carolina, South Carolina, and Tennessee). These southern states are strongly Democratic and, as we discussed above, the Deep South commits fewer resources to environmental protection than do other regions. Consequently, we must view these findings with extreme caution for the reasons discussed previously. Once again, we conclude that economic resources shape state commitment to the environment.

Finally, in Table 9.7, our four theoretically related independent variables are again regressed upon state policy in those states where the pollution problem is less salient (i.e., in less industrialized states). The findings are quite consistent in underscoring the critical importance of state economic resources (betas = .52 to .85), and the insignificance of political/institutional variables.

In summary, within the limitations imposed by our data and temporal domain, we must conclude that a state's budgetary commitment to environmental quality is largely a function of the severity of the pollution problem and the economic resources available to the state; this commitment does not, on the other hand, appear to be encouraged by partisan political activity and/or institutional reforms aimed at increasing governmental performance and effectiveness.

Table 9.6 Multiple Regression Coefficients (Beta Weights) Showing the Relationship between Independent Variables and States' Commitments to Environmental Quality, 1975–1979 (Highly Industrialized States)

Independent variables	1975 (N = 25)		1976 (N = 25)		1977 (N = 25)		1978 (N = 25)		1979 (N = 25)	
	B	Beta	B	Beta	B	Beta	B	Beta	B	Beta
Median income	.0000	.24	.0001	.33	.0001	.20	.0000	.13	.0002	.49
Democratic Party strength	-.0122	-.33	-.0122	-.29	-.0229	-.48[a]	-.0317	-.52[a]	-.0067	-.18
Bureaucratic consolidation	-.0009	-.18	-.0008	-.15	-.0004	-.07	-.0018	-.20	-.0016	-.25
Legislative professionalism	.0011	.13	.0015	.17	.0004	.04	-.0007	-.06	-.0005	-.05
Summary:										
Mean (Dependent variable)	.0073		.0074		.0086		.0088		.0085	
Multiple R	.48		.55		.59		.59		.51	
R^2	.23		.30		.35		.35		.26	
R^2	.08		.16		.22		.22		.12	
F-ratio	1.52		2.13		2.64		2.78		1.87	

[a] p = .05

Table 9.7 Multiple Regression Coefficients (Beta Weights) Showing the Relationship between Independent Variables and States' Commitments to Environmental Quality, 1975–1979 (Less Industrialized States)

Independent variables	1975 (N = 25)		1976 (N = 25)		1977 (N = 25)		1978 (N = 24)		1979 (N = 24)	
	B	Beta	B	Beta	B	Beta	B	Beta	B	Beta
Median income	.0001	.52[a]	.0002	.70[c]	.0001	.85[c]	.0001	.55[b]	.0000	.61[a]
Democratic Party strength	.0041	.18	.0033	.12	.0033	.19	.0035	.21	-.0010	-.09
Bureaucratic consolidation	.0004	.15	.0005	.15	.0000	-.02	.0002	.09	.0003	.16
Legislative professionalism	-.0002	-.09	-.0003	-.11	-.0002	-.11	-.0006	-.23	.0000	.03

Summary:

Mean (Dependent variable)	.0053		.0065		.0053		.0052		.0046	
Multiple R	.56		.73		.84		.60		.67	
R^2	.31		.54		.71		.36		.45	
R^2	.18		.44		.65		.22		.33	
F-ratio	2.29		5.80[b]		12.25[c]		2.65		3.85[a]	

[a] p = .05
[b] p = .01
[c] p = .001

Discussion: Implications for Future State Commitments to Environmental Quality

This study has examined the degree to which state governments have responded to environmental pollution problems. We have extended the analyses of others, conducted within individual state legislatures or the U.S. Congress, to a comparative research design including all the American states. Our findings are relevant for both the comparative public policy literature and research on environmental policy more specifically.

Our findings, based on the above analyses, suggest three important considerations. First, the American states are clearly responding to the existence of the pollution problem. While the level of state commitment is indeed miniscule, it is nevertheless a function of the severity of the problem itself. That is, highly industrialized states have established a higher budgetary priority for environmental protection than have the less industrialized states, as shown by the mean scores for each group of states (confer with Tables 9.5–9.7) and by our results in Tables 9.3–9.4. This is encouraging and perhaps somewhat surprising.[38]

Second, this analysis underscores the general findings that political and institutional factors are without significance in influencing states' policy commitments. This is no surprise to students of comparative state policy research, although it conflicts with the thrust of much of the environmental policy literature. That literature has continued to stress the importance of Democratic partisanship in promoting environmental protection; in addition, some environmental policy literature has emphasized the importance of consolidation of the environmental bureaucracy and/or legislative professionalism in advancing environmental protection activities.[39] It is therefore necessary to reconcile our findings with those of the environmental policy literature.

To begin with, there are very few studies available that are devoted to a fifty–state analysis of environmental protection activities. Of those available, our findings are generally consistent with those who employ *fiscal measures* of the dependent variable. In these studies, economic resources and demographic factors are positively associated with environmental policy outputs, while political factors (such as interparty competition, malapportionment, consolidation of the environmental bureaucracy) are not.[40]

On the other hand, our findings are inconsistent with those studies that employ state *legislative enactments* as measures of environmental protection activities; those few studies found both economic and demographic, as well as political factors (such as state innovation, bureaucratic consolidation, Democratic party strength, and legislative professionalism), to be strongly related to state environmental protection.[41] Perhaps this divergence in findings may be explained by the suggestion that socio–economic factors are more important in explaining *fiscal policy decisions* (or the monies spent by the state for environmental protection), while political factors may be more influential in explaining *substantive policy choices* (or state variation in types of environmental policies enacted).[42] Until the two types of indicators are systematically compared, this possible explanation must remain tentative and await further study.

Finally, our findings suggest that a state's commitment to environmental quality is largely explained by the economic resource base of the state itself. It is this finding that carries significant implications for future state commitments to environmental quality. Assuming for the moment that our analysis has included the primary variables of concern and is empirically valid, the future of state budgetary commitments for the environment is not an optimistic one. As the states face the prospect of federal budget cuts (and undertake cuts of their own), together with increasing state unemployment and a general decline in state income, the amount of state expenditures given to environmental protection will surely decline as well. According to this scenario, we are likely to witness even fewer state resources devoted to environmental protection activities in the future.

However, if we assume that factors outside the scope of this analysis are equally (or more) important, then a more optimistic future may be drawn. For example, as we noted earlier, the level of public support for environmental protection remains high.[43] State decision makers may respond to the politics of environmental concern, especially if there is a public outcry against further budgetary cuts in this area. In addition, the private sector is not unresponsive to the need for environmental protection expenditures; indeed, as we noted earlier, private industries within the states have spent substantial amounts for environmental protection activities in the past.[44] Finally, it is also possible that a more intense commitment of executive leadership (either at the federal or state levels) will increase the amount of public attention given to environmental protection in the future. Given the severity (as well as the complexity) of the issues associated with environmental pollution, social goals and priorities may change on the basis of new knowledge that commands a sufficient consensus to serve as a guide to public policy.

In any case, both political and economic phenomena will likely dictate whether present levels of state environmental commitments are sufficient, insufficient, or overly munificent. That is, such budgetary commitments will likely hinge on the particular mix of political leadership (whether benevolent or hostile to this policy area), changing governmental objectives, and the level of state fiscal capabilities, rather than the type of institutional structure found in various states.[45]

Notes

1. Charles O. Jones, "Regulating the Environment," in Herbert Jacob and Kenneth N. Vines, *Politics in the American States: A Comparative Analysis* (Boston, Massachusetts: Little, Brown and Company, 1976), p. 408.

2. *Id.*, p. 396.

3. According to Daniel R. Grant, this period of the early 1970s was one of "consensus mobilization," or a period when the political process produced policies beyond the political, administrative, financial, and immediate technological capabilities that normally limit decisionmaking. See his "Carrots, Sticks, and Consensus," in Lynton K. Caldwell, (ed.), Environmental Studies (Bloomington, Indiana: Institute of Public Administration, 1976), p. 29. See also Anthony Downs, "Up and Down with Ecology: The Issue Attention Cycle," *The Public Interest* 28 (Summer, 1972): 38–50.

4. Jones, *op. cit.*, pp. 424–25; see also Regina S. Axelrod, ed., *Environment, Energy, Public Policy: Toward a Rational Future* (Lexington, Massachusetts: Lexington Books, 1981), pp. 1–2; and "Reagan and Deregulation," in *Environment and Health* (Washington, D.C.: Congressional Quarterly Press, 1981), pp. 129–33.

5. See testimony of Louis Harris before the Subcommittee on Health and the Environment of the House Energy and Commerce Committee, October 15, 1981.

212 *Selected Natural Resource Issues*

6. Although there is an abundance of comparative state policy research in other functional areas, there is relatively little research focusing on the environmental arena which employs the fifty American states as the units of analysis.

7. Of the few available state level policy studies, almost none examine the possible effects of legislative professionalism, gubernatorial powers, Democratic party strength, or executive reorganization in relation to policy outputs in this functional area.

8. The year 1979 was selected as the end point of our analysis because it is the most recent period for which existing data are presently available. The year 1975 was selected as the beginning point because it was the first year data on *total* environmental quality control expenditures were compiled by the U.S. Census Bureau.

9. See Jones, *op. cit.*; and Lettie M. Wenner, *One Environment Under Law: A Public Policy Dilemma* (Pacific Palisades, California: Goodyear Publishing Co., 1976).

10. Jones, *op. cit.* See also J. Clarence Davies, *The Politics of Pollution* (New York: Pegasus Publishing Co., 1970); Kingsley W. Game, "Controlling Air Pollution: Why Some States Try Harder," *Policy Studies Journal,* 7 (1979): 728–38; and Thomas R. Dye, "Politics Versus Economics: The Development of the Literature on Policy Determination," *Policy Studies Journal* 7 (1979): 652–62.

11. Michael Maggiotto and Ann O.M. Bowman, "The Impact of Policy Orientation on Environmental Regulation: A Case Study of Florida's Legislators," *Environment and Behavior* 14 (March, 1982).

12. James E. Anderson *et. al.*, Public Policy and Politics in America (North Scituate, Massachusetts: Duxbury Press, 1978), pp. 17–18.

13. Thomas R. Dye, *Politics, Economics, and the Public: Policy Outcomes in the American States* (Chicago, Illinois: Rand McNally, 1966).

14. John Sacco and Edgar Leduc, "An Analysis of State Air Pollution Expenditures," *Journal of the Air Pollution Control Association,* 19 (1969): 416–19.

15. In addition to Sacco and Leduc, *op. cit.*, the aggregate (fifty–state) studies of state environmental protection activities include Kingsley Game, *op. cit.*; Susan Clarke, "Determinants of State Growth Management Policies," *Policy Studies Journal* 7 (1979); Lettie M. Wenner, "Enforcement of Water Pollution Control Laws," *Law and Society Review* 6 (1972): 481–507; and James P. Lester, "Partisanship and Environmental Policy: The Mediating Influence of State Organizational Structures," *Environment and Behavior* 12 (March, 1980): 101–131.

16. Riley E. Dunlap and Richard P. Gale, "Party Membership and Environmental Politics: A Legislative Roll–Call Analysis," *Social Science Quarterly* 55 (1974): 670–90.

17. Leonard G. Ritt and John M. Ostheimer, "Congressional Voting and Ecological Issues," *Environmental Affairs* 3 (1979): 459-72; Dunlap and Gale, *op. cit.*; Riley E. Dunlap and Michael P. Allen, "Partisan Differences on Environmental Issues: A Congressional Roll-Call Analysis," *Western Political Quarterly* 29 (1976): 384–97; and Henry C. Kenski and Margaret C. Kenski, "Partisanship, Ideology, and Constituency Differences in Environmental Issues in the U.S. House of Representatives: 1973–1978," *Policy Studies Journal* 9 (1980): 325–35.

18. See Jerry W. Calvert, "The Social and Ideological Bases of Support for Environmental Legislation: An Examination of Public Attitudes and Legislative Action," *Western Political Quarterly* 32 (1979): 327–37; Riley E. Dunlap and Kent D. Van Liere, "The Social Bases of Environmental Concern: A Review of Hypotheses, Explanations, and Empirical Evidence," *Public Opinion Quarterly* 44 (1980): 181–97; and Lester, *op. cit.* There are, however, some important exceptions to the findings that Democrats are more likely than Republicans to support environmental protection. See, for example, Frederick H. Buttel and William L. Flinn, "The Politics of Environmental Concern: The Impacts of Party Identification and Political Ideology on Environmental Attitudes," *Environment and Behavior* 10 (1978): 17–36; and Daniel Mazmanian and Paul Sabatier, "Liberalism, Environmentalism and Partisanship in Public Policy–Making," *Environment and Behavior* 13 (1981): 361–84.

19. Lester, *op. cit.*

20. V.O. Key, *Southern Politics* (New York: Alfred A. Knopf, 1949).

21. See Arthur E. Buck, *The Reorganization of State Governments in the United States* (New York: Columbia University Press, 1938): 14-28; and A.E. Buck, *Administrative Consolidation*

in State Government, 5th ed. (New York: National Municipal League, 1930). See also Committee for Economic Development, *Modernizing State Government* (New York: Committee for Economic Development, 1967). For a brief history and typology of overall executive reorganization activities, see James L. Garnett, *Reorganizing State Government: The Executive Branch* (Boulder, Colorado: Westview Press, 1980): 1–50.

22. See Kenneth J. Meier, "Government Reorganization for Economy and Efficiency: Some Lessons from State Government," paper presented at the 1979 Annual Meeting of the Midwest Political Science Association, Chicago, Illinois. See also Garnett, *op. cit.*, and T.R. Carr, "State Government Reorganization: The Impact of Executive Branch Reorganization on Public Policy," paper presented at the 1981 Annual Meeting of the Midwest Political Science Association, Cincinnati, Ohio.

23. Edward G. Carmines, "The Mediating Influence of State Legislatures on the Linkage Between Interparty Competition and Welfare Policies," *American Political Science Review* 68 (1979): 1118–24; A.K. Karnig and Lee Sigelman, "State Legislative Reform: Another Look, *Western Political Quarterly*, 28 (1975): 548–52; Lance T. LeLoup, "Reassessing the Mediating Impact of Legislative Capability," *American Political Science Review* 72 (1978): 616–21; and Phillip W. Reoder, "State Legislative Reform: Determinants and Policy Consequences," *American Politics Quarterly* 7 (1979): 51–69.

24. See Citizens Conference on State Legislatures, *The Sometime Governments: A Critical Study of the 50 American Legislatures* (New York: Bantam Books, 1971).

25. See sources in note 23 above.

26. Jones, *op. cit.*, p. 396.

27. Id., pp. 397–400.

28. See Wenner, *op. cit.*, p. 499.

29. This point was originally made by Ira Sharkansky. See his "Agency Requests, Gubernatorial Support, and Budgetary Success in State Legislatures," *American Political Science Review* 42 (December, 1960): 1220–31; see also Joyce M. Munns, "The Environment, Politics, and Policy Literature: A Critique and Reformulation," *Western Political Quarterly* 28 (1975): 646–67.

30. Federal environmental aid would obviously affect the amount spent on environment quality by state governments. On this point see J.C. Strause and P. Jones, "Federal Aid: The Forgotten Variable in State Policy Research," *Journal of Politics* 36 (1974).

31. The measure for interparty competition was derived from the index of Democratic party strength. All scores above .500 were subtracted from 1.000 in order to transform the index from one of degree of Democratic control to degree of party competition, with higher scores representing greater degrees of competition. All data are concurrent; no lags were employed.

32. See Appendix A for the source of legislative professionalism scores.

33. The first measure of administrative reform was based on the governor's ability to appoint the head of the agency responsible for environmental protection, with scores ranging from 0–5; the second measure was based on the degree of consolidation of the environmental bureaucracy, with scores ranging from 0–3; the measure of "reformism" is an index combining both these individual measures. It is derived by multiplying appointive powers times consolidation of the environmental bureaucracy. The scores thus range from 0–15. See Appendix A for data sources.

34. The "Deep South" includes the eleven states of the Old Confederacy. These states are: Alabama, Arkansas, Florida, Georgia, Louisiana, Mississippi, North Carolina, South Carolina, Tennessee, Texas, and Virginia.

35. See notes 22 and 23 above.

36. The five variables selected for inclusion in the multivariate model include those selected on the basis of their being the strongest (i.e., highest r values) and most significant representative of each model (e.g., indicators of problem severity and state resources), or because the variables provide strong theoretical interest (e.g., indicators of partisanship, legislative reform, and bureaucratic reform). According to Klein's criterion, none of the five variables selected was judged to be excessively co–linear.

37. Helen Ingram, Nancy Laney, and John R. McCain, "Water Scarcity and the Politics of Plenty in the Four Corners States," *Western Political Quarterly* 32 (1979): 298–306.

38. This finding is particularly encouraging given Charles Jones's earlier argument

which implied that the "need" for environmental protection expenditures by the states was not strongly related to their actual expenditures. See Jones, *op. cit.*, pp. 415–23.

39. See references in notes 17 and 18 above. See also Walter A. Rosenbaum, *The Politics of Environmental Concern* (New York: Praeger, 1977); Cynthis H. Enloe, *The Politics of Pollution in a Comparative Perspective* (New York: McKay, 1975); Paul Sabatier, "Regulatory Policy-Making: Towards a Framework of Analysis," *Natural Resources Journal* 17 (1977): 415–60; John A. Worthley and Richard Torkelson, "Managing the Toxic Waste Problem: Lessons from the Love Canal," *Administration and Society* 13 (1981): 145–60; Lester, *op. cit.*; and Gerard A. Bulanowski, *The Impact of Science and Technology on the Decision-Making Process in State Legislatures: The Issue of Solid and Hazardous Waste* (Denver, Colorado: National Conference of State Legislatures, 1981).

40. See Sacco and Leduc, *op. cit.*, Wenner, *op. cit.*

41. See Clarke, *op. cit.*, Lester, *op. cit.* See also James P. Lester, James L. Franke, Ann O.M. Bowman, and Kenneth W. Kramer, "Technology, Politics, and Hazardous Waste Regulation: A Comparative State Analysis." *Western Political Quarterly* 36 (June, 1983).

42. The primary reason for the expected importance of both legislative professionalism and bureaucratic consolidation of the environmental sphere is that these variables embody a concentration of technical expertise and fiscal resources which are important to policy enactments involving complex issues, such as those associated with protection of the environment.

43. See Robert C. Mitchell, "Public Opinion on Environmental Issues," a national public opinion survey conducted by Resources for the Future in 1980 for the Council on Environmental Quality. This survey is reprinted in Council on Environmental Quality, *Environmental Quality 1980: The Eleventh Annual Report of the Council on Environmental Quality* (Washington, D.C.: Government Printing Ofice, 1981); see also footnote 5.

44. State–level data for the amount of private industry expenditures for pollution abatement are available. See, for example, U.S. Department of Commerce, Bureau of the Census, *Pollution Abatement Costs and Expenditures, 1980*, MA–200(80)–1 (Washington, D.C.: Government Printing Office, 1981): 26–28.

45. Although they are difficult to quantify, both the type of leadership provided and changing governmental objectives (both federal and state) are no less important. On this point, see E.H. Haskell and V.S. Price, *State Environmental Management: Case Studies of Nine States* (New York: Praeger, 1973).

10

U.S. Water Resources Development Policy and Intergovernmental Relations

HENRY P. CAULFIELD, JR.

Introduction

This chapter highlights certain basic features of U.S. water policy—past, present, and future. Also, it relates that past and projected history to concepts of political scientists about *federalism* and *political decisionmaking*. The explanatory adequacy or inadequacy of these concepts and basic views is indicated in relation to the author's perception of the development of U.S. water policy. Finally, the chapter provides a basis for the reader's own consideration of possible future developments in U.S. resource policy.

The Past to 1977

The historic past to 1977 will be discussed in two parts. First, the author's perception of the assumption of federal responsibilities for navigation and irrigation, the initial federal water programs, will be related to the views of political scientists on federalism. Second, the author's perception of the rise and decline of federal dominance in water resources development will be related to three academically conceived processes of political decisionmaking.

Assumption of Federal Responsibilities for Navigation and Irrigation

Early in the nineteenth century, in the humid East, the principal port cities along the Atlantic coast (particularly Boston, New York, Philadelphia, Baltimore, Washington, and Charleston), in cooperation with private business interests, became rivals in development of inland commerce through making river navigation improvements and constructing canals and portage facilities over mountains to the Great Lakes and the Ohio and other inland rivers. Flood control on the lower Mississippi River (but no doubt elsewhere) was then performed by cooperative enterprise of landowners or local public districts. And provision of municipal water supply changed largely from private enterprise to municipal government, notably in Baltimore, Philadelphia and New York.

In the latter half of the nineteenth century, in the arid West, private, cooperative, and local public enterprise undertook construction of diversion, canal, and storage works to provide water for irrigated farming. Before the century ended, local enterprise in eastern Colorado went beyond works to divert and convey water from the South Platte and its tributaries. Collection works high up on the west slope of the Rocky Mountains, with transbasin conveyances to the east slope, were constructed to supplement river flows. In Utah, Arizona, Idaho, California, and in other states of the arid West, local irrigation development also occurred.

These manifestations of private and local governmental enterprise with respect to navigation and irrigation proved to be inadequate. The period between 1817 and 1838 is known to historians as the Canal Era, meaning the period when state governments took a leadership role, along with cities, in developing inland navigation projects in cooperation with private enterprise. Some federal assistance was given in the form of public land grants and army surveying personnel.

The failures of state and private enterprise during the Canal Era (due largely to inadequate financial and technical resources); the strong belief of the new Republican Party in federal constitutional power; the political conviction that development of inland navigation to provide cheap transport of agricultural and other commodities was a key national public means to encourage economic development of the West (the humid West, that is); the availability of financial resources brought about by import tariffs; and the then unique and available technical skills of the Army Corps of Engineers—all resulted in federal assumption after the Civil War of responsibility for planning, financing, construction, operation and maintenance of inland navigation.

Later, and after a short period of reliance upon state initiatives under the Carey Act of 1894, federal provision of greater technical competence and of longer term financial capital for irrigation developments was also perceived in Washington as the key national public means to foster settlement and economic development of the arid West. This perception led politically to assumption of federal irrigation development responsibilities via the Reclamation Act of 1902.

These assumptions of responsibility for inland navigation and irrigation developments need to be seen too in a wider political context. From the earliest years of the Republic, occupation and military defense of the lands claimed and acquired west of the original eastern state boundaries were a major national political preoccupation. Great Britain, France, Spain, and Mexico were seen as rivals in quest for the land that became the United States. The comprehensive inland navigation plan of Secretary of the Treasury Gallatin in his Report on Roads and Canals of 1808 had, as its stated goals, economic development, furthering political unity, and military defense. A similar comprehensive inland navigation plan was produced in 1822 by John C. Calhoun when he was Secretary of War and before he gave up his nationalist predispositions in 1829.

Consistent with its belief in the principles of a strong national government, the Republican Party, in its first platform of 1856, resolved that "appropriations by the Congress for improvements of rivers and harbors, of a national character,... are authorized by the Constitution."[1] After the advent to power of the Republican Party, the navigation program of the Army Corps of Engineers furthered national political goals of comprehensive inland navigation planning and action. Moreover, this inland navigation program to provide cheap bulk transport for the products of Midwest farmers and its tariff policy in aid of eastern manufacturers provided the party with a solid, long–lasting, national political foundation.

The Reclamation Act of 1902 was a logical specific means of furthering the national political objectives of settlement and economic development of the undeveloped arid West. Toward the end of the nineteenth century, gold, silver, and other types of mining were petering out in Nevada and elsewhere. However, extensive grazing of cattle was flourishing. Nevertheless, more irrigated agriculture was then seen by Senator Newlands of Nevada and other leading proponents of the Reclamation Act of 1902, including the new Republican president, Theodore Roosevelt, as a national strategic instrument to further settlement and economic development of the arid West.

At the time of its enactment the Reclamation Act was not justified by argument that more agricultural commodities were needed. There were then no general food shortages. In fact, midwestern farmers saw increased production of western farmers as depressing prices further that were already seen to be too low, so they opposed the Reclamation Act.

In summary, the assumption of primary federal responsibility for internal navigation and irrigation developments was politically animated by widely held broad national goals of settlement and development of the East, Midwest, and then West, and by practical considerations of finance and efficient provision of scarce technical competence. All water resources development was not taken over by the federal government from the states and local governments; neither municipal water supply nor flood control responsibilities were changed. The only water functions then changed were those seen to be useful instruments of broad national political purposes.

Academic Views on Federalism

How does this author's perception of federal assumption of responsibility for inland navigation and irrigation developments square with academic views on federalism?

Both political consideration of federalism, as well as early academic literature, dwell upon constitutional considerations. Prior to the Civil War the political arguments in Congress over federal assumption of responsibility for inland navigation took the form of a constitutional debate. As David Walker has put it, the debate was between a "nation–centered or state–centered theory of federalism."[2] The Supreme Court entered the fray in 1824 by deciding in *Gibbons* v. *Ogden* that navigable waters usable in interstate and foreign commerce were subject to federal authority. But in 1829 the Court also agreed, in *Williams* vs. *Blackbird Creek Marsh Company*, that a state had concurrent authority to dam a navigable creek. After the Lincoln Administration succeeded in giving the Corps of Engineers authority to plan, construct, operate and maintain navigation projects utilizing federal financial resources, the Supreme Court in 1877 decided in *Wisconsin* v. *Duluth* that such authority was constitutional.

In the academic literature of federalism this legal view has been described as "dual–federalism," analogous to a "layer cake."[3] And Walker sees "dual–federalism" as the appropriate operative concept between 1879 and 1930 but with increasing qualifications as time passed.[4] The subsequent realities of federal–state relations were later seen as "cooperative federalism," analogous to a "marble cake."[5] In the late 1960s, federal–state–local relationships, particularly those involving Federal financial grants and Federal regulatory authority, were described as analogous to a "picket–fence."[6] And in 1972, federal–state relations were described by Michael Reagan as "permissive federalism," a sharing of power and authority but with the state's share dependent upon the permission and permissiveness of the federal government.[7]

None of these static conceptions involves a political explanation of change in responsibility for governmental functions from the states to the federal government. However, Glendening and Reeves in their 1977 book, *Pragmatic Federalism*, attempt a general explanation of such a change.[8] Looking at the history of changing federal–state relationships, they see "steadily increasing centralization" via "rampant incrementalism." They identify catastrophe, population shifts, technological developments, and financial considerations as causal factors of change. Broad strategic political objectives, like those involved in federal assumption of responsibility for inland navigation and irrigation developments, are not seen as explanatory variables. Instead, they emphasize the financial and technical variables. Thus, the historic evolution of federal–state relations generally is seen as just "pragmatic adjustment."

No other political theories of functional change between levels of government within American federalism have been identified. If one is concerned about change in responsibility between levels of government one appears to have little academic theory to draw upon.

The Rise and Decline of Federal Dominance

The conservation movement, which became a national political force early in the twentieth century (through the leadership of Theodore Roosevelt, Gifford Pinchot, John Wesley Powell, and others) fully supported national economic development through federal water resources developments as well as nonfederal developments through license under the Water Power Act of 1920. The conservation movement articulated the concept and need for comprehensive multiple–purpose river basin planning and development throughout the nation.[9] The first such comprehensive plan was undertaken for the Colorado River. This plan led to the Colorado River Compact of 1922, the Boulder Canyon Project Act of 1928, and subsequent projects. In a larger national perspective, Section 308 of the Rivers and Harbors Act of 1927 called upon the Army Corps of Engineers to develop comprehensive multiple–purpose river basin plans for the major river basins of the United States. Multiple purposes at that time meant navigation, irrigation, hydroelectric power, and flood control.

Many of the "308" plans, calling for construction of large multiple-purpose dams, were completed by the beginning of the great depression of the 1930s.

Policies to overcome the depression provided the federal financial means for large public works projects. Where regional political support for water resource development was also present, great Federal developments began—for example, on the Tennessee and Columbia Rivers and in the Central Valley of California (by the U.S. Bureau of Reclamation). Where regional political support was denied—for example, on the Connecticut and Potomac Rivers—development did not occur, despite the availability of both plans and funds. Nevertheless, the federal big–dam era prevailed in much of the country, but particularly in the South and West, well into the 1960s.

On the level of broad political purpose and political visibility, underdeveloped river basin regions substituted for "The West" as the geographic object of economic development. Geographic regions, like the ten–state Missouri basin, became political aggregates in support of federal development.

Municipal and industrial water supply, recreation, and fish wildlife enhancement later became authorized purposes of multiple–purpose dams. On the ideological level, water resource developments could thus be seen as more fully utilizing natural resources, and, on the political level, new constituencies were added in support of water projects.

Much of the most aggressive political support for the federal big– dam era came from the supporters of federal public power, which gave preference in the distribution of power to public bodies and rural electric cooperatives. Much political support was also provided by the proponents of navigation, irrigation, and flood control through the lobbying efforts of the National River and Harbors Congress (formed in 1902) and the National Reclamation Association. Membership, particularly that of the former, included many members of both the House of Representatives and the Senate. Before lobbying the executive branch and the Congress for water resource developments, these organizations regularly performed the necessary political function of prioritizing potential authorizations and funds for "new starts" on the basis of geographic and other distributive factors acceptable to their members.

The report in 1950 of President Truman's National Water Policy Commission reflected fully the ideology of comprehensive multiple–purpose river basin planning and development, including the progressive ideology of development of federal public power and preferential distribution of power to public bodies and rural electric cooperatives. It recommended strong action measures.

The Eisenhower Administration, upon coming to power in 1953, announced its "partnership policy" calling for private utility or other nonfederal development of hydroelectric power under license of the Federal Power Commission (now the Federal Energy Regulatory Commission), together with federal provision of funds to nonfederal developers to cover the costs of other project purposes that would be nonreimbursable federal costs if a part of a federal project. Except for this major policy difference, the basic ideology of comprehensive multiple–purpose river basin development, as reflected in the 1955 report of its Presidential Advisory Committee on Water Resources Policy, was consistent with that of the Truman commission's report. Several major partnership projects were proposed, but only one (with the State of California) was licensed, federally funded for the flood storage portion, and built. This reversal of Democratic Party policy with respect to federal power development, together with refusal by the Bureau of the Budget (now the Office of Management and

Budget) to approve many project proposals, particularly by the Army Corps of Engineers, led to the characterization of the Eisenhower Administration as having a "no new starts" policy by the Democratic Party.

At the beginning of the 1960s national political leaders called for renewal of water resource development. The bipartisan Senate Select Committee on National Water Problems unanimously called upon the executive branch to undertake a new era of comprehensive river basin plans on all the major river basins of the United States. President Kennedy immediately acccepted the goal of completing these plans by 1970. Moreover, he repudiated the alleged "no new starts" policy of the Eisenhower Administration and reestablished administrative policies favorable to public and cooperative electric power. At both the political and technical level, it was assumed that that much water resource development still needed to be accomplished and that the political support to get it done was widespread and strong. These assumptions proved, increasingly, to be false as the years passed to the advent of the Carter Administration in 1977. The decline in this federal function became increasingly evident and has now been documented.

The Bureau of Reclamation had had no significant new authorizations since the Colorado Basin Projects Act of 1968, which authorized the Central Arizona Project and a few other projects. From 1950 through 1980 the constant dollar average authorizations for the Corps of Engineers by each Congress (a two–year period) was about $3.1 billion. The last Congress to exceed the average authorization level was in 1968. "The absence of an 'average' congressional authorization since that time is clear evidence of a decline in the Corps' role in water resources development as measured by project authorizations."[10] As measured by constant dollar appropriations, "the lowest average five–year average for [the thirty–year] period occurred from 1975 through 1979 when construction appropriations averaged $1.9 billion."[11]

Why this decline? Among the interrelated factors that help to explain the decline in political support for water resource development projects is the political emergence of a national urban majority. Irrigated agriculture and other traditional resource development concerns are not a major interest of this relatively new national majority; they are foreign to it. The urban majority is primarily concerned with urban problems: housing, transportation, health, welfare, air and water pollution, urban open space and recreation areas, energy, etc. An active, educated segment of this urban majority is also concerned with the rural and natural hinterland, expressed effectively now for some twenty or more years in the environmental movement. The origins of the ideology that motivates these activists are set forth in Roderick Nash's *Wilderness and the American Mind*.[12] These urban people (not rural people) strongly support establishment of wilderness areas, national parks, wild and scenic rivers, and fish and wildlife enhancement.

In 1962, preservation in their natural condition of particular rivers or segments of rivers was authorized for water resource planning within the federal government; in 1964, the Wilderness Act, which precluded water resources development within designated wilderness areas, was enacted; and in 1968, Congress passed the Wild and Scenic Rivers Act authorizing establishment of eight wild or scenic rivers and the study of twenty–seven more. These actions were taken within the context of what became the Water Resources Planning Act of 1965. Under this act, it was assumed by federal policy and planning officials that only particular rivers or segments, not all rivers, would be preserved and removed from the possibilities of multiple–purpose development. However, with passage of the Wilderness Act, the Wild and Scenic Rivers Act, and other manifestations of the environmental movement (particularly passage of the National Environmental Policy Act of 1969), the most ardent environmentalists adopted the view that dams and reservoirs could never be justified for any purpose. As a consequence, the environmental movement succeeded in stopping many water resource developments and in authorizing many wild and scenic river proposals.

Also, since the mid–1960s, the futility of reservoir storage as the primary means of flood control had become widely recognized. "Nonstructural" measures such as flood insurance, flood plain zoning, warning systems, and flood–proofing, etc., came to be seen as the most appropriate means to be employed in flood hazard mitigation. And most of these measures require largely state and local (not federal) nonstructural implementation.

Decline in federal dominance in water resources development also was fostered by federal policy. Implementation of Title II of the Water Resources Planning Act of 1965, which authorized establishment of federal– state river basin commissions, encouraged states to share responsibility. Even more directly, implementation of Title III, which authorized 50 percent matching financial grants to states to encourage them to undertake comprehensive water resources planning, increased state professional capability to undertake greater responsibility. In 1980, these financial grants to states were substantially increased. Increased state professional capability has also been fostered by the Water Resources Research Act of 1964, as amended and supplemented, under which water research institutes have been established at each state land–grant university.

With the achievement of multifaceted economic development throughout the country, water resource development can no longer be cited as the "key" national public means of encouraging general economic development. It has lost its traditional national ideological relevance. Moreover, major resource basin developments have largely been accomplished on the Columbia, Colorado, Missouri, Ohio, Tennessee, Mississippi, Arkansas, and Rio Grande Rivers. Majority votes in the Congress for water projects, East and West, which were increasingly hard to put together before the Carter Administration, have been much more difficult to obtain since 1977. This is not to imply, of course, that no new developments are needed, but most of those really needed are smaller intrastate projects for municipal and industrial water.[13]

An Academic Conception of Political Decisionmaking and the Rise and Decline of Federal Dominance

Three processes of political decisionmaking have been identified by political scientists: ideological adherence, bargaining, and vote-trading (i.e., distributive or pork–barrel politics). The usual context of decisionmaking to which these processes refer is the legislative process involving a particular piece of legislation. In his Inter–University Case Program study of November 1971, Lowi classified acts of Congress into the above types of processes, implying that only one type is involved in the passage of each act.[14] On the basis of the author's role as a participant/observer during the 1960s in the passage of much legislation (including the Wilderness Act of 1964, the Land and Water Conservation Fund Act of 1965, the Water Resources Planning Act of 1965 and the Wild and Scenic Rivers Act of 1968), he would hypothesize that all three processes of decision are involved in the passage of each piece of legislation.[15]

Lowi, also in his November 1971 study, looked at the record of passage of the Omnibus Rivers and Harbors Act of 1950 and classified its decisionmaking process as vote–trading. Looking at the immediate processes of that act alone one might reasonably conclude that the classification is correct. But if one looks at decisionmaking processes involving a subject matter area, such as water resources development, over a long period of time, and notes that the key decisionmakers also have been involved for a long period, then the three process hypothesis, above, appears explanatorily valid in this broader context too.

During the rise of Federal dominance, *ideological adherence* was operative, first in terms of basic policy objectives of economic development and settlement of the West and second as a part of the traditional conservation movement's commitment to comprehensive multiple–purpose river basin planning and development. Through the generations in Congress of, say, Senators Robert Kerr of Oklahoma and Clinton Anderson of New Mexico, as well as Congress-man Wayne Aspinall of Colorado, the author would assert that ideological adherence was an important part of political decision processes.

But *bargaining* too has been involved. The Pick–Sloan Plan for the Missouri River basin, authorized by the Flood Control Act of 1944, was made politically feasible by the striking of several important bargains, including finally the O'Mahoney–Milliken Amendment. The Colorado Basin Project Act of 1968 involved several major bargains; the most dramatic was the ten–year morato-rium on studies of interstate interbasin water transfers struck with Senator Henry Jackson from the State of Washington. But that struck with the Sierra Club, substituting the Mohave steam–electric plant for Bridge and Marble Canyon dams as "cash registers" to subsidize other project costs, was the most wrenching for political traditionalists.

Finally, there is no doubt that *vote-trading* has been importantly involved in political decisionmaking processes. As noted above, the National Rivers and Harbors Congress and the National Reclamation Association played important roles in assisting political leaders in carrying out this type of political decision-making. These interest groups represented individual project interests, asso-ciated, respectively, with the Army Corps of Engineers and the Bureau of Reclamation.

The decline of federal dominance in water resources development can be explained, first, by gradual loss of ideological adherence. New unbelieving political leaders displaced the old political believers, particularly of the West and South, who had constituted an effective coalition. And, as practical interest in developments declined with much of the needed development already built or building, the two major interest groups lost enthusiasm, members, and budgets. Despite attempts to reorganize them, they are now no more than mere shadows of their former selves. Within the context of water resources policy, there are currently only a few bargains to be struck (e.g., over user charges for navigable waters and over cost–sharing policy more generally). Increasingly during the last fifteen years or more, authorizations and appropriations for water resources developments have been widely seen by the public (as Lowi thought he saw them in 1950) as "just pork barrel."

As a consequence, as the author first argued in a National Water Conference in 1975, "the federal water development program is politically dying if not already dead."[16] There was no bargaining with other programmatic interests then (or now) to develop a wider and renewed basis of support. Also, no new ideological foundation for the federal program has been formulated. Officials concerned with high policy issues in the federal government ask: What is the federal public–interest in further water resource development? Without an ideological formulation consistent with contemporary objective factual analysis, they see no reason for federal development, except as a response to "pork–barrel" pressures. Not only are all three processes of political decisionmaking necessary for the "rise" of a program, but also to sustain it or prevent decline.

Clearly the foregoing analysis of the rise and decline of federal dominance in water resource development, utilizing the three decisionmaking processes of ideological adherence, bargaining and, vote–trading, has the potential of providing an academic theory of federalism, now almost nonexistent (as was seen above) for the explanation of change of primary responsibility of a program from the state to the federal level or vice versa.

The Present: 1977–1983

The "present" in U.S. water resource policy affairs refers to the policy change proposals that surfaced in Washington during the Carter Administration and in the years that the Reagan Administration has been in office. During this period nothing has happened yet that gives a clear and definite indication of long–run U.S. water resources policy.

The Carter "hit list," developed by ardent environmentalists within the administration, clearly poisoned the atmosphere for negotiation of policy change. Economists within the administration, in agreement with environmentalists, were able so to tighten the Principles and Standards of the Water Resources Council (WRC), and their related procedural manuals, that economic justification of almost any water resource development project, but especially an irrigation project, would be next to impossible. This action, when combined with the president's proposal that all recommended federal water resource projects be subject to review by the WRC, made fruitless any attempt by Congress to reach agreement with the Carter Administration on U.S. water resources policy.

The only policy proposal of the Carter Administration that appears to be having some carryover value to the Reagan Administration is the federal–state financing proposal. This called for states to supply 10 percent of the front–end funds for projects with vendible outputs and 5 percent of the front–end funds for those without vendible outputs.[17]

The Reagan Administration came into office with the solid support of the western states where water projects still have, at minimum, great political symbolic importance. The chairman of the water policy task force of the Cabinet Council on Natural Resources and the Environment is Assistant Secretary of the Army William Gianelli, the former director of the California Department of Water Resources (which implemented California's $2 billion water resource development plan). The administration soon indicated its desire to abolish the WRC staff and to rescind the WRC's Principles and Standards and related procedural manuals for planning in their entirety. It has abolished the WRC staff and all the federal–state river basin commissions created under Title II of the Water Resources Planning Act of 1965; and it has succeeded in stopping the 50 percent grants to states under Title III for comprehensive water resources planning.

Economic and Environmental Principles and Guidelines for Water and Related Land Resource Implementation Studies were promulgated by James Watt as Chairman, Water Resources Council, on March 10, 1983. They replace completely the WRC Principles, Standards and Procedures of the Carter Administration.[18] Apparently, great difficulties were encountered in their long period of internal preparation and acceptance. The Reagan Administration not only includes officials who believe fundamentally in water resources development, like Mr. Gianelli, but also economists who are even stricter constructionists in their principles of benefit–cost analyses than the economists of the Carter Administration, and who have fewer concerns about the environment.

The new "guidelines" abolished the previous provision for multiobjective planning, involving the objectives of national economic development and environmental quality, and substituted a "federal objective" of "national economic development consistent with protecting the nation's environment, pursuant to national environmental statutes, applicable executive orders or other federal planning requirements." Quick appraisal of the required economic analysis leads to the general conclusion that the "guidelines" are not much, if any, less stringent than those of the Carter Administration. However, the "guidelines" are intended to be such and not "rules" enforceable by federal courts, as were the Principles, Standards and Procedures of the Carter Administration.

In line with the financing views of the Carter Administration, but with emphasis upon "the federal budget crunch" rationale, Mr. Gianelli has called for involvement of project sponsors in up–front financial participation. He also advocates fixed percentages of reimbursement over the usual fifty–year period for several development purposes (e.g., navigation, flood control, irrigation, and recreation) by nonfederal interests.[19]

Since the advent of the Reagan Administration, Congress has directed almost all of its attention to the administration's unprecedented tax decrease and domestic budget–cutting proposals. Nevertheless, at least in the Senate, efforts aimed at water resource policy reform have been made. They have focused on a bill to create a National Board of Water Policy, with an independent chairman, to replace the WRC; on the "Domenici–Moynihan bill," entitled the National Water Resources Policy and Development Act; and on bills to authorize federal funds for grants to states and local governments to rebuild the nation's overall "infrastructure," including water systems.

The Domenici–Moynihan bill, if enacted, would make very substantial changes in U.S. water policy in the direction of substantially increasing state control and financial responsibility (i.e., 25 percent of the cost of planning studies, 25 percent of construction costs and 50 percent of operation and maintenance costs). The bill also would remove the power of Congress to determine the authorization and funding of individual planning studies and individual development projects, except in limited cases of dispute.

These present tendencies for policy change all point in the direction of decentralization of financial responsibility and control of water resource development to states and local governments—a direction that would appear to be in accord with current widely held political beliefs in decentralization. Nevertheless, all present proposals—except the infrastructure rebuilding grant proposals—retain substantial federal executive involvement. For example, both the Army Corps of Engineers and the Bureau of Reclamation would continue as planners, constructors and operators of intrastate as well as interstate developments. None of the proposals—with the exceptions noted above—appear to recognize that there is no longer an effective coalition of national constituencies, possessing a strongly held national ideology, in support of federal water resource development *per se*.

The Future

The author's reading of the future in the light of the past and present calls for a major break with the past—a real decentralization with state and local governments (and, in some cases, private enterprise) taking primary responsibility for intrastate projects. This reading is based on the belief that most water resource developments of the future will be for municipal and industrial water, for some structural developments as a part of urban flood plain management, and for relatively small–scale, single–purpose hydroelectric power developments or multiple–purpose projects. Also, a few new irrigation schemes will be politically supported in states with further irrigation project opportunities and with continuing strong rural legislative representation (e.g., Nebraska and Oregon). Broad state support will also be given in the West to new watersupply developments that relieve the pressure to transfer water from agricultural to municipal and industrial uses. Rehabilitation of existing irrigation projects in the West and of municipal water systems in the East is needed and will be politically supported. The locus of political action will be the states. Federal block grants, at most, will be the desirable, if not essential, means of federal support. A national coalition of water interests could bring about a grant program as well.

Consistent with the above readings of the future, two alternative political scenarios suggest themselves regarding potential intrastate projects: (1) federal political processes that bring about devolution of primary responsibility into nonfederal hands; and (2) state political processes that bring about primary nonfederal responsibility, with federal programs continuing to "die on the vine" as they are now doing.

Scenario 1

Scenario 1 begins with President Reagan's New Federalism proposal to state, city, and county governments. Some forty–eight federal programs are proposed for transfer of basic authority to these governments. As originally proposed, the transfer would not begin until FY 1984 and would be carried out over a period of eight years. A financial trust fund would be established by the federal government to finance these programs during the period of transition to state, city, and county funding. In meetings with the National Governor's Association, National Association of Counties and groups representing cities, the administration made very clear that the proposal is open to extensive and intensive negotiation. Sensitivity to state versus local interests is manifest in the provision that "a hundred–percent pass–through provision to local units of government (is included) for programs which have traditionally been direct Federal–local programs."[20] Water quality and water supply grant programs were reported as included in the proposal, but the full significance of their inclusion in the context of overall federal water resource policy is not known. A major change in the style of federal action for water resource development is implied. Up to now the federal response, as discussed above, has been largely a traditional federalist response of direct public service such as that of the Army Corps of Engineers or the Bureau of Reclamation. The federal response, to which the administration's New Federalism relates, seems to be in line with the modern approach of categorical financial grants–in–aid and loans to states and local governments, or, more recently, block grants and revenue sharing. Federal water resource development assistance of this type has been confined so far to the small watershed program of the Soil Conservation Service, the small reclamation project program of the Bureau of Reclamation, the rural water supply program of the Farmers Home Administration and some assistance provided by the Economic Development Administration of the Department of Commerce.

How could this scenario be developed with respect to federal intrastate water resource projects via the three processes of political decisionmaking: ideological adherence, bargaining, and vote–trading?

With respect to *ideological adherence*, President Reagan is trying to lead those persons, widespread within the political spectrum, whose current political ideology includes decentralization. In speaking to the National Association of Counties on February 22, 1982, he said: "We must seize this opportunity to reverse a trend that has begun to choke state and local initiative and overload the Federal government." And he asserts that "Democracy depends on government being close to the people." Moreover, he identifies his opponents as "representatives of bureaucracies and particular special-interest groups"—by implication, members of the many "iron-triangles," "power clusters" or "sub–governments" in support of continuance of individual federal programs.

With respect to *bargaining*, he is trying to utilize the negotiating forum of states, counties, and cities to build a presidential, state, county, and city coalition to put through Congress an overall program of New Federalism. If there is no longer an effective national constituency for federal water resources development, as this author has concluded, then the federal water resource program would appear to be a natural for decentralization, via New Federalism.

With respect to *vote-trading*, agreement to a water resource development block grant program would provide, presumably, the necessary "pork," even if only for a limited number of years. In this way, the North Central–Northeast coalition, or the "gypsy moths" in Congress, could get federal funds to help rebuild city water supply systems needing substantial rehabilitation (e.g., Boston and New York) and the South and West could get what they determine they want.

Clearly, no one of the three processes of decisionmaking would be sufficient for Scenario 1 to materialize. All three processes would appear to be necessary to provide coalitional support. As of the summer of 1983, however, New Federalism appears to be on the back burner of intergovernmental relations. The political climate of recession, with state and local finances impaired, is not a propitious time. When the recession is over New Federalism presumably will return. Too many influential people believe with David Walker, Assistant Director, Advisory Commission on Intergovernmental Relations, that the present intergovernmental system is "dysfunctional federalism" and calls for a "fiscal, programmatic, and political strategy of decongestion of and some disengagement within the intergovernmental system."[22] Michael Reagan sees the "United States (as constituting) a single national community," with the emergence of national standards in increasing numbers of policy areas in the last twenty years (e.g., in environmental protection, civil rights, and product safety), and he praises what he calls this "permissive federalism."[23] The programmatic field of intra–state water resources development, if decentralized, could lessen dysfunctionalism without necessarily violating ideology embodied in desirable national standards. Within the Reagan Administration there somehow does not seem to be any coordination between its New Federalism program and its present consideration of major policy changes in federal water resources development programs.

Scenario 2

Glendening and Reeves characterize the American experience as "steadily increasing centralization" through "rampant incrementalism" and "pragmatic adjustment." Toward the end of their 1977 study, however, they indicate a belief that the states "today" are becoming increasingly effective instruments of government. They do not share Luther Gulick's 1933 assessment that the American state is finished. However, they conclude that "effective decentralization cannot be legislated by the central government," contrary to Scenario 1. Only insofar, they say, as the "states indigenously maintain their sovereignty and the loyalty of their citizens, (will) meaningful decentralization . . . be possible."

Scenario 2 is based on the premise that Glendening and Reeves are correct in their general views that the states are "becoming increasingly effective" and that meaningful decentralization can occur if states indigenously maintain their sovereignty and the loyalty (i.e., sense of community) of their citizens. What have states done recently with respect to water resources development that suggests the political viability of Scenario 2?

States and local public bodies have allowed themselves for many years to be locked into national perceptions of values, goals and objectives determined largely in Washington, D.C. As long as these national ideological considerations were consistent with state counterparts, traditional federal water policies, when supported financially in Washington, served them well. Increasingly it is becoming clear, however, that individual states themselves are now determining their own values, goals, and objectives for intrastate water resources development.

South Dakota did so in 1981 when its governor, legislature, and people determined that selling a share of the state's Missouri River water out of Oahe reservoir to an interstate coal–slurry pipeline company, on terms that provided low–cost water along the pipeline to several of its towns and a large financial return to the state, was good for the state.

The Governor of Wyoming proposed to the legislature in 1982 that the state appropriate $100 million per year for six years to develop the state's water resources. The legislature responded by authorizing an initial $50 million program. When more specifics as to the goals and objectives of the Wyoming decision become apparent, it will be interesting to note what novel ideas emerge that are perceived to be in that rural state's basic interests. Maybe, indirectly if not directly, state subsidy of irrigated agriculture will emerge as a state objective.

After experiencing frustrations with federal politics in recent years, other states, including Montana and Oregon, are thinking and acting in the direction of taking major responsibility for intrastate water resources development. In 1982, Montana authorized a water resources development program, based upon financial resources derived from mineral severance taxes, that will enable the state to participate financially in federal projects or undertake them itself if this is more advantageous. After the drought of 1977, Oregon's voters authorized the issuance of bonds totalling $600 million to develop water supplies for irrigated agriculture and other purposes in eastern Oregon. California and Texas have long had active water resources development programs of their own, based upon their own financial and technical capabilities; they will undoubtedly continue such development as they perceive the need.

Local governments also possess the capability to undertake water resources developments, particularly to meet municipal and industrial water needs. For example, the Northeastern Colorado Water Conservancy District is presently fully financing, through revenue bonds, the bringing over from the western slope of the Rocky Mountains of some 48,000 acrefeet of water annually at a total investment cost of $80 million, for municipal and industrial use. In this way the pressure to transfer water out of agricultural use for municipal and industrial use is being relieved on the eastern slope.

With repayment of bond interest and principal based upon water and power user charges, local governments, regional public districts and states can finance a large proportion of their future water resource development costs. To the extent purposes of development cannot be financed on the basis of user charges, state appropriations, federal block grants, or federal revenue sharing, if available, could be utilized.

Considering Scenario 2 in more political terms, the author finds the case of Colorado interesting in itself and suggestive of political processes that may be occurring, or could occur in other states, particularly in the arid West. Politically important water leaders of Colorado, both inside and outside of government, appear to believe increasingly that the state, in cooperation with local public bodies, should see to the development of the state's remaining water supplies sooner rather than later. Much of the water available to Colorado for consumptive use is now being utilized. The remaining undeveloped water involves the state's claim to some 700,000 acre–feet of unappropriated Colorado River water plus the water which in the few years of markedly above–average precipitation "wastes" to contiguous states. This "wasted" water provides the contiguous states (e.g., Nebraska) with more water than Colorado is obligated to provide them under interstate compacts and judicial decrees. Within the internal political context of Colorado, presently and in the foreseeable future, state, and local public action could well be taken to develop these remaining supplies for municipal and industrial use, including energy production. Thus the probable adverse effect of expected growth in Colorado's population and industry upon the state's irrigated agriculture could be minimized.

Underlying the internal politics of Colorado, an ideology would appear to be strongly held that involves the goal of preservation of the state's agricultural industry. This industry is strategically based upon irrigated agriculture. Belief in agricultural preservation would appear to be held by a wide array of leaders, including urban leaders, despite the probable fact that purchase of water rights from agricultural interests would be cheaper, in the short run, to meet increasing municipal and industry water needs than development of the state's remaining water supplies. This ideology, in a rapidly urbanizing state, is clearly different than that operative nationally in Washington, D.C. In 1981, with support of both urban and rural political leaders, the Colorado Water Resources and Power Authority was established by the legislature with a $25 million trust fund and authority to issue tax–exempt revenue bonds for the purposes of participating in the financing of federal or non–federal water resources development projects.

To the greatest extent practicable, development of the remaining water supplies of Colorado will be undertaken cooperatively with private industry, particularly energy and other mineral industries. No doubts are expressed that Colorado has all the technical talent that is needed for water resources development without planning by, or technical assistance from, the federal government. Thus the situation of the nineteenth century and the first half of this century, when both financial capacity and adequate technical talent were not available to the state, no longer exists.

Conclusion

The record of the past has been one of steadily increasing centralization. In the case of water policy, that centralization began early and was sustained by a cohesive national ideology and by congressional bargaining and vote–trading. The result has been a century of national water policy projects dependent upon national technical planning and financing.

The underpinnings of support for that national policy began to erode in the 1950s and the programs have been dying ever since. In part, this evolution reflects the cumulative success of national water policy. The West no longer is relatively undeveloped and western states no longer lack technical capabilities. Most remaining prospects for dam construction are intrastate. In part, the evolution reflects dissipation of support from other regions and inroads on the construction ethos by environmental groups.

Future national water policy will depend upon the viability of a new consensual ideology supporting federal financing of a broad range of mainly urban projects. If support can be created, a new round of grant programs may ensue. Even so, the role of states and localities is likely to grow. The alternative is a much greater devolution of initiative, as well as planning and financing roles, to states and localities. Strong initiatives in this direction already are underway. Thus far, the Reagan Administration has been content to observe continued erosion of the traditional national policy. Though water policy seems an appropriate topic for the New Federalism, it has not yet been included.

Notes

1. Kirk N. Porter and Donald Bruce Johnson, *National Party Platforms 1840--1956* (Urbana: The University of Illinois Press, 1956), p. 28.

2. David B. Walker, *Toward a Functioning Federalism* (Cambridge: Winthrop Publishers, Inc., 1981), p. 54.

3. Edward S. Corwin, *National Supremacy* (New York: Henry Holt and Co., 1913), pp. 108–09.

4. Walker, pp. 46–65.

5. Daniel J. Elazar, *The American Partnership: Intergovernmental Cooperation in the Nineteenth Century United States* (Chicago: The University of Chicago Press, 1962).

6. Terry Sanford, *Storm Over the States* (New York: McGraw Hill, 1967), p. 80.

7. Michael D. Reagan, *The New Federalism*(New York: Oxford University Press, 1972), ch. 5.

8. Parris N. Glendening and Mavis Mann Reeves, *Pragmatic Federalism: An Intergovernmental View of American Government* (Pacific Palisades, California: Palisades Publishers, 1977).

9. For the political concept of "movements" in relation to interest groups, see L. Harmon Ziegler and G. Wayne Peak, *Interest Groups in American Society*, 2nd ed. (Englewood Cliffs, New Jersey: Prentice– Hall, Inc., 1972), pp. 73–86. As the author perceives them, "movements" are not likely to be successful in the longer run without the strong support of interest groups. But movements which lose their ideological attraction of large segments of the public, but not members of interest groups, also lose much of their earlier political force: witness the "Labor Movement" in the 1980s in comparison to the 1930s through 1960s.

10. Charles Yoe, *The Declining Role of the U.S. Army Corps of Engineers in the Development of the Nation's Water Resources* (Fort Collins: Colorado Water Resources Research Institute, Colorado State University, Information Series No. 46, August 1981), p. 115.

11. Ibid., p. 86.

12. See Roderick Nash, *Wilderness and the American Mind*, 3rd ed. (New Haven and London: Yale University Press, 1982).

13. The author believes that the federal government will need to continue to have a very substantial direct role with respect to major interstate water developments. The nation's system of 25,000 miles of inland navigation must be maintained and rehabilitated, as necessary by the federal government. The flood control problem of the Mississippi could not possibly be handled except by the federal government. Any large interstate interbasin transfers of water would also require federal responsibility. The new politics of this necessary residual responsibility is not addressed here.

14. Theodore J. Lowi, *Four Systems of Policy, Politics and Choice* (Syracuse, New York: The Inter University Case Program, Inc., ICP #110, November 1971).

15. The author was closely associated with Secretary of the Interior Stewart L. Udall during the 1960s, first as Assistant Director and then Director, Resources Program Staff, Office of the Secretary. From April 1966 through August 1969, he was Executive Director, U.S. Water Resources Council. For a brief analysis of the legislative decisionmaking processes involved in passage of the Wild and Scenic Rivers Act of 1968, arguing that all three processes of decision were involved, see "Institutional Aspects of Water Allocation in the Upper Colorado River Basin, Implications for Fish and Wildlife; A Discussion," by the author, published in *Energy Development in the Southwest—Problems of Water, Fish and Wildlife in the Upper Colorado River Basin*, edited by W. O. Spofford, Jr., et al. (Washington, D.C.: Resources for the Future, Research Paper R–18, 1980), Volume II, pp. 515–31.

16. Henry P. Caulfield, Jr., "Let's Dismantle (Largely but Not Fully) the Federal Water Development Establishment, or the Apostasy of a Longstanding Water Development Federalist," in the *Proceedings of the National Conference on Water*, Washington, D.C., April 1975, sponsored by the U.S. Water Resources Council (Washington, D.C.: U.S. Government Printing Office, 1976; Stock Number 024–001–02798–4), pp. 180–84.

17. President Carter's Message to Congress on Water Policy, June 6, 1978.

18. Because the Reagan Administration found that the Water Resources Planning Act

of 1965 was still the law of the land, it had to utilize the Water Resources Council to abolish the previous Principles, Standards and Procedures and establish its own "guidelines." Consideration of the latter, however, was through the Cabinet Council on Natural Resources and the Environment with lead staffwork performed by the Office of Water Policy of the Department of the Interior.

19. William R. Gianelli, *New Directions in Federal Civil Works Programs*, a speech presented at the Annual Meeting of the Water Resources Congress, Houston, Texas, February 1, 1982. The Administration's position on this matter is not clear even though on June 15, 1983, Mr. Gianelli testified before the Senate Committee on Environment and Public Works in substantially the same vein as his Houston speech in July 1982. Secretary of the Interior Watt, Chairman pro tempore of the Cabinet Council on Natural Resources and the Environment, although invited to testify on behalf of the Administration, did not testify. Senator Laxalt of Nevada, a leader among western Republican Senators, publicly indicated in 1983 his opposition to change in financing policy. Secretary Watt, a political protege of Senator Laxalt, was reported in the *Rocky Mountain News* of June 18, 1983 as disavowing Mr. Gianelli's testimony and stating that financing arrangements will be flexible in accord with state and local government's ability to pay and "to honor prior (federal) commitments."

20. President Reagan's remarks at a dinner honoring Governors, February 23, 1982, *Weekly Compilation of Presidential Documents*, 18, 8 (March 1, 1982): 215.

21. Bill to authorize renewed infrastructure will become involved politically through bargaining as to what types of public works will be included. Ideas about "urban renewal," presumably will provide the ideology. And the proposal of federal funds for particular projects throughout the nation's urban areas will inspire distributive politics. Thus new or renewed urban water systems could be supported by an urban coalition along these lines.

22. Walker, *Toward a Functioning Federalism*, ch. 7–9; and in particular, p. 258.

23. Reagan, *The New Federalism*, ch. 5.

11

Categorizing State Models of Water Management

TERRY D. EDGMON and
TIM DE YOUNG

I. Overview

In the study of water resources policies and programs, the American states and their political subdivisions have been curiously neglected. The traditional centrality of the federal government in the development of the nation's waterways, harbors, and reclamation projects has produced a rich and abundant literature on federal water policy. In contrast, studies of state and local water institutions are few in number as well as limited in scope. Technical and legal studies predominate, but these tell little about the relative administrative abilities of the states in dealing with water resources problems. However, recent attempts to shift governmental responsibilities from the federal to state and local levels suggest that the state role in water management has and will assume increasing importance.

The purpose of this chapter is to develop a conceptual framework whereby the organization of state water management institutions may be fruitfully analyzed. This framework will be based on the development of a typology by which the states may be classified. Case studies of representative states then are presented both to describe each major category and to identify factors that may account for variations in state water resources organizations. Following a brief summary of the changing federal role, the dimensions of water policies that confront state water resource specialists, and the inadequacy of legal typologies of state water institutions to account for recent organizational developments, an institutional model is presented. The model allows for comparison of state water institutions according to both the degree of centralization of water management within a state and the relative integration of water planning and management functions at the state level. The model is then utilized to compare differences in the organization of state water management institutions with findings from a national survey of state water institutions and case materials from selected states.

II. Emerging Roles for the States in Water Resources Policy

Once a dominant force in water resources development, the federal government's influence has declined over the last twenty years. This trend appears to be accelerating since the Reagan Administration has made policy decisions to reduce substantially the federal role and to shift many water resources management responsibilities to the states. The elimination of the U.S. Water Resources Council, reductions in funding for river basin commissions, and increased cost sharing by the states and local entities suggest that the states will continue to assume greater responsibilities in the planning, financing, and development of future water resources.

The trend of less federal and more state responsibility in water resources management first became evident during the 1960s. A shifting balance of power in the Congress, the rise of the environmental movement, and increased pressures from the states for more active roles in the decisionmaking process began to erode the traditional hegemony of federal agencies (Schooler and Ingram 1982). Political resistance to a strong federal role came from critics of federal water projects who objected to these projects' distributive or "pork barrel" attributes and from environmentalists who were concerned primarily with negative effects on the environment. Moreover, economists have argued that the marginal costs of large–scale projects have been increasing rapidly while the number of productive project sites has declined (Howe 1981, p. 17). Increasing congressional opposition to federal water projects has become evident, with the retirement or defeat of many of the traditional water resources policy advocates that, therefore, reduced the strength of the water developmentalist coalition (Caulfield 1982). Finally, the emphasis of water resources policy and management has shifted from distributive, construction–oriented technical programs to nonstructural, regulatory policies and programs. For example, the river basin plan, whose centerpiece often was a large federally funded impoundment, has been superceded by combinations of regulatory programs including nonstructural flood control measures, conservation incentives, and user permit systems.

The complexity of contemporary water resource problems represents a formidable challenge to the states. Even though assistance to both water resource development and pollution control has declined, federally mandated requirements such as the cleanup deadlines of the Clean Water Act remain in force. Similarly, court–ordered compliance deadlines in water quality protection typically cannot be modified. Moreover, each of the states must respond to a wide range of water–related problems including infrastructure decay, increased competition for limited and, in many cases, fully allocated surface water supplies, and emerging problems of ground water quality and quantity.

Primarily because water resources management on the state level has not been subjected to the standardizing influence of a large–scale federally funded grants program, state water management systems exhibit considerable variation. Unlike water pollution control requirements, the traditional federal role has been one of building constituency support among local groups and congressional delegations. As such, the states, while the recipients of relatively small amounts of federal monies to develop water resource management capabilities, also have been spared meeting strict federal guidelines in the development of water management systems. The most significant form of federal grant aid to the states for water resources planning and management was through the activities of the now defunct U.S. Water Resources Council and Title III planning assistance authorized under the Water Resources Planning Act of 1965. Through this assistance, many states had initiated positive steps toward active water management by:

1. centralizing authority of water allocation and planning to the state level;
2. integrating water quality and quantity management within a single state agency;
3. developing strategies for the complementary management of ground water, surface water and land resources; and
4. developing water resource management programs based upon regulatory and nonstructural strategies.

The states exhibit considerable variation in the implementation of these objectives. Some states have not initiated such steps; others have delegated primary water management responsibilities to local governments or water districts. Some have reorganized agencies to integrate water quality and quantity functions, and others distribute water management authority and responsibility among several state agencies or even branches of governments. Permits for water use, once found only in the West, have now become a near universal technique among the states as a water resources management tool. However, the administrative and legal attributes of the permit and its permit systems vary considerably (Water Resources Council 1981).

III. Limitations of Riparian vs. Prior Appropriation Distinctions

Typically, discussions of state organization for water management have been conducted by students of water law or civil engineering, who have based their analysis on either technological criteria or state legal doctrine. Detailing differences in legal doctrines for water allocation from state to state constitutes the bulk of state–level comparative analyses (See Dewsnup and Jensen 1973). Indeed, many apparent differences exist between the riparian doctrine states of the humid East and the appropriation states of the arid West.

Most arid western states developed their water law under the appropriation doctrine, where the right to use water is established as a property right subject to review by some state institutions that possess the responsibility to determine if the private use of water meets the criteria of "beneficial use." Defining the water property right as a usufruct also allows the state to act on behalf of other users when the right to use must be shared. State regulation of an individual's right to use water is based upon constitutional definitions of water as a public good, owned by the state, people, or public (Radosevich 1980, pp. 262–63). Public ownership requires an elaborate system to affect the administration and allocation of water rights. This includes sophisticated water supply monitoring, permit systems, and detailed legal procedures for the adjudication of water rights.

In contrast, riparian states east of the one hundredth meridian, where water has been abundant, have until recently had no need for complicated allocative mechanisms because water was freely available to all, and in most years, allocation decision rules to deal with scarcity were not needed. However, economic expansion, population growth, and threats to ground water quality have forced many eastern states to modify riparian laws in order to cope with increasing conflicts in water use. What has emerged then is a third category, the "permit" state. In these states, riparian doctrines have been altered to allow for administrative procedures to effect greater efficiencies in the management of scarce water resources. Many of these administrative systems resemble those operative in appropriation states. Our survey of state water institutions indicates that thirty–five states administer some form of permitting system to regulate surface and ground water use. Of these, twenty–nine administer permits on a statewide basis, while six do so for designated areas only. Eight have a decentralized permit system, six of which utilize regional offices and two utilize local water districts. These results indicate that administrative regulation of water use, once found exclusively in appropriation states, has been spreading east to the riparian states. Moreover, there appears to be a trend toward concentration of authority or centralization of power at the state level.

The widespread adoption of permit systems in the riparian states has diminished the riparian–appropriation distinction. The distinction is also of limited use in comparing those states where both doctrines are operable. California and Texas, for example, are governed by the riparian doctrine for ground water but the appropriation doctrine governs surface water use (Dewsnup and Jensen 1973). The increasing importance of protecting in–stream uses further suggests the possible utility of adoption of riparian concepts in the appropriation states. Hence, traditional classification of states according to legal doctrine is increasingly obsolete and contributes little to our understanding of administrative variation among the states.

IV. A Comparative Model of State Water Institutions

The increasing salience of water resource problems in combination with declining federal involvement represents a significant political and administrative challenge to the states. The adoption of statewide permit systems in most states suggests the increasing uniformity of state water management systems. Yet, on close inspection, much variation exists from state to state. For example, some states have delegated major water resource management responsibilities to local organizations, others have assigned water management to a number of state agencies, and still others have integrated and centralized such functions into one state agency. Such diversity provides a fertile ground for empirically testing a number of competing hypotheses about the proper organization of water resources decisionmaking. On one side of the debate is Wittfogel (1957), who has hypothesized that water scarcity, and the need to construct and maintain complicated water systems, leads to the centralization of power and authority within government. In contrast, Maass and Anderson (1978) have countered with a comparative study of six hydraulic communities in different regions of the U.S. and Spain. They find that local communities are able to maintain independence from the central government, in spite of receiving subsidies to develop water resources (Maass and Anderson 1978). They deny the contention that aridity and irrigation constitute the necessary and sufficient condition for the centralization of political power and the integration of water management institutions. They argue that:

> The technological characteristics of irrigation agriculture—especially the flow, stochasticity, and singularity of water supplies—create special problems of control to which there are two polar responses, with a relatively small number of alternatives between them: a single leader or leadership group, which may be from outside the irrigated area, can operate the control structures and procedures, or all the irrigators of a water source who live within a defined service area can create and support a user's organization with authority to operate the structures and procedures of water control.

Maass and Anderson, then, argue that local control is inherently more democratic. In direct contrast, Wittfogel's argument is that water management requires authoritarian, centralized decisionmaking.

Both Wittfogel and Maass/Anderson perhaps overstate their respective positions by failing to examine carefully the dimensions of central or local control within the context of contemporary water resources management. Wittfogel focuses upon societies which lack democratic traditions. Hence, his work may have limited application in Western democratic societies. Maass and Anderson's historical case studies in Utah, Colorado and California are also limited since they focus upon once dominant federal–local relationships; little attention is paid to state–local relationships. This is of interest because it is the state government, not the federal government, which possesses the authority to regulate and control water use, even if that authority—as is also the case in land use—may be delegated to local political institutions.

If state–local relationships are taken into consideration on a state–by–state basis, there may exist a wider range of alternatives than Wittfogel's outside, top–down despotism and Maass and Anderson's grass roots democracy. We will argue that water resources management is being achieved through a variety of democratic institutions as diverse as the political and hydrologic environments within which they are found.

Water resource management varies across the states according to two dimensions: (1) the relative centralization of management authority on the state level, and (2) the discretionary authority or autonomy of local water supply organizations. Relative centralization refers to the degree to which management authority is distributed among state-level agencies. For example, water management in some states is the primary responsibility of a single state agency, which conducts water resource planning, develops and designs projects, manages water quality programs, and administers a permit system for water allocation. On the other extreme, some states fragment the responsibility for water management among various state agencies. For example, water quality programs may exist in an environmental protection department, water supply programs in a natural resources department, and water planning may be the responsibility of a state planning agency or commission. In this latter type, some management functions may not be the responsibility of any agency. For example, there may be no permit system to control water use. In consequence, it is assumed that the fragmentation of management functions inhibits state control over local water organizations.

The range of local discretionary authority ranges from states where no local water organizations exist to those where local water organizations possess authority to plan, finance, and implement water projects, and therefore may effectively manage the water resources under their control, with little oversight or direction from state government. In some states, local organizations operate within the context of enabling legislation that has explicitly delegated state regulatory authority to them, or they may operate within the context of a political climate whereby state agencies defer to local policies and decisions.

Combining these two dimensions gives us a four–celled typology as illustrated in Figure 11.1, below:

State organization

		Centralized	Fragmented
	Weak	Unified	Fragmented
Local discretionary authority			
	Strong	Federalist	Home rule

Figure 11.1 A Typology of State Water Management Institutions

The first type is a unified structure, which approaches Wittfogel's authoritarian view, in that there exists a high degree of centralization or concentration of authority on the state level combined with weak or nonexistent local organizations. The lower right cell represents a situation illustrative of the Maass and Anderson model of strong local water organizations that can operate with relative independence of state government because power and authority on the state level is fragmented.

The upper right cell represents an organizational pattern of fragmentation on the state level and weak or nonexistent local organizations. In its extreme form, this is tantamount to the absence of positive water resources management. We would expect to find this pattern in states where water resources are not a significant input into the political process. The lower left cell represents a federalist model because of the balance between centralized state authority and strong local organization. For this type, we would expect to find a two–tiered management structure, composed of centrally directed administration on the state level and comprehensive management districts on the substate, regional level.

V. Patterns of State Water Institutions

While the typology represents ideal types of state water resource organization, states approximate each of the types in actual practice. Based on a limited number of case studies, we shall review the organization of states that are representative of each of the four major categories in our typology.

Unified States

Relatively few states have developed integrated, centrally directed water management structures, but many appear to be moving in this direction. Centralized structures can be found in both the appropriation and riparian–permit states. Such structures are characterized by the integration of water quality and quantity management within one agency, the utilization of permits to regulate water use on a statewide basis, and considerable coordination between water resources planning and management. States that utilize this structure are Georgia, New Jersey, Maryland, and Wisconsin. In addition, there exist "transitional" states, whose structures are in a state of evolution toward integration and centralization. These states have centralized water management functions within a single regulatory agency, but state regional offices only possess authority to regulate water use in specific designated "capacity use" regions. These states include New York, North Carolina, Virginia, and New Mexico. We shall discuss the organization of two states as representatives of the centralized type: Georgia and New Mexico.

Georgia represents a highly centralized set of institutional arrangements for water resources decisionmaking. The state was riparian until the mid–1960s. Prior to that time, Georgia water law protected landowners adjacent to lakes and streams from unreasonable uses that would diminish water quality or quantity (Sellers 1978). At the state began to expand its population and economic base, municipal and industrial water use began to overshadow traditional uses. The existing riparian doctrine, adopted to protect domestic use and mill dams, was soon to be considered outmoded. A system of water rights that would lead to a greater level of certainty for users in terms of the extent of the right and the conditions under which these rights could be revoked without resorting to court proceedings was needed. Kates (1961) has noted:

> It has been observed that the administration of water rights by courts does not lend itself to the same efficiency in a state wide, comprehensively planned development of water resources as does such administration by an administrative board composed of experts.

This observation captures the thrust of the development of Georgia water law since 1964, as each successive act has led toward the development of a statewide allocation process which, as Sellers has observed, "tended to extract managerial discretion from the individual and place it in the hands of the state governing body" (Sellers 1978, p. 129).

The path toward state regulation of water use began with regulations to protect water quality. The Water Control Act of 1964 established the principal that there exists a public interest in all water uses, and therefore the state possesses the responsibility and authority to regulate waste discharges into

Georgia waters. The Ground Water Use Act of 1972 served as the mechanism for the state to assume managerial control of its ground water resources. This act established "reasonable beneficial use" as the basis for allocating ground water supplies in designated capacity use areas in excess of 100,000 gallons per day. In 1973, the General Assembly amended the law to extend the ground water permit system to the entire state. Finally, the Georgia Water Quality Control Act was amended in 1977 to establish limits on surface water use and a permit system. However, both the ground water and surface water permit systems allow for the exemption of agricultural uses, and withdrawals of less than 100,000 gallons per day (Kundell and Breman 1982, pp. 20–21).

An example of centralized water management in an appropriation state is New Mexico. Unlike Georgia, the integration of water management functions in New Mexico has been achieved through organizational linkages rather than by legislation and governmental reorganization. For example, water supply and water pollution control conditions are conducted by two different agencies, but the Office of the State Engineer, responsible for water supply regulation, is an ex officio member of the Water Quality Commission, which is responsible for establishing water quality policy, standards, and the Environmental Improvement Department's management strategies. In addition, the Office of the State Engineer staffs the Interstate Stream Commission, established to represent the state in negotiations and court cases on interstate and international water matters.

The appropriation, adjudication, and distribution of all surface water rights have been the continuous responsibility of the State Engineer since 1905. As early as 1910, New Mexico courts have upheld the right of the State Engineer to evaluate all facts and circumstances related to competing proposals for water use in the determination of the public's interest based on state constitutional provisions whereby unappropriated water is declared public property. Moreover, the New Mexico Constitution provides that beneficial use is the "basis, the measure, and the limit of the right to use water." This is insured by a permit system of water allocation administered by the State Engineer, a closely regulated water market for water rights transfers, and statutory procedures for the adjudication of all water rights. Ground water is regulated within designated basins that include most, of not all, major aquifers within the state. With the authority to regulate both surface and ground water, the State Engineer has developed and implemented a system for conjunctive management of interrelated surface and ground water supplies (Dewsnup and Jensen 1973).

New Mexico possesses a relatively weak structure of local water organizations. The traditional local water organizations, the acequias (community ditch associations), are extremely small in size and influence. Artesian conservancy districts were authorized in 1931, but the Ground Water Act of 1935 gave the State Engineer identical powers in areas where no district had been formed. Although reclamation projects and districts do exist within the state, their influence is limited because they are relatively small in size and scope. A notable exception is the Middle Rio Grande Conservancy District, which serves as a linkage between federal, state, and local interests in the management of an intensively developed segment of the Rio Grande Basin.

As in Georgia, the State of New Mexico has undergone successive executive branch reorganizations that have led to integration of natural resource management functions. However, New Mexico's State Engineer has successfully resisted attempts to shift statutory and constitutional powers to a departmental secretary. For example, in 1978 the Office of the State Engineer was exempted from the authority of the newly created Department of Natural Resources after persistent and persuasive lobbying (Cline 1978).

Federalist Structure

The federalist structure is essentially a two–tiered system, incorporating a relatively centralized administration and a second level of substate management authority. While many states have decentralized water management functions to field offices to effect some degree of regional water resource management, federalist organizational structure goes beyond the field office by establishing distinct regional or basin water management organizations with substantial management authority and separate budgets and tax bases. Florida and Nebraska's water management structure exemplifies this form.

Florida's hydrology is a study in contrast. For much of the state, excess water and drainage pose significant management problems, while other areas are subject to periodic drought. Prior to World War II, the major thrust of water management activity was the disposal of excess surface waters, especially in southern Florida. With the construction of drainage works, hundreds of small drainage districts were created along with the large Everglades Drainage District. Two other large multipurpose districts were formed, the Central and South Florida Flood Control District in 1949 and the Southwest Florida Water Management District in 1961. The latter district covers about one–fifth of the state (Blake 1980, pp. 166–94).

In line with the evolution of water districts were changes in state water law and administration. The state considered, then rejected, a change from common law riparian to a doctrine of prior appropriation. Then, in 1957, the Florida Water Resources Act established a statewide administrative agency responsible for the development of the state's water resources. The agency was empowered to issue permits for surface and ground water use and to generate rules for water conservation in areas of severe water withdrawals (Maloney 1978, pp. 25–26).

During the 1960s and 1970s, Florida was the scene of rapid growth and development as well as environmental conflicts. Controversies included the Cross–Florida Barge Canal, the threatened over–drainage of the Everglades National Park, and the endangerment of critical water recharge areas, such as the Green Swamp near Tampa. These crises, and other serious water problems created by droughts in 1970 and 1971, brought on the realization by many that the state had neither policy nor administrative procedures to manage effectively the severe resource problems facing a rapidly growing state with a fragile ecosystem (Blake 1980, pp. 195–222).

In 1971 Governor Askew called a Governor's Conference on Water Management for South Florida. The conference soon broadened its mandate to consider land and water management for the whole state. New actions were recommended and the governor established a Task Force on Resource Management that developed two notable legislative acts: the Environmental Land and Water Act of 1972, and the Comprehensive Planning Act of 1972 (DeGrove 1978, p. 69).

The Water Resources Act granted broad and sweeping powers to state and regional agencies to plan and to regulate virtually all surface and ground water activities in the state. The administrative framework called for the coordination of state planning and management activities with those of five substate water management districts, which also were authorized by the legislation. The districts covered the entire state and policymaking functions were assigned to nine–member citizen boards, with members appointed by the Governor (DeGrove 1978, p. 72). While all districts have a local tax base, only four regulate water use through a permit process.

In 1975 the Environmental Reorganization Act consolidated water quality and quantity functions on the state level. Three agencies previously had responsibilities for the regulation of surface and ground water activities. Water use functions were moved from the Department of Natural Resources to a new Department of Environmental Regulation, which effectively integrated water quality and quantity planning, management, and regulation. The five substate water management districts operate under control of the Department of Environmental Regulation, which has regional offices co–located with the water management district offices. This allows for a "one stop shop" for both water use and quality permitting on the regional level.

Florida's water management districts, therefore, offer a unique mix of water management functions. In respect to water law that governs the allocation of surface and ground water, they act as instruments of state government, but they also function as local sponsoring agencies for federal water projects. Because they operate on the basin level, they have been able, through research and development programs, to produce appropriate expertise to manage the highly variable and complex water problems within their jurisdictions. The policy board members are appointed by the governor, and they are therefore to a certain degree insulated from local political pressures. However, because their revenues are generated by a local tax base and policy board members must reside within the districts they represent, the water districts are able to maintain a degree of independence and responsiveness to local concerns. As DeGrove has stated, the water districts are neither fish nor fowl. They are too independent and diversified to be categorized as instruments of the Department of Environmental Regulation or the Office of the Governor. But, because their policy boards are not locally elected, neither can they be classified as local governmental entities. They can only be defined as regional institutions, which occupy, in a political and administrative sense, the middle ground of intergovernmental relations between state and local government (DeGrove 1983).

Nebraska's water management institutions possess similar regional characteristics. As in Florida, executive branch and environmental reform legislation of the 1970s led to the creation of autonomous regional organizations with some measure of state agency control. As one of the top three states (with California and Texas) in terms of dependence on ground water for irrigated agriculture, the Nebraska State Legislature responded to increasing problems of overdraft and ground water quality with the Nebraska Ground Water Management Act of 1972. The activities and jurisdiction of 150 single–purpose water districts were assumed by 24 Natural Resource Districts (NRDs) organized according to the major hydrological regions of the state. These districts are served by locally elected policy boards, whose rulemaking is subject to review and oversight by the state (Aiken 1979). However, while state enabling legislation allows for the establishment of ground water control areas by the NRDS, only three are engaged in regulatory activities.

Other states have attempted to regionalize on a more limited basis. For example, New York and Wisconsin have divided their states into regions, with agency regional offices providing water management functions. North Carolina, South Carolina, Virginia, Arizona, and Kansas have enabling legislation for the creation of regional management entities, either legislatively or on the basis of local initiative, but none of these states appear to allow for the degree of autonomous regional management that can be found in Florida or Nebraska.

Fragmented Structure

The third organizational type is characterized by a fragmented state level organization. There are many examples of fragmented systems, particularly in regions of the United States where water resources do not play a critical role in state politics. However, two examples exist in the arid West where water resources are politically controversial: Arizona and Colorado. Colorado is illustrative of policy and administrative fragmentation among the branches of state government and Arizona represents a fragmented structure undergoing reorganization toward centralization.

In Arizona, administrative responsibility for water management has been shuffled among various state agencies. The State Water Commission was enabled in 1912 upon statehood, but water resource functions were transferred temporarily to the State Land Commission from 1923 to 1925— then permanently in 1942. A Division of Water Resources was created in the Land Department in 1956 and this office eventually became a cabinet level department in 1980. Citing the transfer of duties of the Underground Water Commission to the State Land Department in 1954, Mann concluded that administrative reorganizations of water institutions in Arizona are often a facade for the destruction of undesirable programs (Mann 1960, p. 119). Cortner (1977) has observed that state water policy and administration has largely played an instrumental role for dominant federal, private, and local interests. In both Arizona and Colorado, the power of state water institutions is compromised by the authority of the courts to adjudicate disputes and allocate water resources. In Arizona, court–appointed water masters administer the distribution of surface water in certain areas, and in Colorado, the rights of appropriators are determined by district–level water judges. Recent state legislation in Arizona has led to the development of state authority to regulate ground water through the creation of capacity–use areas called Active Management Areas (AMAS). To date, four AMAs exist, which are managed by area directors appointed by the Director of Water Resources and advised by unpaid councils appointed by the governor. This provides for a degree of central control, but to date the implementation of ground water management by the AMAs is limited by the state legislature's reluctance to fund them adequately. Thus, on paper Arizona appears to be a fragmented state in transition, but that has yet to be realized in day–to–day management decisionmaking.

The absence of a measure of central control to temper the vagaries of local management is especially evident in Colorado, where no system was devised to keep statewide records of decreed priorities until 1969 (Dewsnup and Jensen

1973). A review of litigation also indicates that Colorado courts have tended to limit the discretion of state water management agencies. For example, the Colorado Supreme Court has held that the attempt by the State Engineer to shut down wells presumed to be responsible for depletion of the Arkansas River amounted to arbitrary and capricious action (Chalmers 1974, p. 65). Further dominance of the judicial branch in water management is reflected by the fact that the state is divided into regional court districts, with judges acting to settle water allocation disputes between local users.

Home Rule Structure

The last category in the typology is the home rule structure. This system approximates the type of community–state relationship embodied in the Maass-Anderson model of strong, independent local organization and a fragmented state organization. A state that best fits this example is California. Water resources policy and administration functions are shared by several agencies on the state level. The Department of Water Resources has general responsibility for water resources policy and administration. It has constructed and manages the State Water Project and sells water on a long–term contract basis to local water districts. However, the State Water Resources Board serves as the regulatory arm of the state. It administers water rights and water quality management programs, and provides oversight of state and federal wastewater treatment facilities, grants, and other water quality programs. The California Water Commission acts in an advisory role in respect to the Director of Water Resources on policy matters and approves all rules and regulations of the Department. It also provides a budget review process, reports annually to the Department and state legislature on the operation of the State Water Project, participates in eminent domain procedures, and presents its position on federal project appropriations (Goodall and Sullivan 1979, pp. 210–212).

Underlying this state organization of multiple–agency water management is a vast and complex water industry composed of numerous municipal, private, and mutual water organizations. Approximately one thousand special purpose districts were created as political subdivisions of the state by either general or special legislation. Most of these districts have been granted the power to levy taxes, exercise eminent domain, issue bonds, enter into contracts, construct works, and charge for services (Goodall and Sullivan 1979, p. 208). The result is that state and local organizations manage water resources within the context of separate spheres of influence. The state is responsible for the regulation of water quality, custodial management of surface and ground water rights, and large–scale water projects, while local organizations have been free to develop water supply management practices. The result has been a lack of uniformity in surface and ground water use that has made it difficult to coordinate water quality, surface, and ground water management except in localized areas where local interests have allowed such integration to occur.

VI. Analysis

Investigation of the status of water resources management among the states has provided data that renders the Maass/Anderson–Wittfogel debate of limited relevance. Neither authoritarian central control nor grass roots democracy is the norm for water resources management among the American states. Critical surface and ground water problems, the influence of federal water quality programs, historical development of the state role in active water resources management, and the decline of the traditional federal role with its emphasis on developing local, rather than state–level constituencies, have been factors that have led to considerable experimentation among the states in balancing the necessity for a measure of central control with political expediency for direct local participation in water resource management decisionmaking.

While much variation does exist, the overall direction of state water resources management since the early 1970s has been toward integration and centralization of managerial control. This process appears to be evolving at a faster rate among those states that do not possess an infrastructure of local water management organizations. Indeed, we would argue that the unified structure can only evolve in states that either have no local water organizations, as exemplified in the southeastern region of the United States, or in states with very small and highly specialized districts. In states with an infrastructure of local organizations, federalist structures have evolved largely because water problems were perceived to be of critical statewide concern. However, in some states, local organizations have retained sufficient political influence to resist attempts to develop federalist structures.

Contrasting California with Florida is illustrative. Local water management districts have played important roles in the development of both California and Florida, albeit for different water purposes. In California, the development of extensive irrigation and municipal water supply systems was essential for the economic development of the state. In Florida, water abundance was the initial problem and the creation of numerous local water drainage organizations by land speculators was the immediate response. In the ensuing years, Florida's local organizations were gradually consolidated into a few large areawide systems, which provided the foundation for a two–tiered federal system of water management. This change in water management structure in Florida was the consequence of a sweeping environmental reform movement that led to comprehensive changes in the manner in which land and water resources are managed in the state. The environmental movement in California, however, has not focused on water management as an issue critical enough to reform its management structure, and agricultural interests have continued to be successful in maintaining a highly decentralized water industry.

The importance of local water organizations is also illustrated in those states that are going through a transition from one type of water allocation system to another. For example, Arizona combines a relatively decentralized management structure for surface water with a recently centralized ground water management system. Hansen and Marsh (1981) attribute this change to several key factors including pressure from the federal government and the gradual loss of political influence of agricultural interests in the face of increasing political importance of cities and industries.

The Arizona pattern is similar to processes that are occurring in transitional states in the Southeast. For example, both North Carolina and Virginia have been experiencing an alteration of their basic economies from agriculture to industry. Like Arizona, these states have experienced a shift away from traditional forms of water management toward a centralized administrative process.

However, unlike in the past, agriculture may promote, rather than hinder, this process in the future. This is due to recent consolidation of agricultural lands into large "superfarms," and an increasing reliance upon irrigation. The necessity for certain, long–term water supply sources in combination with limited prospects for new federal reclamation projects has given agricultural interests the motivation to support legislation that will broaden state authority to regulate water use.

It is anticipated that Georgia–type systems will evolve on the state level in the absence of viable local water management organizations. In those states where water resources have been developed primarily on the local level through the creation of water districts, state water management systems will resemble either a federal or home rule–type structure. In their absence, a state will either remain fragmented or, if significant demands emerge, evolve into a centralized structure.

A second factor that accounts for variations in water resources management is the level of competition for existing water supplies. Our comparison of humid eastern states with arid western ones partially supports the Wittfogel hypothesis, but with some necessary modifications. It is not water scarcity per se that leads to the concentration of state power. Rather, the concentration of the power of the state appears to be a function of competition among different user groups over existing supplies. In New Mexico, the centralization of water allocation decision-making was instrumental in the reallocation of water resources from the northern part of the state to the south, or from one set of users (Hispanic) to another (Anglo). In Georgia, centralization of water allocation was a response to conflicts among different classes of users. In the absence of a viable local network of agricultural water users, the state assumed reallocative responsibilities. The Maass/Anderson hypothesis about the strength of grass roots democracy via decentralized water management appears to have been operative in states where irrigated agriculture is a dominant water use and farmers have organized into water districts to promote federal water projects. Water districts provide local interests with an organizational base to successfully oppose attempts to integrate water resource management functions on the state level.

A third factor appears to be the extent of external pressure on the state's water resource base. This is evident in at least four states: New Mexico, Arizona, North Carolina, and Virginia. In New Mexico, the State Engineer has played a strong leadership role in maximizing the flow of water resources into the state and minimizing the flow out (Ingram 1969). Pressures from the federal government were instrumental in creating the conditions for the centralization of ground water management in the state of Arizona. In North Carolina and Virginia, significant interstate water conflicts may not be resolved until each state centralizes its water management system sufficiently to allow for the development of an interstate regional water resource plan.

VII. Conclusion

This analysis indicates that great diversity exists among the American states in respect to the organization of water resource management systems. This diversity is attributable to the great variety of water resource conditions, problems, and political histories among the states. Yet within this diversity, distinct patterns emerge. First, water resources decisionmaking is increasingly agency based. Administrative rules and rulemaking are being substituted for political and judicial decisions. Second, a trend among the states is toward centralizing and integrating water resource management functions. This process is multidimensional. Water resource management functions related to water quality and quantity control are being concentrated within single agencies on the state level. These agencies either implement administrative rules governing pollution control and water use directly, or possess authority to supervise the decisionmaking processes of a variety of regional or local management organizations.

The process of centralization can be attributed to a number of factors. These include the ever–increasing complexity of contemporary water resource problems and changing political climates within the states. Urbanization, industrialization, and the decline of federal leadership in water resources development have eroded the political base of once dominant interests that benefited from decentralized water resources policymaking. Contemporary water problems, such as ground water pollution, conservation, and the necessity to allocate dwindling supplies among a variety of competing uses has necessitated the development of decision structures that optimize efficiency and promote water quality standards, rather than maximize the distribution of the benefits that large–scale, federally sponsored water resource development projects of the past have promoted.

References

Aiken, J. D., and Supella, R. J. "Groundwater mining and western water rights law: The Nebraska experience." 24 South Dakota Law Review 656, Supp. 1979, pp. 607–48.

Blake, Nelson M. 1980. *Land into Water—Water into Land: A History of Water Management in Florida.* Tallahassee: University Press of Florida.

Caulfield, Henry P., Jr. 1975. "Let's dismantle (largely but not fully) the federal water resources development establishment, or the apostasy of a longstanding water development federalist." Paper delivered to the National Conference on Water, Washington, D.C., April 22–24.

Chalmers, John R. 1974. *Southwestern Groundwater Law: A Textual and Bibliographic Interpretation.* Arid Land Reserve Information Paper No. 4, University of Arizona, Office of Arid Lands Studies.

Cline, Dorothy. 1978. *Reorganization of the Executive Branch of State Government.* Albuquerque, New Mexico: Division of Government Research.

Cortner, Hanna J., and Berry, Mary P. 1977. *Arizona Water Policy: Changing Decision Agendas and Political Styles: Hydrology and Water Resources in Arizona and the Southwest,* vol. 7: 7–14.

DeGrove, John M. 1978. "Administrative systems for water management in Florida." In W.E. Cox, ed., *Legal and Administrative Systems for Water Allocations and Management.* Proceedings of a regional conference held at Va. Polytechnic Institute and State University, Va., pp. 68–90.

DeGrove, John M. 1983. Personal interview, 24 March.

Dewsnup, Richard L., and Jensen, Dallin W., eds. 1973. *A Summary Digest of State Water Laws.* Washington, DC: USGPO.

Goodall, Merrill R., and Sullivan, John D. 1979. "Water District Organization: Political Decision Systems." In Ernest A. Englebert, ed., *California Water: Planning and Policy.* Davis, Ca.: California Water Resources Center, pp. 207–228.

Hansen, Scott, and Marsh, Floyd. 1981. "Arizona ground–water reform: innovations in state water policy." Paper for the Groundwater Management Symposium, Western Social Science Association Annual Conference, April 23–25.

Howe, Charles W. 1980. "The coming conflicts over water." in *Western Water Resources: Coming Problems and the Policy Alternatives.* A symposium sponsored by the Federal Reserve Bank of Kansas City, September 27–28, Boulder, Colorado: Westview Press.

Ingram, Helen M. 1969. *A Case Study of New Mexico's Role in the Colorado River Basin Bill.* Albuquerque, NM: Institute for Social Research and Development, University of New Mexico.

Kates, Robert C. 1981. *Georgia's Ground Water Law.* Athens, Georgia: Institute of Government, The University of Georgia.

Kundell, James E., and Breman, Vickie A. 1982. *Regional and Statewide Water Management Alternatives.* Institute of Government, University of Georgia, Athens, Ga.

Maass, Arthur, and Anderson, Raymond L. 1978. . . . *and the Desert Shall Rejoice: Conflict, Growth and Justice in Arid Environments.* Cambridge, Mass: MIT Press.

Maloney, Frank E. 1978. "Florida Water Law." In Cox, supra. pp. 22–67. Dean Mann 1960. *Politics of Water in Arizona.* Tucson, Arizona: University of Arizona Press.

Radosevich, George E. 1980. "Better use of water management tools." in *Western Water Resources,* pp. 253–289.

Schooler, Dean, and Ingram, Helen. 1982. "Water resource development." *Policy Studies Review.* Vol. 1, No. 2, pp. 243–253.

Sellers, Jackie. 1978. "Alternatives for Managing Georgia's Water Resources," in Cox, pp. 123–153.

U.S. Water Resources Council. 1981. *State of the States: Water Resources Planning and Management.* Washington, DC: USGPO.

Wittfogel, Karl A. 1957. *Oriental Despotism: A Comparative Study of Total Power.* New Haven, Conn.: Yale University Press.

12

The Pacific Northwest: A Regional "Soft Path" Experiment

LAUREN McKINSEY

Introduction

The first phase of an exhaustive self–analysis of the Pacific Northwest's electricity future was completed in the spring of 1983. The Northwest Power Planning Council adopted a twenty–year demand projection and resource acquisition plan in fulfillment of a congressional mandate to guarantee adequate supplies of electricity at the lowest cost to power customers of Washington, Oregon, Idaho, and Montana. The preceding two years of Regional Council deliberations had seldom escaped the public spotlight, especially because of the simultaneous collapse of the ambitious nuclear supply system (WPPSS) under construction in the state of Washington.

There are two primary reasons why this attention–grabbing planning process should be of interest to analysts and policymakers outside the region. First, the Regional Council is an innovative multistate body which plans for a federal agency: the joint creature of Congress and the affected states, its authority extends to the Bonneville Power Administration (BPA). Constitutional lawyers and intergovernmental relations specialists will probably find the Council an interesting and fertile subject for years to come. Second, the Pacific Northwest Electric Power Planning and Conservation Act of 1980 (PL 96–501) that created the Council extends a strong institutional preference to conservation and renewable energy sources over thermal plants in planning new sources of power.

This article features the Northwest Power Plan as an energy transition experiment informed by the "soft path" paradigm of Amory Lovins. (Lovins 1976). Lovins introduced a lively dialogue to U.S. energy policy in the wake of the first oil shock. Debaters were the "soft path" proponents partial to conservation/renewable energy sources and the "hard path" defenders of conventional energy production systems. Lovins's principal message is that society must choose between the soft path and the hard path because they are incompatible in terms of both cost and ideology. Rarely, however, does the complex decision-making structure in the United States provide the opportunity for an explicit choice to be made. Policy changes are seldom more than incremental shifts that result from counterbalancing and compensating decisions by governments. The past decade of energy policy in the United States has witnessed equal measures of subsidizing traditional energy development, promoting synfuels and renewable resources, and encouraging conservation in the face of the plurality of interests (Goodwin 1981). The Northwest Power Plan might be an exception — an opportunity for such a clear choice to be made. The acid test is whether it can be accomplished through the complex intergovernmental mechanisms of American federalism.

Within this perspective, several other themes are interlaced. One is the chronic quarrel of public versus private power as the Pacific Northwest became one of the chosen battlegrounds. A second can be labeled the political economy of electricity production as it concerns the principles of costing and pricing power and how these affect in turn the demand for and supply of power. A third is the question of technocratic versus democratic participation in the decisions to build the system of power plants and distribution lines.

The Pacific Northwest: A Region in Transition

The Pacific Northwest is a natural region that coincides largely with the Columbia River basin. Among the people there is a sense of region engendered by the many blessings that arose from taming the river. The federal government provided institutional parameters for the region when it committed resources for Columbia River development in the New Deal. Ever since, the Pacific Northwest has been supplied with the nation's cheapest electricity. For better or worse, this cheap electricity has shaped the region's character.

The classic struggle between private and public power was rooted in the federal presence in the Pacific Northwest. Comparisons with the TVA are instructive. The Columbia's development was premised on multiple–use objectives to include flood control, irrigation, navigation, recreation, and electricity that would help an inland desert to bloom. The TVA also involved a complex river development with New Deal origins. BPA and TVA came from "different branches of the public power family" (Energy News, I, 5:12).

TVA became the model, the experiment to show that public power could provide vast amounts of cheap electricity to rural and urban dweller alike. This was easy to do because the public utilities were entitled to federal power at wholesale rates, rates held down because of the federal subsidies to electricity. TVA grew into a public monopoly in the production, transmission, and sale of power to more than a hundred local public authorities and became the largest system in the nation. BPA remained limited to wholesaling federal power in a mixed public/private system: both types of utilities build plants and sell power via the BPA transmission grid. Public utilities, however, have always had first crack at the wholesale federal power and are known as "preference" customers.

In practice, the real power of the two entities converged in the decades following their New Deal origins. This resulted from the growing dominance of the electricity mission—the "dam mentality"—in both regions (Davis 1982, p. 183). In the TVA, power production developed a life of its own and came to dwarf the other functions which the TVA was empowered to perform.

Likewise, hydropower production came to dominate multiple–use planning in the Pacific Northwest because cheap electricity was the key to steady economic growth. Sustained prosperity was linked to the continued availability of electricity in the minds of planners and consumers alike. The dams for hydropower provided the benefits most easily measurable, while costs were disguised, deferred or not easily agreed on. The electricity grid came to dominate the perceptions of the region.

The construction of each new dam was followed by rising demand for power. This pattern lasted for several decades and the power planners came to believe that it was a natural law—even after the river system had yielded its suitable hydro sites by the end of the 1950s. An ambitious Hydro–Thermal Power Program was introduced in the late 1960s in the State of Washington. It called for construction of twenty nuclear plants and two coal plants over two decades. In the absence of this program, it was widely believed that electricity demand would eventually exceed supply, thereby imperiling the link between electricity growth and economic prosperity.

TVA also moved from a hydro to a nuclear phase although it developed a coal system in between. Because of its unity of command and its autonomy from either federal or state agencies, TVA could act more decisively. In the Pacific Northwest, a great deal of orchestration was required among the public and private utilities, the BPA, and the many state, local, tribal, and federal actors. Nonetheless, this complexity did not prevent the region from developing equally grandiose nuclear system plans.

Expensive coal or nuclear plants can be acquired if the costs to consumers are disguised and deferred by using average or "rolled in" rates. This means that customers are shielded from paying the high cost of the more expensive power at "the margin" for which their demand creates the need: it is a rather elaborate pyramid game of pushing these costs onto future users. BPA provided "net billing" agreements to the sponsors of three nuclear plants (WPPSS 1, 2, and 3) that allowed their costs to be melded with the costs of cheap hydropower from existing dams. A large base of cheap hydro allowed for the addition of high marginal cost plants because average costs rose slowly. Only this subsidy afforded by using average rates made these expensive thermal plants seem to be viable. The need for some new thermal plants in the region was self–fulfilling because underpriced electricity induces additional demand.

The real cost of the new plants eventually would be felt just as it was with the lagged effect of higher oil prices. Reduced demand gradually undermined OPEC and so it was with the nuclear power plants. Unlike gasoline, however, the nuclear plants might have to be paid for whether or not they are ever used. Declining load growth meant not only that the new plants would not be needed when expected but that their construction would have to be reflected in higher rates as well. Consumers faced a vicious conservation paradox: using less electricity was going to result in higher rates because of the redistribution of fixed system costs.

Soon it appeared that the Hydro–Thermal Power Program had come to the rescue too late. Construction of the large nuclear plants would not be completed in time to meet the steadily rising customer demand. In 1976, Bonneville issued "notices of insufficiency" to its preference customers: it could not guarantee to meet their aggregate demand by as early as 1983. Even worse, the cost of the new plants was beginning to soar and concerns arose over how these costs would be distributed among BPA customers. Project sponsors turned to Congress to provide a method for guaranteeing financing. They were followed to Washington, D.C., by a chorus of voices that argued for three years over the shape of the Regional Act (Lee 1980).

The original legislation was introduced under the general impression that electricity was becoming "scarce" in this well–endowed region. This perspective largely ignored the distinction between shortages because of physical limitations on supply and shortages because of high levels of consumption spurred by low prices. This concept of scarcity was entirely consistent with prevalent thinking in the mid–1970s regarding the oil and natural gas problems that constituted the "energy crisis." Shortage of electricity in the Pacific Northwest meant a fear of denial to accustomed cheap electricity. Shortage per se could have been alleviated by redesigning rates; instead, the focus was on creating new supplies in the spirit of making the United States "independent" or "self–sufficient" in energy.

Customers of public and private utilities alike had become addicted to cheap electricity. Energy is consumed excessively if its price is set lower than its cost and federal sponsorship through loans helped the Pacific Northwest acquire electricity that in the Depression would not have been otherwise forthcoming. The low price encouraged consumption and the economy flourished as a result. The boom of cheap and increasingly cheaper hydropower made it easy for producers and consumers alike to overlook the unpaid costs of the cheap power.

The hydrosystem was expanded to nearly full capacity with virtual indifference to the deleterious impact on the fish runs. The loss of fish has a direct economic impact that was seldom explicitly compared to the benefits of the hydropower system expansion. The loss of fish and fisheries habitat is also an indirect measure of the ecological health of the entire river system. These externalities of hydropower production remained hidden, in part because of lag time in cause and effect.

The contest between private and public power is a complex and colorful story told elsewhere (McKinley 1952). One aspect important here is that the BPA provided room and common incentives enough for both types of utilities to coexist and to support the mixed system that evolved. BPA managed the multiplicity of power exchanges, both across state lines and in sales or purchases conducted with other regions. During the halcyon days of steady hydro growth, there was always expansion of consumer demand across public and private districts alike.

The private utilities were forced to augment their generating capacity with thermal power ahead of the public utilities. But when the time came, BPA created the net billing arrangements that allowed the public utilities to participate despite their lack of capital. For several decades the tension between the public and the private utilities remained relatively subdued and their differences held subservient to the common interest in expanding power sales—through the creation of self– fulfilling demand forecasts. Here was an experiment in selective socialism.

Now, with the tremendous cost overruns of the WPPSS nuclear system, the issue of who should pay has revitalized the debate over public versus private power. The so–called scarcity that inspired a plan to guarantee adequate electricity now is widely recognized as in great part excessive use of federal hydropower by preference customers. Beneath the rhetoric, the debate is not over total costs, but rather cost sharing. Today there is a widening consensus that the broad question of preference power will have to be reexamined if the region is to escape its dilemma.

The Regional Act and Regional Council

Complex motivations lay behind the compromises achieved in PL 96–501. The Regional Act meant many things to many people. It was supposed to: guarantee that the region would not "run short" of electricity; ensure that the costs of expensive new power plants would be underwritten by BPA; guard against overbuilding power capacity and allow only the development of cost–effective electricity; ensure to user groups continued access to electricity; expand access to "preference power" to some new users; provide a way for the states to rein in the powerful BPA; give citizens a greater say in policy; and account properly for the side effects of power production on fish and wildlife.

The set of choices constitutes a hard path/soft path dichotomy. One group of proponents in the congressional battle, led by WPPSS sponsors, wanted "business as usual": to strengthen the BPA mandate and obtain assurances that the nuclear plants would be built and markets would be guaranteed. An alliance of environmentalists argued that soft path principles ought to guide selection of incremental supply alternatives. They were concerned primarily with restraining BPA's proclivity to build a bigger power distribution complex and successfully sought a mandate to conserve.

The interlocking utility network (publics and privates) had its own substantial prerogatives to protect but it realized that federal authority—and BPA resources—would be necessary for an acceptable solution. Downstream states had different motives than upstream states and Washington (public) and Oregon (private) were divided by their differences in utility systems. But the important point is that all parties recognized that the solution transcended any one state: there was no avoiding a federal role.

On the other hand, there was no interest in telling Washington, D.C., that the states could not manage their problem or in giving BPA a longer leash to solve it. The common interest in avoiding federal preemption resulted in agreement to the idea of the Regional Council. And despite the absence of equality in stakes among the four states, this common interest led to the egalitarian solution of equal sized delegations for the four states.

Several compromises emerged. The rates for long–term, interruptible power to direct service industries (DSIs) would rise in exchange for guaranteed access to power. Residential and farm customers of private utilities would receive some benefits of preference power through exchanges made possible by the higher DSI rates. BPA would now be mandated to acquire all the power required to meet the demands of public power customers.

In essence, the goal of the legislation was to guarantee adequate supplies of electricity but without upsetting drastically the pattern of actors in the regional system. The Regional Act was a compromise to this end.

The Regional Council is composed of eight members, two each appointed by the governors of Washington, Oregon, Idaho, and Montana. It was charged to create an independent forecast of electricity demand and a plan to develop the resources to meet the demand. BPA must follow the council's resource acquisition plan or return to Congress for special permission to depart from it. Senator Mark Hatfield called the act creating the Regional Council "the most important piece of legislation to affect the Pacific Northwest since the 1937 Bonneville Project Act" (Congressional Record, November 19, 1980).

The Regional Act gave explicit advantages to the soft path in selected respects. It instructed the Council to give priority to conservation first and renewable energy sources second ahead of coal and nuclear in the selection of resources to be acquired in the plan. In fact, conservation is to be given a 10 percent cost–effective premium against any other acquisitions. In addition, the Council is supposed to consider environmental costs whose exclusion in the past has given thermal plants a hidden advantage.

The Council adopted four load growth forecasts from a 1981 base of more than 15,000 megawatts. The high forecast assumes a booming economy and calls for average annual growth of 2.5 percent by the year 2002. This growth rate would necessitate the acquisition of an additional 10,000 megawatts of capacity by the year 2002. The low forecast assumes a modest economic recovery and calls for average annual growth of only .7 percent. Only 660 new megawatts of capacity would need to be acquired over the next twenty years. The medium–high forecast is set at 2.1 percent and the medium–low forecast at 1.5 percent. This range forecast approach is touted as a vast improvement over the single point forecasts used by utilities which, by implication, produced the mess.

The plan stacks different resources or combinations of resources to meet the expected range of forecasts. Conservation meets virtually all of the load growth in the low scenario and 50 percent in the high. Only some small–scale hydro is additionally required under the medium–low scenario. Some cogeneration and coal are required under the medium–high. Coal plants become significant only in the medium–high, accounting for about 33 percent of the load growth. Combustion turbines provided about 1,000 megawatts of "planning reserves" in the high and medium–high. Nuclear power is not included in the plan, not even the partially completed WPPSS 4 and 5 plants counting only the remaining costs to completion.

The Council's treatment of conservation would be the key to understanding how it viewed its own range of authority to make electricity policy in the region. It recognized that conservation could be achieved through a variety of means: regulations, subsidies, information programs, and prices or redesigned rates. These choices are surrogates in the chronic debate over planning versus the market. Reliance on prices or rates enhances consumer freedom of choice and eliminates having to choose between government regulations, incentives, and penalties to achieve supply and demand balance. Reliance on governmental tools allows targeting of benefits to preferred institutions and groups.

Higher rates are the self–evident market solution: incremental costs should be assigned to incremental users. This would defer the "need" for more power plants. But, to induce conservation by rates is to ration supply and this approach goes against the grain of the philosophy that the region should have all the electricity it "needs" for economic growth. Conservation incentives to encourage the substitution of energy efficient measures eventuate in higher rates just the same.

Rate design changes would rearrange the current distributive outcome in the region. The motivations behind the Regional Act need to be recalled: imputed shortages of electricity rationalized congressional action but the real fears were directed at expected shortages of traditionally cheap electricity. Rate design changes could have alleviated the problem of shortage: higher rates could have been absorbed because the Pacific Northwest has enjoyed very cheap rates compared with other regions. But users, municipalities, and public utilities vied for continued access to the cheap electricity.

The formidable question of "deregulating" electricity utility industry pricing, therefore, was abandoned for a more constrained set of alternatives: "The Council finds that three primary tools are available for achieving conservation: incentives, regulatory standards and rate designs. To give utilities and state and local governments as much freedom as possible this plan favors incentives" (Draft Plan, pp. 2–4).

Rates and regulations are to be used selectively. The Council recommended that rates be used to help balance supply and demand in extraordinary cases. One case is during critical water years when power might have to be imported for the region. Another is if programs of conservation fall short; then, those users making the least progress are to be penalized by a surcharge. As for regulations, building codes are to be required for new construction on the grounds that the greatest potential for conservation is in new homes and businesses and it cannot be captured fully once they are built.

How is the compromise philosophy of the Northwest Power Plan to be regarded in terms of the "soft path"?

The Soft Path Paradigm

Lovins criticized the predominate supply side emphasis in energy planning and focused on the electrified economy as the classic example of mismatching energy sources and needs. Lovins and others demonstrated that economic growth does not depend on ever greater consumption of finished energy (Stobaugh 1977; Ross 1981). Elements of the soft path paradigm relevant in the subsequent analysis of the Northwest Power Plan follow.

DEMAND. Projections of energy "requirements," especially electricity, have been inflated. Future energy demand is almost certain to be far lower than the consensus of self–serving industry and government estimates. Lovins has demonstrated that the "high" estimates consistently turn out to be "low" estimates within the span of just a few years (Marshall 1980).

CONSERVATION. An expected "shortfall" in premium energy supplies was used to rationalize the preoccupation with energy "security" during the 1970s. Conservation could not make up this shortfall and was repressive as well according to hard path proponents. But improvements in energy efficiency can greatly reduce the magnitude of the shortfall and they already have undermined the case for developing the preferred supply side alternatives of nuclear, coal, and synfuels (Bethell 1980).

POWER PLANTS. Renewable energy sources supplant conservation to close the window on the need to plan for a major expansion of the fossil fuel plant system. Some coal systems will continue to be built during the transition period—as the "bridge" to a society using renewable fuel flows instead of depletable fuel stocks.

MARKET CHOICE. Conservation and renewable energy sources are competitive against fossil fuel systems in a completely unsubsidized market. Some subsidies for renewables are required during the transition until the time when all fuels should be made to stand on equal footing and none should be given the artificial support that made nuclear attractive (Norman 1981).

EXTERNALITIES. The soft path has environmental impacts but they are more "benign" than hard path choices. The failure to quantify or price fully the externalities of hard path choices has helped to give them a preference in the energy market (Ramsay 1979). Soft path decisions bring together the region of benefits and the region of costs.

SCALE. Investment in small–scale renewable projects is preferable because it induces a greater rate of capital turnover within the local community. Large–scale domestic fossil fuel plants have been touted as the key to "energy security" but they actually are more vulnerable to sabotage and system failure: a variety of decentralized, small–scale plants could prove to be more secure in this respect.

SOCIAL IMPLICATIONS. The soft path is superior to the hard path on technical, economic, and environmental grounds alone. But it is further preferable because it promotes greater popular participation, social diversity, and individual self–reliance. Lovins calls for a resurgence of the conserver spirit in America but is willing to defend his proposition solely as a technical fix.

SYSTEM INCREMENTS. The soft path does not call for dismantling the current energy supply system. Instead, it recommends those additions or increments that take maximum advantage of the considerable investment in the present electricity grid, pipeline network, and other parts of the energy distribution system. The soft path values sunk costs and is an incremental least cost strategy.

MUTUAL EXCLUSION. Proponents of fossil fuel systems portrayed the soft path position as unreasonable because its major premise is that the two paths are mutually exclusive (Forbes 1977). But it was easy, patriotic, and misleading to argue that there is room for—indeed, need for—any and all energy developments that can help to solve the energy problem. Lovins argued against the mistaken luxury of promoting all energy developments equally, because of limited revenues for both research and development and capital construction: every energy dollar should be invested in the cost–effective source.

Conservation is the soft path cornerstone. A nearly universal criticism of the soft path is that it endorses and even praises a forced austerity: conservation means reduced energy consumption or, what is worse, doing without. There is frequent allusion to a return to the dreaded days of washboards and wood piles. Lovins responded that the end uses of energy do not have to fall in the soft path:

increased energy efficiency means accomplishing the same jobs with reduced energy consumption. Although rising prices play some part in raising the efficiency of energy conversion and use, conservation is still cheaper than most replacement fuels. Because increasing energy efficiency is a market response, the lowering of consumption by conservation is simply not equivalent to a socially engineered constriction of need.

The Plan as a Soft Path Experiment

The Conservation Cornerstone

Conservation was defined in the Regional Act as improvement in the efficiency of energy conversion or end use. It could be treated as independent of demand forecasts and, therefore, as a source of supply by displacement. Divorced from demand, conservation was capable of being compared along with other supply choices such as hydro, coal and nuclear to meet the cost–effectiveness test.

Treating conservation as an independent supply source is the way to avoid the soft path pitfall. Conservation seen as supply, rather than as reduced demand, is philosophically preferable because it conveys the impression that electricity will be made available to anyone who wants it as a result of displacement by anyone who wants to conserve. And, only cost–effective conservation is to be acquired in the plan. Conservation for its own sake is not a philosophical objective: cost–effectiveness is the only relevant property of conservation.

The plan skirts reference to forced conservation and tries to avoid having to acknowledge explicitly the conservation which comes from higher prices. "Conservation actions as a result of specific programs must be distinguished from conservation actions motivated by higher electricity prices" (Draft Plan: 3–36). But semantic ambiguity in the legislation sometimes forced the Council over the fine line between treating conservation as a price–induced reduction in demand and treating conservation as an increase in the efficiency of conversion and use. For example, the law refers to "conservation and efficient use of electric energy" as though conservation could sometimes be different from efficient use. And the Council had to deal with the question of implementing retail rate designs that would encourage conservation: the response to higher rates may reduce consumption without necessarily increasing the efficiency of electricity use.

How much is conservation worth? Within the present configuration it is worth more to the utility than to the consumer. Conservation may allow the utility to avoid the cost of expensive new generating capacity (since the utility is required to meet any load demand). So, conservation is highly valuable when it reduces load growth. But consumers, who pay lower average costs instead of avoided costs, do not value conservation as highly. How can conservation be made to be worth as much to the consumer? The Council's answer is by paying the consumer the avoided cost to conserve. The plan calls for acquiring 75 percent of all conservation that is cost–effective up to four cents per kilowatt hour equivalent. This is the "avoided cost" of the next resource acquisition after discounting for the 10 percent premium accorded conservation.

A major issue was raised over the reliability of conservation programs. As noted, conservation is to account for all of the load growth under the low–

growth scenario in the plan. Critics argued that the plan overestimates the rate of penetration expected of untested conservation programs. Success of the conservation program depends on orchestrating the activities of hundreds of thousands or even millions of consumers responding to a variety of programs: this kind of program is allegedly highly unpredictable compared to the reliability of centralized power systems to which planners are accustomed.

Another crucial issue was how much to invest in conservation incentive programs. A debate between "full" and "partial" financing emerged late in the Council's deliberations. Proponents of full financing argued that the test of cost–effectiveness legally and sensibly required the Council to acquire all of the available conservation at the avoided cost of new generating capacity.

Opponents were concerned about several issues. One was the equity of further reimbursing consumers for making rational, money–saving decisions like weatherizing their homes. It was argued that consumers will take a more responsible approach if they have some economic stake of their own in the project. Another issue was the logic of having the BPA conservation at a time when the combination of higher rates and reduced economic activity had turned shortage to surplus: why raise rates further (creating cash flow problems for BPA) to fund more conservation when reduced consumption has emerged as the villain in the plot?

Critics also argued that full funding is not necessary to achieve the desired penetration rates. This last argument is incongruous with the assertion that the plan's expected penetration rates and hence its conservation goals are too optimistic. But anything less than full financing at the level of the region's avoided cost would be a rejection of the soft path principle of treating all potential energy sources evenhandedly.

The Council chose to fund fully all conservation measures for low– income households. This approach may reinforce individual initiative and reward those most in need, but it is not certain that it is the most economically efficient approach to fund all conservation efforts as part of the price of electricity in the region.

The Logic of the Planning Strategy: Resource Stacking and Banking

The Council was bound by law to provide for the acquisition of all of the resources which are needed to meet BPA's load. At the same time they faced the mandate to acquire only cost–effective resources: by implication, only what is needed. So the Council was left with the technically complex and politically vulnerable task of balancing the risks against overbuilding and underbuilding. More accurate demand forecasts are one way; flexibility in resource acquisitions is another: "Certainty about the future does not come from the technical sophistication of a forecast. Instead, it comes from the flexibility and confidence one has in the array of resources available to meet any given conditions" (Draft Plan: 1–4).

The plan calls for no acquisitions of active solar, wind, geothermal or biomass in the next twenty years: only small–scale hydropower among renewables has been included. The Council recommended a few demonstration projects and further evaluation research. But at present these energy flow resources do not meet the cost–effectiveness test in a region with such a large base of cheap hydropower. Soft path proponents are disappointed with the plan's passive approach to renewable energy sources, failing to credit them with any intrinsic values: "no resource is inherently better than any other—they all produce electricity" (Draft Plan, 2–2). Nonetheless, there are some soft path overtones in the selection and stacking of resources in the plan.

The concept of the resource option is proposed to guard against overbuilding by incorporating risk management strategies. The region pays promoters the costs of the long but relatively inexpensive licensing process. A number of projects would then be "shelved" at regional expense until they might be needed. They could be completed expeditiously when the need for them has been triggered. It is hoped that half–built projects will not lie idle using this method. Monitoring load growth projections is supposed to guard against unnecessarily exercising any option. Options are an insurance policy in the event that the high load growth forecast is correct. This is risk avoidance common to private enterprise.

To address risks of underbuilding, the plan favors resource acquisitions that can be brought on quickly and/or in small increments. This practice reduces dependency on long–term demand forecasts. Conservation and low–head hydropower are favored because they can be acquired incrementally, in pieces as needed. Conservation also is "load following" because a greater amount will be available through incorporation in new houses, production techniques, et cetera if the economy booms. Cogeneration has the same quality: it becomes more feasible when industrial output expands. A new power plant cannot be built one megawatt at a time, but conservation can.

Flexibility and system compatibility, in addition to cost and lead time, are the main criteria in stacking resources to be acquired. Decisions on resource suitability were made keeping the nature of the hydro system in mind: "economics point toward getting maximum use out of the hydropower system while planning new resources that complement [it]" (Draft Plan: 1–3). Coal was given the nod over nuclear at the bottom of the list partly because of its shorter construction time and partly because it is more compatible with the hydropower system. This decision reverses the logic of the WPPSS planners.

In order to make nuclear plants viable they must be run as base load systems, turning the hydroplants into peaking systems and reversing their original purpose. In the Council's plan, one of coal's advantages over nuclear is that coal plants are fuel cost driven and can be used more selectively, not only as base load plants. The preference for coal over nuclear is consistent with Lovins's premise that new resources should be selected keeping in mind the maximization of the current system.

If renewable energy sources are regarded as the heart of the soft path, then the Northwest Power Plan seems to fail the test. On the other hand, the plan may spell the end of the nuclear era, which opens up long– term possibilities for renewables. More important, the criteria adopted in the plan—flexibility, scale, and timing—all lend themselves to eventual use of more renewables. With today's projected needs it is simply the case that conservation with a little help from hydro better suits the criteria. Once conservation is used up and constraints have stabilized energy consumption growth, it will be easier to find the necessary small increments in growth from a variety of small, decentralized sources. Putting nuclear, coal, solar, biomass, et cetera on the same footing is what the soft path is about.

The Fish and Wildlife Plan: Paying the Real Costs

Hydropower production dominated development of the Columbia River and the other New Deal purposes receded. The Regional Plan seems to begin a process of working backward from this one–dimensional river basin planning. Most notably the legislation required giving equal status to the fish runs and hydropower. A fish plan and a water budget were adopted ahead of the power plan. This is one of the crucial dimensions of the soft path approach: taking environmental externalities into account in energy production.

The Fish and Wildlife Plan affects the contribution that hydropower can make to the total system. It requires that fish access up and down the system be improved.One aspect is to spill more water over the dam to help flush fingerlings to the ocean, thus reducing the amount available to generate power by 550 MW. Some observers argue that this concession was made possible by the current surplus and is a nominal gesture. Nonetheless it could establish an important precedent. Operating the hydrosystem to minimize water fluctuations will improve the fish runs and other wildlife habitat as it reduces both power production and peaking ability.

Consumers are in for a shock whenever high–cost new power sources are mixed with low–cost old power sources to raise their rates. There is a proclivity, therefore, to buffer the impact by disguising the costs of the new power. Coal and nuclear have constituted the alternative resources in the Pacific Northwest since the 1960s and, as elsewhere, the negative externalities went largely unconsidered and unmeasured. These externalities—air quality, water quality, et cetera—were different from those of the hydropower system but neither were given adequate attention. The Council's plan appears to break no new ground in this regard despite a mandate to include those comparative environmental costs that are measurable in its resource portfolio. It has not been able to add externality costs for existing coal or nuclear plants.

The Grass Roots: Underlying Soft Path Values

The choice of mechanisms for implementing the Northwest Power Plan also is relevant to the soft path paradigm. Though critics often charge that conservation—as forced reduction of consumption—would be repressive, Lovins has argued that soft path decisionmaking instead could easily foster pluralism. The underlying values of the soft path are decentralization, social diversity, and individual involvement, which are contrasted with the technocratic elitism behind the expansion of the electric utility network.

There was widespread opportunity for participation in the formulation of the plan. The plan should "offer an orderly public process for deciding how the region will meet its power needs for the next 20 years" (Energy News, I., 9: 25). Ordinary citizens are to be brought into a process previously the exclusive domain of utility executives. A statistical and scientific advisory council was established with membership from the public sector, industry, the utilities, universities, and public interest groups. Meetings of the council were rotated among the four states of the region. Newspaper coverage of the council's work was extensive. Hearings were held on the draft plan and considerable written testimony was submitted.

Most of the public involvement in practice came from the interested public, not the average consumer or citizen. Representatives from the utilities, the direct service industry, builders, contractors and others with a direct economic stake had their say throughout. And the council's assumptions and choices were hidden deep in layers of reports and printouts from consultants using sophisticated econometric models. The actual process was one of interest group pluralism with a heavy aura of public technocracy.

The plan also makes a concerted effort to merge grass roots populism with private sector management practices:

> The range forecast and the portfolio of flexible resources is the Council's attempt to bring a more private sector–like management method and better control to the Northwest's troubled electric energy scene (Energy News, I., 9: 31).

The plan is replete with this duality: "Resource acquisition programs should use existing market mechanisms and organizations as much as practicable" (Draft Plan: 2-4). Business practices, markets, freedom, private sector management—these are to be the tools of a more open, public process. How reminiscent this is of President Roosevelt's goals for TVA as "a corporation clothed with the power of government but possessed of the flexibility of a private enterprise" (Derthick 1974: 32). Ironically, the call for more public involvement was directed primarily at the public elements of the system—the federal BPA and the Washington Public Power Supply System, which together inspired the fatally flawed hydro–thermal system.

The promotion of conservation through market incentives raises a major implementation challenge: promoting and coordinating the efforts of millions of "citizen generators" who have been legitimized as home–based power plants by the classification of conservation as a power source. One interesting irony of the plan is that to achieve an emphasis on smaller, diverse, and dispersed resources has necessitated a regional or more centralized planning and approval mechanism.

The Regional Dimension

Lovins argues that society cannot afford to walk both energy paths simultaneously. By implication a discrete choice must be made, one possible in a unitary system with a centralized government. But plural, incremental decisionmaking prevails in the United States, especially in energy policy (Rosenbaum 1981: ch. 3). There tend to be countervailing choices, compounded by the checks and balances of divided governments and further complicated by layers in the federal system. Little attention has been given to the question of how a society collectively makes a soft path choice through the dispersed institutions of federalism (Washington 1980). This is important because the soft path decisions in fact are made by millions of people in the aggregate through proper structuring of market choices.

Is the switch to the soft path made easier or more difficult by the existence of federalism? It should be more difficult in the sense that so many voices of authority are involved, but easier in the sense that some parts or regions are prepared to move more quickly than others. Federalism is an appropriate soft path conduit according to the social laboratory principle that one region can learn from either the negative or the positive experiences of another. The Regional Council provides a model of shared jurisdiction by federal and state agencies in which a new institution is meant to facilitate decentralized choices. It may show that an intermediate collectivity can exist—below the layer of the national government yet above the level of the neighborhood, community, or state. How does the Regional Council measure up as a regional organization?

Regional organizations in the United States arise from a mutual interest by two layers of government in managing conflict over jurisdiction. Various taxonomies have been created by analysts. One approach is to divide the state inspired from the federally inspired organizations. The most common regional entities are those spun from Washington, D.C., as a method of focusing or enhancing federal authority. Infrequently, neighboring states have organized themselves as regional entities under a provision of federal rules or laws, either to capture federal financial benefits or as a defensive measure against perceived federal encroachment.

Top–down regional organizations generally do not require state approval, although the federal government often legitimizes its initiatives by inducing participation. Bottom–up regional organizations uniformly require federal approval, although the degree of ongoing structural involvement by federal agencies may vary widely. There are few true regional organizations in the confederal sense, whereby power is conferred jointly by both layers of government and, more importantly, selectively binds them both (Derthick 1974).

Regional organizations also can be classified according to the extent of their authority. They range from organizations with only informational or advisory functions to ones with management duties or regulatory powers. Entities with vague planning mandates fall in between. Size of budget and independence of revenue sources are vital measures of the organization's power. In practice, only a very few regional organizations exercise real power. The Regional Council has the following characteristics.

First, it is state–inspired, although as a defensive measure against the proposed strengthening of a federal agency, the BPA. A compromise acceptable to the states, to the federal government, and to the many interests in the region was forged in Congress and ratified by an interstate compact.

Second, it seems to be a true regional or confederal body, simultaneously accountable to constituents at both the state and federal levels. Unlike some regional organizations that are constituted of federal and state representatives on a joint management team or planning task force— such as the joint coal leasing program in the West—the Regional Council is a separate formal entity jointly constituted by authority of both states and the federal government. History may accord the Council a special place in American federal experience for this reason alone.

Third, the Regional Council's scope of power remains to be tested by experience. BPA can deviate from major aspects of the Regional Plan only by returning to Congress for an exemption. The Council also has some financial independence because its budget is tied to BPA rates. On the other hand, the law imposes few action–forcing regulatory mechanisms on either the BPA or the states. The four states have to be persuaded independently that it serves their interests to adopt the plan's features in such areas as retail rates, building codes, and plant site banking.

Further comparisons with the TVA are instructive. TVA is usually regarded as a regional organization while the BPA is not: "By any tangible measure such as revenue, employment, or capital assets, the TVA is by far the most important regional organization in the United States" (Derthick 1974, p. 18). Yet Derthick's comprehensive treatment makes no mention of the BPA even though it is comparable in the measures she suggests.

The TVA is a creature of the federal government, which encompasses parts of several states and interacts with state and local governments in many of its activities. It is a superagency, not a regional government that supercedes state authority. Its distinguishing feature is its status as a public corporation, which provides a great deal of autonomy from the federal government and its line agencies. Moreover, it is a single agency with multiple–purpose management missions and regulatory powers. Thus, it is able to coordinate in one region the activities belonging to a multiplicity of federal agencies in other regions: navigation, flood control, reforestation, erosion control, industrial development, and hydropower. This incorporates the principle of scientific management stemming from the Progressive era: an authority extending to all uses of the water, providing national planning for an entire watershed.

The BPA differs in principle and practice. The difference between "authority" in TVA and "administration" in BPA is revealing. BPA also covers several states and interacts with many state and local institutions. Because, like TVA, it generates most of its own operating revenues, it is regarded by many as a superagency, but it is, in fact, accountable within a federal agency—the DOE. Most important, however, it is unlike TVA because it has a single function—on paper at least—which is to market hydropower from federal dams. The BPA mission, therefore, is far more circumscribed in law.

In the Pacific Northwest, the centrality of dam construction heightened the BPA's authority even though it was meant to have no say over companion functions such as navigation or flood control. And the BPA's role in building a complex distribution grid for power sales and exchanges together with creative financing arrangements meant that it did not require TVA's powers such as bonding or power plant construction to accomplish virtually the same end. The BPA's role as a marketing agency evolved into a strong de facto regulatory presence, especially because federal dams provide half of the region's electricity.

The mixed public/private system in the Pacific Northwest helped inspire the idea of a Regional Council. The cleavages of public/private power are superimposed on state boundaries. That institutional complexity may be a major impediment to the implementation of the council's plan in the next two years. In contrast, the simplified system of the TVA enabled a quicker recognition of the problem. TVA used unity of command and its multiple–purpose powers to implement conservation rapidly once it had received direction from the president to do so (Durant, et al. 1983). No new regional framework was required. There was no shift either toward a greater or lesser centralization of authority.

The creation of the Northwest Power Planning Council as a regional organization resulted from fears that the BPA either could not or would not be able to implement an energy transition. Advocates of bailing out the nuclear supply system wanted to broaden BPA's authority to guarantee purchase of the power. Advocates of conservation wanted a check on this tendency and succeeded in gaining the Council in place of a stronger BPA.

Major statutory responsibility for formulating a plan has been completed but the task really has only just begun. The plan now triggers an elaborate, dispersed planning process that includes state and local governments, utilities, industry, and citizen groups. Success of the plan will depend on the orchestration of efforts by actors who need not always go along.

The states hoped to use the Council as a counterweight to the BPA. But with completion of the plan, the states are likely to become increasingly concerned that implementation of the plan unduly affects decisions normally made by them in areas such as facility siting, rate regulation, and building codes. The Council's power to plan may be squeezed by BPA from above and by the states from below. It may be years before the four states alter their regulations to be mutually consistent with the plan. By then conditions may have made the plan moot.

There is a precedent already. The planning process in the Pacific Northwest has already dramatized the liability of lag time with institutional mechanisms of this scale and complexity. The multiplicity of actors and crosscutting motives in the struggle for legislation delayed creation of the plan until it had been overtaken by events. The manifestations of the problem had shifted as the Council was being formed. Spiraling rates had turned shortage to surplus. Instead of a conservation plan, the region's most pressing need is for a plan to market the surplus that no one in the region will now buy at these higher prices.

Should the Regional Act be viewed as a centralization or a soft path decentralization of electricity planning for this region? Either interpretation is arguable. Through the Regional Council the states have thrown a few ropes over the semi–autonomous federal agency that has dominated development of the system. On the other hand, BPA's powers are enhanced in some respects. It is no longer limited to being a wholesaler of federal power; its de facto power broker role has been legitimized through extensive new authority to acquire electricity resources of all kinds. And while the states together have some clout over the federal agency through the Council, their perspective is likely to be reversed when they are individually requested to revise their laws to conform to the Regional Plan.

The final irony is that the proclivity to accommodate both federal and state presumptions of jurisdiction in superimposing new authority on a natural region has produced an elaboration instead of a simplification of the planning process. Despite allusions to greater market choices and freedom, there is more central direction in this new approach: government involvement begets government involvement. The paradox of the soft path is the apparent need to confirm a new, encompassing layer of authority in order to manage the diversified response entailed in the soft path.

Conclusion

On conservation criteria, it would seem that the Regional Plan is a major step down the soft path. If that is true, it might also be argued that the soft path is not a divergent path after all and that Lovins is no visionary. In retrospect, the wisdom of energy efficiency should have been self–evident. Conservation has moved quite rapidly from a drastic to a conventional, consensual choice. As a result, the Lovins contribution tends to be minimized, forgetting the great debate of the 1970s. In fact, in a time of surplus (Tucker 1981), the plan's stress on paying for conservation must be seen as quite remarkable. The Pacific Northwest power planning experience cannot simply be copied by other regions. In an important sense, it is tied to the unique hydropower properties in the region. But many of the lessons can be generalized for other regions and a variety of suggestive implications for federalism have been noted.

References

Bethell, Thomas. 1980. The energy crisis, how to keep it going: Synfuels. *The Washington Monthly* (October): 11–20.

Bonneville Power Administration, Daily Newsclips.

Cook, James. 1982. The great conservation fallacy. *Forbes* (May 10): 156–57.

Davis, David Howard. 1982. *Energy Politics*. 3rd ed. New York: St. Martin's Press.

Derthick, Martha. 1974. *Between State and Nation*. Washington, D.C.: Brookings Institution.

Durant, Robert F.; Fitzgerald, Michael R.; and Thomas, Larry W. 1983. When government regulates itself: The EPA/TVA air pollution control experience. *Public Administration Review* 43 (May-June): 209–19.

Forbes, Ian. 1977. Energy strategy: A time for realism. *Electric Perspectives* 77: 27–35.

Goodwin, Craufurd N., ed. 1981. *Energy Policy in Perspective*. Washington, D.C.: The Brookings Institution.

Lovins, Amory. 1976. Energy strategy: The road not taken? *Foreign Affairs* 55 (October): 65–96.

Lee, Kai N. and Klemka, Donna Lee. 1980. *Electric Power and the Future of the Pacific Northwest*. Seattle: University of Washington Press.

Marshall, Elliot. 1980. Energy forecasts: Sinking to new lows. *Science* 208 (20 June): 1353–56.

McKinley, Charles. 1952. *Uncle Sam in the Pacific Northwest*. Berkeley: University of California Press.

Norman, Colin. 1981. Energy conservation: The debate begins. *Science* 212 (24 April): 424–26.

Northwest Power Planning Council. 1983. *Northwest Energy News*, vol. 1, no. 1–no. 9; vol. 2, no. 1—no. 2.

———. 1983. *Regional Conservation and Electric Power Plan*, draft.

Ramsay, William. 1979. *Unpaid Costs of Electrical Energy*. Washington, D.C.: Resources for the Future, Johns Hopkins University Press.

Rosenbaum, Walter A. 1981. *Energy, Politics and Public Policy*. Washington, D.C.: Congressional Quarterly Press.

Ross, Marc, and Williams, Robert H. 1981. *Our Energy: Regaining Control*. New York: McGraw-Hill.

Stobaugh, Robert, and Yergin, Daniel, ed. 1979. *Energy Future*. New York: Random House.

Tucker, William. 1981. The energy crisis is over. *Harpers* (November): 25–36.

Wade, Nicholas. 1979. Synfuels in haste, repent at leisure. *Science* 205 (13 July): 167–68.

Worthington, Richard. 1980. The politics of energy self–sufficiency in American states and regions. In Gregory Daneke and George K. Lagassa, eds. *Energy Policy and Public Administration*. Lexington, Mass.: D.C. Heath, pp. 37–61.

Yergin, Daniel, and Hillenbrand, Martin, eds. 1982. *Global Insecurity: A Strategy for Energy and Economic Revival*. Boston: Houghton Mifflin.

13

Regulatory Management of Multistate Energy Projects: The Case of the Allen/Warner Valley Energy System

RICHARD GANZEL

Introduction

Americans have been debating the components of an optimal energy policy for the past decade. Public argument has for the most part remained at an excessively general level, rarely focusing upon the role of existing institutions that make many crucial energy decisions. Two judgments seem to underlie the emphasis upon broad policy contours. The fact that significant parts of the energy production and distribution network are international suggests that a national, security–oriented, approach is essential (Stobaugh and Yergin 1979; Deese and Nye 1981). In contrast, choices have been depicted in terms of broad sets of value choices that will shape widely differing alternative futures. Amory Lovins has christened these alternatives as "soft and hard paths," the former entailing an increased reliance upon decentralized technology and also decentralized, allegedly more democratic, decisionmaking structures and the latter dependent upon centralized, massive technologies and elitist decisionmaking structures (Lovins 1976 and 1979; Morrison and Lodwick 1981). Others have portrayed the choice as governmentally directed intervention versus the goals that are dictated by the marketplace (Reagan 1983).

During the Carter Administration, debate repeatedly shifted back and forth among these perspectives. Jack N. Barkenbus has delineated four federal energy policy paradigms in an effort to capture key aspects of the debate— National Energy Policy, Energy Mobilization, Nationalization, and Free Market. Barkenbus suggests that the Carter Administration should be seen as the proponent of a systematic National Energy Policy approach (1982). Perhaps it should. Carter's two attempts at formulation of a National Energy Policy included an effort to be comprehensive and a fetishlike emphasis upon data gathering seemingly inherent in a "systems" approach. But in four short years, the Carter Administration embraced both a soft energy path and federal sponsorship of a massive synfuels industry. It sought expediting authority to override state and local objections to "crucial" national interest energy projects in an Energy Mobilization Board, and deregulated prices set by an oligopolistic, cartelized market. Amidst these "contradictions," it seemed comfortable with the structure of decisionmaking embodied in the Tennessee Valley Authority (Durant, et al. 1983).

Since the inception of the Reagan Administration, debate has shifted toward the Free Market paradigm. It remains excessively abstract, instead of refining the limits of decisions to be left to consumers, enterprises, and markets. The desire for symbolic victories, such as dismantling the Department of Energy while shifting its functions to other agencies, is partly responsible. But rhetoric rooted in the long–standing commitment of many liberals to strengthening central authority vis–à–vis the states as well as the central planning and redistribution (Reagan and Sanzone, 1981) contributes greatly to the problem. All administrations oversimplify arguments to promote even modest reforms. It is far too soon, and policy changes since 1981 have been much too modest, to justify denunciation of the administration's energy policy as immoral in perspective or to charge that

it is in some degree *uncivilized* (perhaps out of frightful naivete) to suppose that so basic an ingredient of modern society as national and world energy supply and usage patterns can be simply entrusted to a particular set of market institutions whose compatibility with long–run societal needs is itself in continuing need of "course corrections" through public sector interventions (Reagan 1983, p. 383).

The reality of intergovernmental energy policymaking is far different than implied by abstract paradigms or ideological rhetoric. State and local governments make many relevant energy related resource policy decisions. Some state policy and implementation roles have been strengthened by federal decisions over the past decade (Sylves 1982). Moreover, as much relevant debate indicates, the federal government necessarily exercises many energy policy functions by virtue of its role as proprietor of the public lands and an enormous mineral estate, and as the trustee of the national system of land preserves.

State regulatory commissions exercise some control over the prices at which electricity can be marketed. Perhaps more importantly, they rule on whether new energy conversion plants and transmission facilities are in the public interest through issuance of certificates of "convenience and necessity." In exercising these functions, they have shown a capacity to reflect both independent judgment and shifting public values (Anderson 1981). Some commissions, including the California agency discussed in this analysis, exercise much more directive planning functions by examining the relative merits of alternative ways of meeting energy demands. Other state and local agencies exercise additional functions that shape utility behavior. State agencies administer environmental regulations or delegate that role to local government. States in the West approve beneficial water uses. Local governments in many cases exercise land use planning and permitting authority, though many state governments are reassuming some of this authority (Nelson 1977; Popper 1981; Healy and Rosenberg 1979). Finally, state tax policy has significant energy policy implications (*Natural Resources Journal*, 1982).

This enumeration of state and local powers that affect energy choices demonstrates the importance of subnational regulatory roles even if the exercise of those powers falls short of formal authority. In practice, much has changed since state energy regulatory mechanisms were created. Today, only the smallest utilities operate wholly within a state's territorial jurisdiction. Many of those are dependent upon power wholesalers. The crucial practical question, therefore, is whether these regulatory institutions can be adapted sufficiently to meet contemporary and future challenges. In particular, can they meet the challenges associated with approving and siting multistate power projects? If not, or if the process of coordinating local, state, and federal roles proves unbearably inefficient, the pressure for a more aggressive federal role will become even more persistent.

In seeking to answer that question, this study examines the case of the proposed Allen–Warner Valley Energy System. The case is rich in nuance, involving a variety of decisions made by local and state governments in three states as well as by several federal agencies. It also involves both state and federal lands and thereby is prototypical of many resource conflicts in the West. The environmental, preservationist, and developmental conflicts are sharp, raising issues of broad principle. Finally, inconsistencies and gaps in the regulatory process, stemming mainly from differences among local, state, and federal regulators regarding values to be pursued through regulation, reveal the difficulty of viable reform.[1]

The case study is approached through an adaptation of classical political economy, which emphasized the effect of incentives on behavior. But whereas classical political economists emphasized maximization of economic efficiency through contracts and exchangeable property rights, governmental intervention introduces a variety of goals as well as means for their achievement. Consequently, the political economist must be concerned with a variety of incentives and with the question of whether a system as designed "works."

But what does "work" mean in a complex, federal, political context? Inspired in part by Theodore Lowi's effort to place decisionmaking and implementation within a context of democratic theory (Lowi 1979), a growing body of descriptive and analytical political economy is emerging. It examines ways in which laws and institutions shape as well as reflect producer, consumer, environmental, and regional interests (Sanders 1981). The approach is dynamic. Consequently, as interests evolve or the relative influence of contending interests shifts (sometimes as a result of an effective appeal to a more general public), existing institutions undergo stress. Definitions of legitimacy change. Politically viable institutional reform therefore entails design of changed roles serving a modified constituency. The issue in a societal sense is whether such modifications permit reformed institutions to perform assigned tasks in useful and efficient ways, as judged by major participants—old and new.

Though submerged, economic incentives retain importance. The general market situation, expressed in costs of real options, sets effective parameters for energy regulators. Within that context, imposed constraints and costs will shape the nature and location of production activities. Similarly, the existence of a complex web of governmental factors at local, state, and federal levels allows interacting interests to pick and choose among alternative governmental forums. Consequently, even though regulatory policymaking seems to take the form of discrete decisions, one decision may in fact significantly determine a wider set of decisions not under discussion. Such, in fact, has been the result of decisions made in response to the proposed Allen–Warner Valley Energy System.

The Allen–Warner Valley Energy System: General Aspects

Notification was given to the Bureau of Land Management in March 1973 of plans to construct the Allen–Warner Valley Energy System (AWVES) and the Warner Valley Water Project (Arlidge 1983). That notification culminated a decade of effort by the Las Vegas-based Nevada Power Company (NPC), whose subsidiary, Nevada Electric Investment Company, had obtained leases in the Alton Coalfield in the early 1960s. NPC serves the booming southern Nevada district as a private utility. It has prior experience not only in cooperative endeavors with California utilities but also in operating a slurry pipeline. The Nevada utility was a determined entrepreneur, eager to use its leased coal to establish a role as a major electrical power wholesaler. It also was anxious to take advantage of technological efficiencies of scale feasible if it could capture a (wholesale) share of the California market as an extension of its own district. Those desires led logically to construction of a large, state-of-the-art, coal-generating facility, much larger than needed for NPC's service district.

The components of the full system as proposed centered on a 2500 megawatt (MW) electrical generating capacity (Arlidge 1983) and typify the complexity of western energy projects:

1. a mining operation in the Alton Coalfield adjacent to the southern boundary of Bryce Canyon National Park, involving both surface and underground mining and facilitated by pooling of leases on private, state of Utah, and federal lands by the NPC subsidiary and by Utah International, Inc.;
2. deep wells and a pulverizing plant near the mine site to prepare the coal for slurry transport;
3. slurry pipelines to the proposed generating sites in the Warner Valley of southern Utah and the Allen Valley of southeastern Nevada;
4. an off–stream storage reservoir to impound 55,000 acre/feet of water diverted from the Virgin River, with 6–10,000 acre/feet intended as cooling water for the Warner Valley generators (Arlidge 1982);
5. dehydration facilities plus twin 250 MW electrical generating units in the Warner Valley;
6. a pipeline to transport treated Las Vegas and Clark County effluent from an existing secondary treatment facility (that was discharging the effluent into Lake Mead) to the Allen Valley site for use as coolant;
7. a dehydration facility and four 500 MW electrical generating units in the Allen Valley; and

8. transmission lines connecting with the nearby (for both selected generating sites) Navajo–McCullough 500 kV transmission artery that links the Navajo generating station with California consumers.

Target date for operation of the initial generating unit was set as 1979, with other units going into service over a period of years.

Nevada Power Company was the driving force behind the proposed project, assembling partners and negotiating with local governments and with the state of Utah to obtain needed water supplies. In a more fundamental sense, however, the driving force came from California consumers seeking clean (for them), cheap electrical power from sites outside the state. Massive projects in Arizona and New Mexico had preceded the AWVES proposal and additional projects in those states were underway. The stillborn Kaiparowitz project for a mine site generating complex to the north of Zion and Bryce Canyon National Parks was a close contemporary. So was the Intermountain Power Project in western Utah, which was successful in obtaining approvals for the generation station and transmission line. Several additional projects, including an important competitor proposed for siting in White Pine County, Nevada, are at varying points in the process of obtaining regulatory approval or are under construction.

These regional projects follow a common pattern. All are on a large scale. All depend crucially upon substantial sales of electricity to California utilities. All are responsive to similar technological, political, and environmental forces, even though prospective impacts and other factors affecting siting at any specific location in Utah and Nevada vary considerably.

Technological advances during the 1950s and 1960s affecting both generation and transmission capabilities created the possibility of substantial large-scale system efficiencies. But crucial prerequisites must be met. A network of high voltage transmission lines is essential to deliver the electricity and especially to permit "wheeling" of electricity among utilities that participate in the system. The existing Navajo–McCullough transmission line is designed to handle high-voltage, low-loss, transmission. It was linked into a large exchange network. Wheeling, the routing of electricity around an intertied system, becomes most efficient when power to be consumed is drawn from the nearest available source. Such efficiency, in turn, requires a negotiated system of billing rates.

If these prerequisites are met, pooling of power creates three distinctive possibilities for enhanced efficiency: (a) reduced need for backup capacity to cover units shut down for maintenance or repair; (b) sharing lowest–cost options to even out seasonal disparities; and (c) sharing lowest–cost options to even out daily use disparities. Building plants close to an existing low voltage line would maximize these prospects and minimize economic costs for construction.

In practice, small utilities that wish to take advantage of technological possibilities frequently purchase power from industry leaders or agree to assume a small share of the construction costs of a large generation/transmission facility. Even large utilities often enter partnerships to make financing burdens more manageable. But the problem of coping with large incremental additions to generating capacity is especially crucial for small utilities. In 1981, Nevada Power Company's service demand reached 1,423 megawatts (MW). A 250 MW or 500 MW increment, available under initial plans in 1979, obviously would be too large to absorb for that company. Technological possibilities, therefore, inspired negotiated partnership approaches.

Complex national and regional political factors also contributed to an electrical supply strategy based on large regional projects. A well-established network of public and private leaders had cooperated on numerous irrigation and hydroelectric projects, which in turn spurred population growth (Wiley and Gottlieb 1982; Fradkin 1981; Kahrl 1982). The Federal Power Commission had been working quietly for many years to facilitate regionwide planning and transmission interties. Along with increased opportunities for efficiency noted above, the Commission thereby would have obtained greater regulatory authority because many wholesale exchanges would have crossed state lines. Related to these promotional efforts, and partly dependent upon their success, were the well-known activities of the Atomic Energy Commission and its successors. No less important, though only infrequently analyzed in terms of the underlying politics or the implications for intergovernmental relations, has been federal funding of planning studies, research, and development (Rosenbaum, 1981). A pointed example affecting the case study was the funding of the Great Basin of Utah search for suitable power generating facility sites (Utah Consortium for Energy Research and Education 1980). It was intended as a prototype study to be replicated in neighboring states.

A variety of state and local interests also worked to promote large scale projects. A complex coalition of dynamic, development–oriented interests allied with promoters seeking to eliminate rural stagnation yielded support from legislators well beyond levels endorsed by public opinion (Ingram, et al. 1980). This success, in turn, underlines a fundamental reality of the West. The region's dynamism exists mainly in the cities. It only episodically is shared via mining or energy booms with the countryside and only in selected locations. The prevalence of rural stagnation occasionally is broken by successes in marketing peculiar local features for their recreational or retirement merits. Within this context of limited options, an energy generating facility with an expected life of forty to sixty years promises stable, comparatively lucrative payrolls and tax bases. Consequently, the attitudes of state legislators who must balance environmental degradations and the objections of local farmers or ranchers against these opportunities are not surprising (Hayes 1980).

In summation, the potential for a relationship of *perceived* mutual benefits among local, state, federal, and private interests is common. Perceived benefits for any particular project must be substantial enough and clear enough to overcome general attitudinal and site–specific objections. (Lamm and McCarthy 1982). Among these intrinsic and site–specific environmental considerations, the water issue is especially fascinating. Colorado Governor Richard Lamm and his collaborator Michael McCarthy articulate the competitive ethic of western water development in a most revealing language. In terms of necessities inherent in the Colorado River Compact, water in rivers must be developed or it will be lost to downstream claimants. In an interesting variation upon traditional conservation principles, the capture of escaping river flows has long been considered to be water conservation (Lamm and McCarthy 1982, pp. 160–207). Consequently, this "use it or lose it" ethic neatly counteracts localist sentiment relying upon slogans challenging "export of water by wire." The practical consequence is that the region's aridity fails to generate solid resistance to new projects. Instead, projects are judged on a case-by-case basis.

Regulation in Action: Specific Issues

Two components of the AWVES—The Warner Valley generating facility and the slurry pipelines—involve water exportation. NPC was able to avoid this issue for the Allen Valley facility in Nevada by proposing a solution to a local problem. Local governments in the Las Vegas area faced the prospect of installing a tertiary treatment facility to meet stiffened discharge standards for effluent flowing into Lake Mead. NPC suggested that the fluids be used instead as coolant and offered to pay for their transport to the Allen facility. The situation for the slurry lines and the Warner facility depended upon Utah water law and the judgments of Utah's State Water Engineer, and promised to be far more controversial.

The very fact that NPC decided to construct a generating facility in the Warner Valley, near the pristine air of Zion and Bryce Canyon National Parks, was certain to inflame environmentalists (Leydet 1980; Martin 1980). Subsequently, that decision and the related slurry lines were challenged on economic grounds by the staff of the California Public Utilities Commission (CPUC) (Knecht, et al. 1980, Appendix 7). The staff argued that coal shipped by rail from Wyoming or central Utah would be cheaper, a conclusion sharply challenged by the power company and difficult to accept on the basis of the approach taken and evidence offered by CPUC staff. But this argument involved extraordinary regulatory reductionism in judging a proposal on small economic differences subject to dispute and inherent estimating problems (Cawley and Griffin 1983).

In practice, NPC assumed that *its* coal would be used. If it failed to develop existing leases, or at least to make a good faith effort, the leases might be forfeited. And NPC knew that regulatory approval for a right–of–way and a construction permit for a new railroad through the rugged, scenic terrain of southern Utah was out of the question. The lone option was for a slurry line. Its approval was by no means certain, under Utah law or in terms of political considerations that would be on the mind of the State Water Engineer as he weighed the benefits of slurry pipeline use of water. When the project was proposed, Utah law required a finding of "reciprocal" benefits to Utah before stream flows could be approved for export. Authority regarding exportation of ground water was hazy (Utah State Water Engineer 1980; State Sen. Ivan Matheson 1981). NPC's strategy to secure regulatory approval provides the rationale for the Warner Valley facility and for impoundment of Virgin River waters.

A key step involved securing participation of local interests. Residents and promoters from the St. George area had sought federal or state assistance to dam the Virgin River for many years, believing that the locality could become a major retirement haven. NPC proposed a partnership. It would bear a substantial portion of costs for an offstream impoundment, in exchange for the rights to 20 percent of the "conserved" water. That water would serve as coolant for the Warner Valley generating facility. The St. George municipal utility could become a minor partner and thereby assure an adequate future supply of electricity to handle the demands of a burgeoning community. Moreover, the partnership to construct the dam would not be contingent upon regulatory approval of the generating facility (Arlidge 1982). That handled the question of reciprocity of benefits for the export of stream water, but left the question of ground water for the slurry lines.

In 1979, at the behest of southwestern Utah legislators, Utah water law was revised to ensure that exportation provisions applied to ground water. This "clarification" in effect instructed the Water Engineer on the content of the public interest, leaving only technical questions of impacts of large water withdrawals for regulatory judgment. That judgment was to be made jointly (at least in effect) with the National Park Service, which had partial responsibility for assessing effects upon Bryce Canyon water tables and stream flows (U.S. Department of the Interior, Office of Surface Mining Reclamation and Enforcement 1980).

While Utah facilitated energy development projects, California pursued a different path. National air quality legislation had made it virtually impossible to site fossil fuel fired plants near the state's population centers. In some respects, California environmental quality laws went even further. In the early 1970s, at the height of the energy crisis, utilities faced a choice between nuclear facilities and participation in joint projects sited outside the state.

The California regulatory framework became more complex with the creation of the reform–oriented California Energy Commission (CEC). Two aspects of the CEC role are directly relevant to this case study. First, it was given the task of formulating energy demand forecasts (hitherto the domain of utilities and their coordinating associations). During the years AWVES was being actively pursued, demand estimates have been revised drastically downward. The twenty-year projection now calls for annual increases of only 1.29 percent (California Energy Commission 1982). That compares with approximately 7 percent annual increases during the 1960s. Utilities are *required* to use the CEC forecast as a basis for planning. Second, the CEC actively promotes conservation, cogeneration, and alternative energy sources, and is aided in this effort by modifications of California tax law. Its aggressive efforts have reinforced the movement of CPUC into the role of evaluating the alternative ways of meeting energy demands that utilities are required to formulate whenever expansion is proposed.

At the beginning of the 1980s, regional utility plans called for the addition of 12,000 MW of generating capacity during the 1980s, with most projects on the list to be sited in Arizona, New Mexico, Utah, and Nevada, and most electricity to be consumed in California (Western Systems Coordinating Council 1980). However, the effect of reduced growth rates is to make many of the projects on the list into competitors. With both public and private utilities involved, and with greatly varying prices charged to consumers in different utility service districts, the stakes in this competition can be substantial. The experience of the Los Angeles Department of Water and Power (LADWP) is illustrative.

In the initial formulation of AWVES, the LADWP was to be the major partner of NPC. Along with its insatiable need for electricity, LADWP as a municipal utility brought the prospect of tax–free bonds to finance construction at below market costs. However, that partnership ran into an inconsistency between Nevada law and Internal Revenue Service regulations. In order to retain some control over large power projects, Nevada law requires that externally owned shares of a project revert to Nevada utilities at the end of their planned service life. Actual service life of a facility usually is considerably longer than the accounting fiction used by planners and regulators. The probable result was transfer of a valuable asset, financed largely by municipal bonds, to a private utility. LADWP was advised that its status as a municipal corporation would be jeopardized by participation in AWVES. It withdrew in February 1978.

Within the year, LADWP was back in Nevada working on a substitute project to complement its participation in the 3,000 MW Intermountain Power Project (IPP) near Delta, Utah. IPP was organized as a public agency, freeing it from crucial elements of Utah regulation as well as avoiding taxation of bonds that finance construction (*Wall Street Journal*, September 15, 1982). LADWP sought an analogous arrangement in Nevada. Working with representatives of depressed White Pine County, it outmaneuvered Nevada utilities and obtained modification of the County Economic Revenue Development Bond Act (Ganzel, 1979). As a result, the county technically will "own" a large generating facility financed by its own bonds collateralized by guaranteed purchases (mainly by LADWP).

The loss of its California municipal partner was painful for NPC. By April 1978 a new partnership had been formed with Southern California Edison (SCE) and Pacific Gas and Electric (PG & E). One major cost had been delay. A second was that, because its new partners were private utilities, the project now required a certificate of convenience and necessity from the CPUC. In short, approval of the project now rested with California regulators and with federal agencies implementing environmental laws and the new Surface Mining Control and Reclamation Act.

The initial difficulties arose with federal regulators. From 1975 through 1981, the AWVES partners were forced to deal with both Utah state office and national BLM officials, and with both District VIII (i.e., Denver, for the Warner Valley, Utah site) and District IX (i.e., San Francisco, for the Allen Valley site) offices of the EPA. A detailed company summary of these interactions indicates that it never was able to obtain a "check list" of what information would be required to complete its submission. Not only were new requests forthcoming; they were used as a rationale to delay decisions (Arlidge, 1983). In retrospect, some aspects of this frustrating process can be attributed to the newness of and changed standards reflected in relevant federal law—the 1977 Clean Air Act amendments and the new mining law are key examples. But not all. Fuzzy federal legislation was coupled with delaying tactics to avoid difficult decisions. Out of that context, repeated many times, came the proposal for expedited review through a federal Energy Mobilization Board. Ironically, its main target was state and local regulators, while the main culprits seem to have been federal regulators and Congress itself.

Decisions finally began to emerge in 1980. In September, EPA's District VIII office concluded that the Warner Valley plant could not meet Prevention of Significant Deterioration standards and recommended denial of a permit. In October, NPC offered to drop planned capacity to a single 250 MW unit. At the end of the month, District VIII replied that it would not reconsider its decision unless the original application was withdrawn and a new one submitted. This was not merely another delaying tactic because NPC was challenging the validity of the computerized models of air flows toward the national parks used by EPA to judge its application. To comply was to concede their validity and start over. In 1981, District VIII denied the permit for a Warner Valley facility; a subsequent appeal has been denied by EPA's national headquarters.

During the EPA review process, the BLM withheld action regarding the public lands involved in the proposed project. However, as the lead review agency it conducted an exceptionally thorough resource planning review that included realistic alternatives to the proposed AWVES (United States Department of the Interior, Bureau of Land Management, 1980). It is clear that the BLM worked closely with the CPUC, which in turn worked closely with the CEC and private interest groups (Knecht, 1981, 1982; Curtiss 1982). However, BLM did not attempt to use these alternatives to challenge company projections nor to draw its own conclusions on relative economic merits. The thrust of the review was directed toward mitigation strategies that might be employed to soften impacts of AWVES. Consequently, much of the data collected and published was superfluous to the BLM regulatory mandate. Concurrent with the AWVES EIS, a Southern Utah Regional Coal EIS also was prepared. Together with the analysis by the CPUC, these studies provide reviewers with a solid data base and independent review of utility company planning.

During the interlude within which the CPUC staff was reviewing the merits of the AWVES proposal and the CEC was in the process of revising downward projections of increases in energy demand within California, the Secretary of the Interior faced a narrower regulatory issue that involved intense emotions. Mining the Alton Coalfield unquestionably would have both visual and noise impacts on the visitors to Bryce Canyon National Park. Environmentalists in coalition with local residents invoked Section 522 of the Surface Mining Control and Reclamation Act of 1977, and petitioned the Secretary to deny a mining permit for substantial portions of the Alton Coalfield (United States Department of the Interior, Office of Surface Mining Reclamation and Enforcement 1980). In December 1980, Secretary Andrus agreed in part with the petitioners. He banned mining of 9,000 acres of the Alton Field closest to scenic Yovimpa Point, affecting approximately 18 percent of the strippable coal in the field on leases held by Utah International.

The decision of Secretary Andrus pleased no one. Utah International and the state of Utah appealed the ban on mining coal resources—the latter because some of the affected coal was leased from Utah. The Sierra Club, Friends of the Earth, and other environmental groups appealed the decision to allow *any* mining, arguing that the entire Alton deposit was legally unsuitable for mining. These challenges continued into 1983.

Other aspects of federal regulation presented fewer obstacles. Secretary Andrus had approved mining of adequate coal resources, especially since EPA decisions had restricted the project to its Allen Valley generating facility. Utah had given a green light to the use of ground water for slurry transport. The BLM was moving toward approval for the slurry's right–of–way over public lands. Local governments in Nevada were moving toward approval of using effluent as coolant for the Allen Valley facility and state regulators were likely to approve emission permits. That left California with the crucial decision—was the project in the (or its) public interest?

Much of the analysis of the CPUC staff seems to have been inspired by objections to the proposed Warner Valley facility and to the use of ground water for slurry transport. As noted above, the latter objection struck at the core of the project's logic. In contrast, the Warner Valley project's main virtue was its political role in soliciting crucial cooperation in Utah—hardly a pressing concern to California regulators with quite different interests. Rumors regarding the negative thrust of the CPUC staff analysis slowly became common knowledge, suggesting an open fight before the Commission if approval was to be obtained. On the eve of the hearing, SCE and PG & E announced their withdrawal from the project. The impression left to the public was that careful independent analysis had shown that the coal project could not compete on its own merits with conservation, congeneration, and alternatives such as solar power or wind generators. The reality was somewhat different, especially in the causal linkages.

Changes in federal law enacted after the decision by the private utilities to join NPC in 1978 sharply affected effective costs of alternatives, altering direct subsidies and tax advantages (Murray 1982). California legislation reinforced this restructing of after–tax comparisons. Three other factors contributed to the decision to withdraw from AWVES. Sharply rising energy prices in the wake of round two of the international energy "crisis" in 1979 stimulated voluntary and involuntary conservation. PG & E, already under attack from consumer groups and critics of private power generation (Roberts and Bluhm 1981), could ill–afford a fight with the CPUC. Soaring credit costs, coupled with the expected deployment of the MX missile system within the general vicinity of the Allen Valley site, threatened to push costs far beyond acceptable limits (MX Energy Review Team, 1981). None of these factors derived from the intrinsic economic costs of alternatives—though one might argue that new subsidies merely compensated for existing subsidies that favored coal (Kalt and Stillman 1980). They were no less real for the reason that their roots were political or external to the proposed AWVES.

A year later, NPC perserveres. It continues to finalize needed permits and seeks construction delays in order to retain existing permits. It continues to fund mandatory studies, e.g., ambient air quality at the proposed Allen Valley site. And it continues to challenge some, though not all, unfavorable regulatory decisions. As time drags on, it becomes clear that the AWVES will not be built until economic recovery has generated substantial new energy demand. Even then, subsidies to public power will work against it and technological progress should make alternate energy sources even more competitive. But the project probably will be built at some point. The key question concerns which utility will become the crucial partner, and hence whether the project must again undergo CPUC scrutiny.[2]

Meanwhile NPC has moved to satisfy its own needs. Its service district demands are now growing 3 percent annually, down from 10 to 12 percent in the past. As it pursues new AWVES partners, the shortterm solution has become participation as a junior partner in the White Pine project (Arlidge 1983). Unable to develop its own coal or to act as a project manager, the utility will not forgo the opportunity to acquire publicly subsidized power.

Concluding Analysis: The Real World of
Energy Regulation

A number of general observations may be offered on the basis of this case study. Some relate to the regulatory process; others are of a theoretical or methodological nature and are addressed to fellow researchers.

1. Implementation of Air Quality Regulations

The existence under current federal law of more than two hundred quality control districts reflects a desire to obtain administrative flexibility and thereby achieve more realistic implementation of the law (Gordon 1982). Delegation of much implementation responsibility to the states exemplifies Michael Reagan's "permissive federalism" (Reagan and Sanzone 1981). Though pristine areas were to be preserved and standards exceeding existing performance were set for many districts, the law was not intended to shut down existing industries, halt traffic, or stop future development. Instead, the goal was to stimulate technological innovation while nudging polluters to improve existing facilities (Ackerman and Hassler 1981).

In practice, it appears that multiple districts with differing standards provide a "window of opportunity" through which utility executives gaze in search of available increments of pollutable air. Certainly that is the case for California utilities, which sustain growth in that state by generously sharing their externalities with neighboring states. The Great Basin has become California's energy hinterland, providing coal, water, and unpolluted air as exportable commodities.

District–by–district implementation creates promotional opportunities and incentives for state and local interests. This study has emphasized the efforts of St. George and the state of Utah. It also could have analyzed the willingness of Nevada to settle for Class II air quality for its Valley of Fire State Park, thereby ensuring approval of permits for the Allen Valley plant. Additionally, one might emphasize the decisions of local regulatory districts in smog–filled southern California, which consider new power sources outside the districts as equivalent to emission reductions. But as new generating capacity is displaced toward cleaner air, it encroaches upon pristine air in parks, wilderness areas, and Indian reservations. How much encroachment will be tolerated?

A key unresolved question concerns the extent of protective corridors on the periphery of federal land withdrawals and reservations. A recent mapping exercise (Enviro–Map, Inc. 1980) probably overstates the geographical magnitude of the question, but does so in a suggestive way. It assumes twenty–mile protective corridors for Class I areas, assumes that Class I protection will be selected whenever possible, and then shades a map of the western United States to reflect the existence of actual and potential Class I areas together with their protective barriers. The resultant shaded area theoretically precluding industrial development is enormous.

The projection cannot be dismissed out of hand. It should be noted that the rejected Warner Valley facility lies *outside* the twenty–mile band, suggesting that the assumptions might even be conservative. Moreover, debate regarding the purposes of special land withdrawals seems to be moving toward a preservationist perspective (Sax 1976 and 1980; Suniville 1979; Martin 1982). The Reagan Administration has been unable to reverse this momentum. Clearly, in the future utilities will undertake more selective consideration of Great Basin generating sites, avoiding preserved areas in favor of districts undistinguished in any sense. Worth pondering, however, is the question of significant local political allies. Can they be marshalled for truly remote sites? Can outsiders secure permits without such allies? NPC needed allies in Utah; LADWP has managed with less visible help in Nevada.

2. Problems with State Energy Regulation

Effective regulation requires a clear, feasible mandate. Wide administrative latitude often erodes the legitimacy of agency decisions and sometimes pushes value choices into the courts (Stone 1982; Breyer 1982). Current law, federal and state, falls far short of a fully protectionist mandate. The mandates to states, dutifully incorporated in state implementation plans for air quality, may protect air quality from a prospective major new stationary source of pollution. They are less effective for existing pollution sources and face a major challenge in the West in coping with cumulative effects of small, dispersed sources of pollution. A notable case is pollution from woodburning stoves, which has pushed some districts into noncompliance with air quality standards.

Conventional public service commission regulation of new generating facilities clearly has become inadequate. On the one hand, allowing a single state commission to veto a project that involves utilities from several states working in partnership cannot yield legitimacy in the minds of participants. On the other hand, permitting public utilities to avoid such regulatory scrutiny mocks the process. The fact that both the Intermountain Power Project and the White Pine Power Project are proceeding under public sponsorship is doubly troubling. Not only do they escape rigorous scrutiny regarding their necessity, on the dubious assumption that what a public entity does is in the public interests (Durant, et al., 1983), even when bonding laws are manipulated by state legislatures. They also involve large subsidies for centrally generated electricity, a strategy that presumably has been rejected nationally—but apparently only for private suppliers of power (Murray 1982).

Though not analyzed in detail in this paper because the withdrawal of California utilities from AWVES made its assessment moot, comparative project review by public service commissions faces a difficult task. Though the data underlying CPUC assessments was far from satisfactory, the real problem is more fundamental. Weighing alternatives is exceptionally difficult in the best of situations. It is virtually impossible when a commission staff must compare a proposed project that has taken years to assemble with nonexistent hypotheticals. Is the fact that a public alternative involves a subsidy to be taken as an advantage or as a disadvantage? What price is assigned to the risk that coal leases issued by the federal government and by Utah may be lost by the substitution of coal from elsewhere? What weight is to be given to tax advantages that may accrue to an entrepreneurial utility based in another state?

Public service commissions traditionally have been staffed by accountants and engineers assigned to exercise oversight. If comparative project review is to be the new role, a full staff equivalent to that employed by corporations must be acquired. Even then, a realistic consideration of opportunity costs will require expansion of jurisdiction to a regional scale. The scope of the region must encompass the elements to be regulated—i.e., the Great Basin and perhaps also the coal fields in the Rocky Mountains.

3. A Reform Proposal for State Public Service Commissions

Those who believe that regulation must encompass comparative project review need to follow the "experiment" undertaken through the federally sanctioned Northwest Power Planning Council very carefully (Pelham, et al. 1981). Its success remains to be demonstrated. But there is an alternative to ever–expanding public planning that may be superior in practice. The alternative is to reform the state energy regulatory process, and in the process partially deregulate the production of electrical energy.

State regulatory commissions ordinarily respond to single requests for permission to construct new capacity. As noted, recent events have undercut the logic supporting that approach. Conservation and a wide variety of alternatives *do* need to be considered; otherwise approval of a project will foreclose development of many alternatives or will create unneeded generating capacity. In essence, the natural monopoly on energy production no longer exists. Contemporary regulation continues to assume its existence and expanded project review compounds the error. In the absence of such a production monopoly, commissions need to review alternative *bids* to supply electricity to a distribution system, not the production of that supply.

A reregulated system would include the innovation undertaken by the CEC. Demand must first be projected for future years. Then, bids could be solicited for delivery of X megawatts of power at a specified future date. Prices and other pertinent considerations could be weighed by the regulatory commission. Undoubtedly, suppliers would resort to insurance to cover contingencies and a futures market for electricity would be created to allow risks to be shared. Meanwhile, commissions would continue to regulate the distribution end of the electrical supply system. They would need to become more involved in administering the wheeling of power—a role that entities such as the Bonneville Power Agency have filled by default (McKinsey 1983). But land use regulation and environmental review would remain with federal, state, and local planning agencies.

4. Energy Regulation and Intergovernmental Relations

Several aspects of the relationship among local, state, and national regulatory responsibilities need to be reconsidered. Most urgent is the need to address the question of protective air corridors around parks and wilderness areas. That question must be answered by Congress because it involves the very purposes for which federal public lands have been set aside. New amendments to the Clean Air Act should specify the form of protection, so that the rules are known and the prolonged, uncertain review process undergone by the Warner Valley element of AWVES can be avoided. It is irresponsible as well as extraordinarily wasteful to delegate broad responsibility to the Environmental Protection Agency and have that responsibility exercised through debates over the proper computer simulation of pollutant dispersal. If federal law is clear, moreover, more administrative responsibility can be delegated to state agencies from EPA districts.

For those who have pushed long, hard, and successfully for intertied electrical transmission systems, a crucial choice arises. Should the federal government now assume many of the regulatory functions previously exercised by state commissions? The trend has been for slow accretions to federal authority (Goodwin 1981), though Congress rejected President Carter's proposal for substantial preemption in 1978 (Pelham, et al., 1981, pp. 195–206). A more useful role would be facilitating state regulatory reform, stimulating voluntary regional planning and regulatory organizations by devolving some responsibilities to them, and policing state practices. One example of such useful policing was the mandate to eliminate previously accepted restraints on access to energy customers by enforcing "wheeling" of electricity. That mandate, enacted as part of P.L. 95–617 in 1978 (Pelham, et. al. 1981, p. 201), makes possible the reform proposal outlined above. However, administrative implementation of wheeling, which was assigned to the Federal Energy Regulatory Commission, could be devolved to regional or even state commissions under cooperative agreements. A second example warranting attention is the question of limits on municipal bonding authority for power projects. The sense of a system in which LADWP might lose its exemption by participating in AWVES but can instead guarantee county bonds for a project that also sells power to private utilities is obscure.

Finally, the virtues of the current federal approach to land use planning for siting and right–of–way seem apparent. BLM rightfully concludes that its role is to facilitate and mitigate, not block, projects. That is perfectly consistent with the aggressive and obligatory effort by the National Park Service to prevent damage to Bryce Canyon National Park. Any effort to go beyond those roles should take the form of designating in advance preferred zones for energy projects (Ganzel 1983). State planning agencies might do the same thing. But when agencies begin to mix planning and regulatory functions, they are likely to encounter the difficulties experienced by the California Public Utilities Commission staff. That should provide ample warning of quagmires to be avoided.

Notes

1. Numerous telephone interviews and informal checks with key participants underlie this analysis. Officials in the Utah State Water Engineer's Office and State Senator Ivan Matheson aided with the section on Utah. Several Nevada Power Company officials were helpful. Three individuals warrant special thanks. Bill Curtiss from the Rocky Mountain office of the Sierra Club Legal Defense Fund answered questions and critiqued an earlier draft. Ron Knecht, who organized the staff analysis for the California Public Utilities Commission and did an excellent job despite the criticism of one element of the analysis offered here, shared his knowledge, judgments, and time fully. Finally, John Arlidge, a model project manager and corporate official, went far beyond duty and was indispensable. Each no doubt will wish that he had been more persuasive in shaping my interpretations.

2. This judgment refers to the system *minus* the Warner Valley generating component and the slurry line that would have served it. Most permits have been obtained, and many underlying motivations have not changed for Nevada and Utah participants. The crucial outside market could be in Arizona or New Mexico instead of in California.

References

Ackerman, B.A., and Hassler, W.T. 1981. *Clean Coal/Dirty Air.* Cambridge: Yale University Press.

Anderson, D.D. 1981. *Regulatory Politics and Electric Utilities.* Boston: Auburn House.

Arlidge, J.W. 1982. The Allen-Warner Valley Energy System. Formal text of presentation to the author's seminar on April 29th.

Arlidge, J.W. 1983. Allen-Warner Valley Energy System History. Nevada Power Company (May).

Barkenbus, J.N. 1982. Federal energy policy paradigms and state energy roles. *Public Administration Review* 42: 410–18.

Breyer, S. 1982. *Regulation and Its Reform.* Cambridge: Harvard University Press.

California Energy Commission. 1982. California Energy Demand 1982 to 2002. State of California, Report PS105–82–001.

Cawley, R.M., and Griffin, K. 1983. Regional equity: The politics of severance taxation. In R. Ganzel, ed. *Resource Conflicts in the West.* Nevada Public Affairs Institute.

Deese, D.A. and, Nye, J.S., eds. 1981. *Energy and Security.* Cambridge: Ballinger.

Durant, R.F.; Fitzgerald, M.R.; and Thomas, L.W. 1983. When government regulates itself: The EPA/TVA air pollution control experience. *Public Administration Review* 43: 209–19.

Enviro–Map, Inc. 1980. Impacts of the Clean Air Act on the West. Bonneville Associates.

Fradkin, P.L. 1981. *A River No More: The Colorado River and the West.* New York: Alfred A. Knopf.

Ganzel, R. 1979. Energy decisions for Nevada. *Nevada Public Affairs Review* (1978/79): 20–23.

———. 1983. Maximizing public land resource values. Chapter 6, this volume.

Goodwin, C.D. 1981. *Energy Policy in Perspective.* Washington, D.C.: The Brookings Institution.

Gordon, R.L. 1982. *Reforming the Regulation of Electric Utilities.* Lexington, Mass.: Lexington Books.

Hayes, L.R. 1980. *Energy, Economic Growth, and Regionalism in the West.* Reno: University of New Mexico Press.

Healy, R.G., and Rosenberg, J.S. 1979. *Land Use and the States.* Baltimore: Johns Hopkins University Press for Resources for the Future.

Ingram, H.M.; Laney, N.K.; and McCain, J.R. 1980. *A Policy Approach to Political Representation: Lessons from the Four Corners States.* Baltimore: Johns Hopkins University Press for Resources for the Future.

Kahrl, W.L. 1982. *Water and Power.* Berkeley: University of California Press.

Kalt, J.P. and Stillman, R.S. 1980. The role of governmental incentives in energy production: An historical overview. In J.M. Hollander, et al., eds. *Annual Review of Energy.* Annual Reviews Inc.

Knecht, R.L., et al. 1980. The Summary Report of the Allen-Warner Valley Project Team. California Public Utilities Commission. October 3.

Lamm, R.D., and McCarthy, M. 1982. *The Angry West.* Boston: Houghton Mifflin.

Leydet, F. 1980. Coal vs. parklands. *National Geographic* 158: 776–803.

Lovins, A. 1976. Energy strategy: The road not taken. *Foreign Affairs* 55: 65–96.

———. 1979. *Soft Energy Paths: Toward a Durable Peace.* New York: Harper &Row.

Lowi, T.J. 1979. *The End of Liberalism.* 2d ed. New York: W.W. Norton.

Martin, J.B. 1982. The interrelationships of the mineral lands leasing act, the wilderness act, and the endangered species act: A conflict in search of resolution, *Environmental Law* 12: 363–441.

Martin, T. 1980. How can we protect southwestern national parks? *National Parks and Conservation Magazine* (March): 4–9.

McKinsey, L. 1983. The Pacific Northwest: A regional "Soft Path" experiment. Chapter 12, this volume.

Morrison, D.E., and Lodwick, D.G. 1981. The social impacts of soft and hard energy systems: The Lovins' claims as a social science challenge. *Annual Review of Energy.* Annual Reviews Inc.

Murray, A. 1982. Congress cuts off funding for study regarded as threat to low–cost public power. *Congressional Quarterly Weekly Report* 40 (October 9): 2630.

MX Energy Review Team, 1981. Informal discussion of probable impacts of interrelated projects. Arlidge and Ganzel were members of the Nevada review committee on the DEIS.

Natural Resources Journal. 1982. Symposium on the taxation of natural resources. *Natural Resources Journal* 22: 525–687.

Nelson, R.H. 1977. *Zoning and Property Rights.* Cambridge: M.I.T. Press.

Pelman, A., et al. 1981. *Energy Policy.* 2d ed. Congressional Quarterly.

Popper, F.J. 1981. *The Politics of Land Use Reform.* Madison: University of Wisconsin Press.

Reagan, M.D., and Sanzone, J.G. 1981. *The New Federalism.* 2d ed. Cambridge: Oxford University Press.

Reagan, M.D. 1983. Energy: Government policy or market result? *Policy Studies Journal* 11: 365–85.

Roberts, M.J., and Bluhm, J.S. 1981. *The Choices of Power.* Cambridge: Harvard University Press.

Rosenbaum, W.A. 1981. *Energy, Politics, and Public Policy.* Congressional Quarterly.

Sanders, M.E. 1981. *The Regulation of Natural Gas.* Philadelphia: Temple University Press.

Sax, J.L. 1976. Helpless giants: The national parks and the regulation of public lands. *Michigan Law Review* 75: 239–74.

————. 1980. *Mountains Without Handrails.* Ann Arbor: University of Michigan Press.

Stobaugh, R., and Yergin, D., eds. 1979. *Energy Future.* New York: Random House.

Stone, A. 1982. *Regulation and Its Alternatives.* Congressional Quarterly.

Suniville, G. H. 1979. The national park idea: A perspective on use and preservation. *Journal of Contemporary Law* 6: 75–91.

Sylves, R.T. 1982. Nuclear power and the states: A typology of state regulatory instruments and how they have been used. American Political Science Association.

United States Department of the Interior, Bureau of Land Management. 1980. Allen–Warner Valley Energy System Environmental Impact Statement. Two Volumes.

United States Department of the Interior, Office of Surface Mining and Reclamation Enforcement. 1980. Southern Utah Petition Evaluation Document.

Utah Consortium for Energy Research and Education. 1980. Utah Energy Facility Siting Study Phase I: Great Basin. The University of Utah and Utah State University Consortium.

Wall Street Journal. 1982. Power plant is held by Utah to be exempt from state regulations. (September 15th): 29.

Western Systems Coordinating Council. 1980. Ten-Year Coordinated Plan Summary, 1980–1989. Western Systems Coordinating Council.

Wiley, P., and Gottlieb, R. 1982. *Empires in the Sun.* New York: G.P. Putnam's Sons.

14

The Supreme Court and Resource Federalism: Commonwealth Edison Co. v. Montana

JAMES J. LOPACH

I. Introduction

A battle is presently being fought out in the name of federalism over natural resource taxation policy. The coal–producing states and their allies are pitted against the coal–consuming states and their allies. The battlefields have been the courts and legislatures of the states and the Supreme Court and the Congress of the United States. The fight over the state of Montana's coal severance tax has been a significant skirmish in this battle. In their attack on the tax in various political and judicial arenas, opponents of the measure have used arguments based on the theory of federalism.

Federalism lies at the heart of the American governmental system and stems from the historic compromise between confederationists and centralists. The concept has neither an inherent nor a consistent justification. Arguments abound, however, that are grounded in political, legal, and administrative principles and conclude that this or that level of government should act in a given set of circumstances. It seems, though, that the actual functioning of federalism can best be explained by the observation that an interest seeking a response from government can approach either the state or national arenas; the source of support is not as important as its reliability. The following pages illustrate that the attack on Montana's coal tax policy was essentially an interest group effort using such an eclectic strategy.

Having exhausted their state legislative and judicial remedies and lost, the opponents of the Montana tax took their arguments to national forums. The Supreme Court ruled on the constitutional issues surrounding state taxation of natural resources in the 1981 case of *Commonwealth Edison Co. v. Montana*. The parallel struggle in Congress is still in progress. The resolution of the taxation and jurisdictional issues by the Supreme Court provides a contemporary commentary on the constitutional doctrine of federalism. The legal framework used by the Court is applicable to other areas of natural resource policy in which the national and state governments have a legitimate role.

II. Coal Politics

The last decade witnessed efforts by all sectors of society to decrease the nation's dependency on foreign sources of energy. Coal has figured in these plans for national energy independence, and this is especially true of low sulfur coal, which more easily permits compliance with current air quality standards. Reliance on coal makes good sense when viewed from the perspective of the size of the nation's coal deposits. The United States has demonstrated coal reserves of 438 billion tons, which represent about 31 percent of the known recoverable coal in the world. Thirty–one states have coal deposits and twenty-six states mine coal.[1]

The state of Montana is especially coal rich because of its total coal reserves and low sulfur coal deposits. Of the nation's reserve base of 438 billion tons of coal, Montana has 120 billion tons, or 27 percent of the total. Included in this deposit is more than 50 percent of the low sulfur coal reserves in the United States. Montana's neighboring states also contain vast coal fields within their boundaries. North Dakota has a reserve of 10 billion tons of coal and Wyoming has 55 billion tons. Montana and Wyoming together have 40 percent of all the coal in the United States and 68 percent of the nation's low sulfur coal.[2] Western states other than Montana and Wyoming possess low sulfur or compliance coal as a substantial part of their total reserves. For example, the low sulfur proportion of coal deposits for Colorado, New Mexico, and Utah are, respectively, 71 percent, 98 percent, and 76 percent.[3]

Mining, manufacturing, and consuming interests have looked to Montana as a major and long–term source of coal for firing electricity generating plants. The interest in developing Montana coal stems from its low cost as well as its abundance. In 1979, Montana coal sold for the average price of $9.76 a ton, including the severance tax, while the national average was $23.75. Colorado coal, in comparison, sold for $16.72, and the price of Pennsylvania coal was $29.97.[4] Utilities are well aware of this fact. In 1975, Detroit Edison negotiated a long–term contract for Montana coal and estimated that the agreement would realize a savings in fuel costs of over one billion dollars.[5] Montana coal is substantially cheaper than other fossil fuels as well.[6] Substituting compliance coal for gas and oil could result in major savings because the investment in capital conversion, from a technological perspective, would not have to include scrubbers to remove sulfur at the point of generation.

Helping to supply the natural resource needs of the nation is not a new role for Montana. For more than a century railroads have carried wheat and logs, cattle and sheep, and copper and zinc to distant points. The river basins of the Columbia and the Missouri have their birth in the state's mountain ranges and carry their water to much of the nation. Similarly, Montana and its neighbors have not held back their coal. In 1976 the nation's coal production amounted to 679 million tons. Montana's share was 26.2 million tons. In 1977 Montana and Wyoming together were responsible for meeting 10 percent of the nation's demand for coal, and by 1990 it has been estimated that this figure will be 33 percent. In late 1981 Montana was the site of nine coal mine expansions, and there were plans for four additional mines. One of these mines was designed to produce 8 million tons a year, which compares to Montana's 1980 production of 30 million tons.[7] The coal mining activity in Montana, therefore, has been largely directed toward serving out–of–state needs. In recent years the mining companies in Montana have shipped as much as 90 percent of their product to utilities located principally in the Midwest and Southwest.[8]

Although Montana citizens have been accustomed to seeing the natural wealth of their state pass before them in long rows of railroad cars, some eyes have not been closed to the damage often done by natural resource exploitation. Boom towns that had turned to ghost towns affected lives equally as long as the ruins remained a part of the landscape. Amidst the conservation–development politics that long have divided the state, the Montana Legislature enacted a coal severance tax in 1975 that was intended to level out the complex and largely unforeseeable costs of coal mining.[9] The rationale of this measure was that the state would receive something in exchange for the depletion of its irreplaceable mineral wealth.

The 1975 severance tax provided that surface–mined coal with a heating quality of under 7,000 Btu per pound would be taxed at 20 percent of its contracted sales price. A tax of 30 percent of value would be imposed on stripped coal in excess of 7,000 Btu. These tax rates are higher than the taxes imposed on energy sources by any other state. The act also permitted counties to levy a tax on the gross proceeds from coal[11] just as on other forms of property, and it left intact the Resource Indemnity Trust Act[11] which placed a tax of one–half of one percent on the gross value of nonrenewable natural resources, including coal. Income from the trust account was intended to improve and reclaim the natural environment. In addition to these taxes, coal companies pay a property tax on their mining equipment,[12] a corporate license tax,[13] and a reclamation bond[14] to make certain that land will be adequately reclaimed after mining operations. The attacks on the coal tax in 1978 and later were based on the rate of the severance tax, the uses to which its revenue was put, and the fact that coal companies had to pay multiple taxes whose earnings were devoted to similar purposes.

Montana legislators who adopted the severance tax were undoubtedly motivated in various ways. They were concerned about damage to the environment and loss of a natural resource, but they also were completely aware of the revenue bonanza that coal presented. A statement attributed to one of the principal sponsors was "in the energy crunch today, Montanans should remember that the Arabs have the oil but Montana has the coal."[15] Legislators had firsthand knowledge of captive consumption because of the state's relationship to Alberta. They chose to imitate that province's pricing policy for natural gas. By raising the royalty on natural gas, Alberta was placing on Montana consumers the burden of its "universities, hospitals, reduction of other taxes, etc."[16] The conclusion of the Conference Committee of the Montana Legislature was: "While coal is not as scarce as natural gas, most of the Montana coal now produced is committed for sale under long–term contracts and will be purchased with this tax added to the price."[17] The probable impact of the severance tax on other states and their citizens, therefore, did not escape Montana legislators.

There was more to Montana's tax policy than "OPEC–like revenue maximization."[18] Alleviating the impact of coal development, as well as increasing the flow of revenue to the state's general fund, was a legislative consideration. The distribution formula for the revenue gained under the coal severance tax was the subject of much debate during the 1975 legislative session, and the plan that emerged was balanced in its composition. The Local Impact Assistance Grant Program, administered by the Montana Coal Board, initially was given 13.25 percent of all coal tax receipts. From 1975 to 1980, the Coal Board received more than $32 million, made ninety grants for coal area governments in the amount of $30 million, and denied $23.7 million worth of proposals. Three Montana counties—Rosebud, Treasure, and Big Horn—have received the brunt of the social, economic, and environmental impact of coal mining. From 1975 to 1980, sixty–three of the ninety Coal Board grants went to the three coal area counties and twelve of their local governments.[19]

The establishment of a permanent fund was another vehicle for alleviating the problems associated with coal development. A constitutional amendment[20] was proposed in 1975 and adopted in 1976 that dedicated 25 percent of total annual collections to a trust fund whose principal would "forever remain inviolate unless appropriated by vote of three– fourths of the members of each house of the legislature." In 1980 the trust fund began to receive 50 percent of the tax collections. The Montana Legislature to date has resisted attempts to spend the principal for major state needs, such as highway expansion and maintenance and large capital projects. The prevailing philosophy is that the fund is for the time when the true social, economic, and environmental costs of coal mining can be known.

Since 1975 the coal severance tax distribution formula (but not the trust fund provisions) has been altered by subsequent legislation.[21] The current allocation of severance tax revenues, which in fiscal years 1981–82 were $86 million, is shown in Table 14.1.

It was the coal severance trust fund that created the strongest opposition to the Montana tax program. In 1980, when the opponents of the tax were asking the United States Supreme Court to hear their arguments, the principal of the trust fund had grown in excess of $56 million. The opponents' position was that the trust fund was not needed because all of the problems in the coal mining

Table 14.1 Allocation of Severance Tax Revenues (in percents)

Trust fund	50.0
General fund	19.0
Education trust fund	10.0
Impact area grant program	8.75
School equalization	5.0
Parks, art, and aesthetics	2.5
Alternative energy research	2.25
Renewable resource development	.625
Water development	.625
Libraries	.5
Land-use planning	.5
Conservation districts	.25
TOTAL	100.0

areas had been remedied. They charged that the coal tax had become a ruse for current and future tax relief.[22]

Montana is one of eleven states that have established a trust fund from mineral tax collections.[23] The typical purposes of these funds have been one or more of the following: compensation for extraction of a non-renewable resource, remediation of the effects of mineral exploitation, or development of a diversified economic base. The projected expenditures, therefore, are in the future. These funds, which number sixteen (Montana has four and Minnesota and North Dakota each have two), have been provided some degree of permanence or inviolability by state constitution or statutory law. The states that have such funds, in addition to Montana, Minnesota, and North Dakota, are Alaska, Colorado, Florida, Michigan, New Mexico, South Dakota, Wisconsin, and Wyoming. In mid-1981, the Alaska Permanent Trust Fund was the largest with a balance of $1.8 billion. The New Mexico Severance Tax Permanent Fund's balance of $530 million was the second highest, followed by the $231 million balance in the Permanent Wyoming Mineral Trust Fund. The June 1981 balance of $75 million in Montana's Permanent Coal Tax Trust Fund was the fourth highest. Of the sixteen trust funds, nine have a principal that is inviolate, two had no balance, five had a balance of under $10 million in mid-1981, and six provided that interest on the principal must go to the general fund.

The attack on the Montana tax singled out a state that had the appearance of political vulnerability. To urban America Montana can seem as provincial and distant as Saudi Arabia, its presumed economic cousin. Montana's political presence in the United States Congress is weak because of its sparse population and lack of senior leaders in its delegation. Montana, however, is only one of many states that has an energy tax as a major revenue source. Consequently, the legal and political challenges to the Montana tax had implications for states throughout the nation.

Severance taxes on mineral production are nothing new or rare; the first was adopted in 1876. Thirty–three states today employ a severance tax[24] which, according to the definition of the Census Bureau, is a tax "imposed distinctively on removal of natural products from land or water and measured by value or quantity of products removed or sold."[25] In 1980 these thirty–three states received $4.1 billion in severance tax collections, and the receipts of more than twenty states were in excess of $20 million. Figures for the top ten states, in millions of dollars, are shown in Table 14.2.

Montana's share was not very significant. In 1980—as previously in 1979—Montana's collections placed the state tenth in the nation. Of the total $4.1 billion, Texas, Alaska, Louisiana, and Oklahoma received a combined share of 72 percent.[26]

Severance tax income is commonly disbursed in four ways.[27] The most significant recipient is the state general fund. The provisions of twenty–six of the thirty–two largest severance taxes in the nation call for at least 50 percent of the revenues to be used in this manner. Thirteen of these taxes dedicate 75 percent or more of the income to the general fund. The second common employment of severance tax revenue is financing a trust fund, a fiscal device that was discussed above. The third use of severance tax income is funding state financial assistance to local governments. Twenty–two states have some type of revenue–sharing program. Finally, there are fourteen states that use their

Table 14.2 Receipts from Severance Tax Collections
for Top Ten States (in millions of dollars)

Texas	1525.1
Louisiana	525.3
Alaska	506.5
Oklahoma	436.1
New Mexico	213.6
Kentucky	177.2
Florida	121.1
Wyoming	105.7
Minnesota	83.5
Montana	74.6

severance tax revenue for purposes other than the three already mentioned. The most common miscellaneous uses are economic development, local impact assistance to mining communities, land reclamation, and research projects.

Many of the severance tax states realized from the outset that they were naturally Montana's allies, but Texas officials were motivated apparently more by the state's coal consuming status than by its role of natural gas and oil developer. The Attorney General of Texas submitted an *amicus curiae* brief to the United States Supreme Court siding with those challenging the Montana tax, and the Texas congressional delegation supported legislation limiting production taxes on coal mined on federal land to 12.5 percent. About three–fourths of Montana coal lies beneath federally owned property.

Regardless of the maverick behavior of Texas, it has increasingly become clear that coal politics is merely a skirmish in the larger battle of energy politics, and the stakes involved in this interstate or federalstate tussle are enormous. It is projected that Texas, a state with no corporate or personal income taxes, will derive $2.6 billion from its severance tax in 1983.[28] Alaska eliminated its income tax in 1981 and receives 90 percent of its revenue from petroleum taxes and royalties.[29] In eight states, severance taxes account for at least one-fifth of gross revenue.[30] The role of severance taxes in the revenue picture of coal states is modest in comparison. Montana's 1980 coal tax receipts represented 13.5 percent of the state's total revenue for that year.[31] In North Dakota, 70 percent of coal tax receipts are needed to handle coal development impacts on school districts and local governments.[32] West Virginia in 1980 produced more than 110 million tons of coal (that year Montana's total was thirty million tons), and imposed a severance tax of only 3.85 percent of gross sales.[33] Wyoming in 1980 collected $42 million from its coal tax, which was 25.5 percent of the state's total revenue, and used the justification that the tax "internalizes both the present and the future costs of rapid coal development by accounting for social, economic, environmental and regulatory costs."[34]

In the attack on energy taxes, however, the amount of tax revenue and the uses to which it is put are not the principal points of contention. Rather, the significant issues are that some states have the revenue to spend, other states do not, and the haves apparently get their additional revenue from the have–nots. The thesis here is that energy consuming states are attacking energy producing states, not just coal states, with theories of federalism as the most convenient weapons of battle. The Montana coal tax is a tale of energy politics that repeats old lessons about our federal form of government.

III. The Framework of Federalism

Although federalism was a compromise struck in 1787, arguments were presented during the course of the constitutional convention that have been given the mantle of framers' intent. Three times delegates adopted the language of Edmund Randolph's Virginia Plan to allow the

> national legislature . . . to legislate in all cases for the general interests of the union, and also in those to which the states are separately incompetent, or in which the harmony of the United States may be interrupted by the exercise of individual legislation.[35]

Included in this formulation are two possible approaches to resolving the puzzle of federalism. The sorting out of what activities are federal and what are state arguably can be determined by the nature of the activity. "National" problems are the responsibility of the general government, and "local" problems are within the province of the states. This approach would appear to be grounded in simple observation and at times the United States Supreme Court has adopted such a mechanical test. The convention's more subjective approach to federalism involves making a judgment about the competence of states or the potential for their activities to disrupt the Union. The court at various times has performed this type of analysis, usually under the label of "balancing test," but the more empirical "subject test" certainly admits of this type of weighing or adjusting, too. The framework of federalism, consequently, has been flexible, ever taking new shape because of lessons of history, contemporary pressures, and the practical demands of administration.

The descriptive phrase, "umpire of federalism," has been given to the United States Supreme Court because so often this body has construed and applied constitutional rules in federal–state conflicts. An analysis of its federalism decisions reveals four categories of controversies and resolutions. These will be discussed to make the point that attitudes of the justices have had a role to play in their decisions. Especially instrumental have been personal judgments concerning the institutional roles of Congress and the Supreme Court, the degree of national urgency, the existence of local provincialism, and the significance of the states as governmental entities. The prevailing jurisprudence is captured well by Judge Learned Hand's view of constitutional interpretation: "in such matters everything turns upon the spirit in which [the Justice] approaches the questions before him. The words he must construe are empty vessels into which he can pour nearly anything he will."[36]

The first category of federalism cases arises when state action is challenged as being in violation of the United States Constitution, even though Congress has not involved itself in the particular subject area. In voiding the policy or program of the state, the Court at various times has warned against state insularity. A state cannot treat out–of–state individuals or businesses in an arbitrary or discriminatory manner or isolate itself from problems confronting the entire nation. The interstate commerce clause is the basis of much of this litigation, and the Court has said that the clause itself as judicially interpreted, and not congressional enactments based on the clause, prohibits what the state is attempting to accomplish. An early statement concerning excessive provincialism was by James Wilson, a founding father and member of the first Supreme Court:

> Whatever object of government is confined to its operation and effects within the bounds of a particular state, should be considered as belonging to the government of that state; whatever object of government extends in its operation or effects beyond the bounds of a particular state, should be considered as belonging to the government of the United States.[37]

Wilson's rule would make a state's boundary a strict jurisdictional barrier; any activity with interstate implications would be dealt with by the national government. *Cooley* v. *the Board of Wardens of the Port of Philadelphia* was a mid–nineteenth century expression of the dangers of narrow state action. Here the court subjected a Pennsylvania system of pilot regulation to commerce clause scrutiny and laid down the "Cooley Doctrine": "Whatever subjects of this power are in the nature national, or admit only of one uniform system, or plan of regulation, may justly be said to be of such a nature as to require exclusive legislation by Congress."[38] This formulation of a subject matter test for resolving federalism questions is reminiscent of the thrice–adopted resolution of the 1787 convention. Another half century later, Justice Holmes declared that the continued existence of the nation depended upon the exercise of judicial review regarding state laws: "One in my place sees how often a local policy prevails with those who are not trained to national views and how often action is taken that embodies what the commerce clause was meant to end."[39] And up to the present day there is a continuation of this type of case that evokes from the Court the sentiment that state action, even in the absence of congressional action, runs afoul of the doctrine of nationalism incorporated in the Constitution.[40]

A second category of cases represents a segment of the law of federalism called the doctrine of absention. Here the set of circumstances includes state activity, the absence of federal activity, and judicial deference to the local program. The underlying philosophy has been that Congress, and not the Court, is the proper institution for determining the extent of the Constitution's check on the states. This doctrine of restraint was advocated by James Madison, the chief architect of the Constitution. Madison believed that Congress, because of its political process and scheme of representation, was the most legitimate formulator of the boundaries of state and national authority.[41]

Following the Supreme Court's early era of nationalism under Chief Justice John Marshall, a philosophy of states' rights became instrumental under Chief Justice Roger Brooke Taney. Taney's view was that no check on the states proceeded from the interstate commerce clause itself. Rather, the states could exercise their taxing and regulating authority in whatever manner they chose until Congress decided to intervene.[42] Years later another period of expanding national powers under the Roosevelt Court and the Warren Court was tempered by new appointments and philosophies. A court dominated by Nixon appointees began to emphasize the other side of the Cooley rule, that a subject of governmental regulation can be so local in nature that state action alone is proper. (The Court in *Cooley* found that the regulation of pilotage required a diversity of approaches which could be supplied only by local attention.) The abstention doctrine of the Burger Court was clearly expressed in its early days when Justice Black, speaking for the majority, gave his view of "Our Federalism" in *Younger* v. *Harris* (1971). The critical concept according to Justice Black is "comity," which includes

> a proper respect for state functions, a recognition of the fact that the entire country is made up of a Union of separate state governments, and a continuance of the belief that the National Government will fare best if the States and their institutions are left free to perform their separate functions in their separate ways.[43]

Both the abstention doctrine and the doctrine of excessive insularity are derived from a local-national subject area analysis. The basic ideas, however, are much easier to formulate than they are to apply to the facts of a state-federal controversy. As the states matured into a nation, it increasingly became clear to Congress and the Court that the local-national dichotomy was simplistic. A third category of federalism cases and the doctrine of economic nationalism emerged as the Court upheld congressional programs to remedy commercial ills, even though the states already were involved.

The expansive rule of national powers had supporters in the founding period, gained the status of constitutional doctrine under Chief Justice John Marshall, fell from favor and then reached full fruition in the court fashioned by Franklin Roosevelt. Several times the case was made for the extreme principle that the national government possessed inherent power to tackle any problem of national scope. The exercise of power by the national government, under this interpretation, would not have to be grounded in a provision of the Constitution. The Supreme Court however, has consistently ruled that the theory that Congress has national legislative powers not expressed in Article I conflicts with the fundamental doctrine of enumerated powers. Even though the Supreme Court rejected the concept of inherent powers, the doctrine's objective was achieved through the Court's generous interpretation of Congress' delegated powers. In *United States* v. *Darby* (1941),[44] the Court ruled that Congress was free to regulate local activities that so affect interstate commerce as to make their control a legitimate means of implementing the delegated power. Subsequent decisions have implied that the commerce power is defined by national needs. The modern approach has been a doctrine of economic nationalism: the Court will respect "any attempt in good faith by Congress to cope with a national economic problem."[45]

In recent years the Supreme Court has recognized a narrow range of exceptions to Congress' broad regulatory power and created a fourth category of federalism cases. The judiciary's brand of "new federalism" has paralleled a decade of congressional concern for decentralizing governmental power and strengthening the states. "Judicial new federalism" was inaugurated in *National League of Cities* v. *Usery* (1976),[46] a case in which a divided court declared that Congress' extension of federal minimum wage and maximum hour standards to state and local governments was unconstitutional. Five justices found that states and their subdivisions possess a residuum of sovereignty that cannot be diminished by the national government. The Court used a variety of expressions to describe this invulnerable sphere of state activity, including "functions essential to separate and independent existence,"[47] "integral operations,"[48] "traditional governmental functions,"[49] and "States as States."[50] The discretion of local governments to determine their employees' wage structure was protected under this view of sovereignty.

The Court's new federalism rests on an interpretation of the states' reserved powers found in the tenth amendment, but the philosophy is not a return to "dual federalism."[51] That theory was far more ambitious in that it postulated that Congress' enumerated powers may not be used to carry out regulatory functions reserved to the states; the very existence of the states placed a limitation on the use of Congress' delegated powers. Consequently, Congress could not regulate local activities of commercial enterprises. The new doctrine of federalism merely holds that certain attributes of sovereignty, activities of "States *qua* States," are free from invasion by the federal government. This is not to say that private businesses may escape the regulatory authority of the national government when they are also subject to the regulatory authority of a state.

The Court in *National League of Cities* gave some guidance as to what internal governmental functions were protected under its version of new federalism. The majority opinion of Justice Rehnquist used a two–step analysis that consisted of (1) identifying an "integral" or "essential" governmental function that (2) was linked to the provision of "traditional" state services. Federal regulation of this highly discretionary area of governmental activity was proscribed because it impaired the "States' integrity or the ability to function effectively in a federal system."[52] In *National League of Cities* the Court found that determination of employment policies was an "integral" function that was essentially linked to providing such "traditional" or nonproprietary services as fire prevention, police protection, sanitation, public health, and parks and recreation.

In subsequent cases, *National League of Cities* has not had a dynamic career before the Supreme Court.[53] Justices have heard arguments in at least twenty cases calling for the application or extension of the new doctrine of federalism. In only one case since 1976 has a majority relied on *National League of Cities* as grounds for its decision, ruling that the commerce clause did not prohibit South Dakota from preferring its residents in selling state–owned cement.[54] Justice Powell found the majority's interpretation of *National League of Cities* so strained that he left their company and joined the dissenters. The continuing appearance of *National League of Cities* in Supreme Court jurisprudence, therefore, has been almost exclusively in minority opinions.

It is evident that the United States Supreme Court since its beginning has settled disputes that were presented to it within the framework of federalism. Four categories of cases have been identified that are distinguished according to the presence of the following variables: whether the governmental activity being scrutinized is state or congressional and whether the Court decided to void or defer to the governmental activity. The doctrines that have evolved from these sets of cases are excessive insularity, abstention, economic nationalism, and judicial new federalism. In the judicial challenge to the Montana coal severance tax that found its way to the Supreme Court as *Commonwealth Edison Co.* v. *Montana* (1981), rulings in all four categories were presented as authoritative or instructive precedent.[55] The appellants argued, for example, that the severance tax was subject to commerce scrutiny because mining affects interstate commerce, that Congress had preempted the tax when it addressed the national problem of energy independence, and that Montana was isolating itself from the nation by exploiting its monopoly position on coal. The appellees, on the other hand, responded that the Court should exercise restraint and defer to the state's power to tax, that the Congress and not the Court was the proper branch of the national government to check the state, and that any invasion by the national government of the power to tax would constitute substantial interference with an essential state function. The substance of these positions and the Court's resolution of the natural resource taxation controversy will be discussed in the following section.

IV. *Commonwealth Edison Co. v. Montana*

The United States Supreme Court during its 1980 term considered an appeal from a July 17, 1980, decision of the Montana Supreme Court upholding the constitutionality of the Montana coal severance tax.[56] The appeal was brought by eleven midwestern and Texas utilities and four coal companies. In addition to the briefs of the appellants and appellees, the Court had before it a considerable array of arguments presented by *amici curiae*. The Attorney General of the State of Texas submitted a brief in support of the appellants. Arguing as *amici curiae* on behalf of the appellees were the states of Wyoming, Colorado, Nevada, Idaho, Washington, Oregon, North Dakota, West Virginia, and New Mexico; the Western Governors' Policy Office; the Western Conference of the Council of State Governments; the Environmental Defense Fund, Natural Resources Defense Council, and Sierra Club; and fourteen United States Senators and Representatives, organized by Max Baucus, United States Senator from Montana. In addition, the United States Solicitor General filed an *amicus curiae* brief upon the invitation of the Court. On July 2, 1981, the Court ruled in a six–to–three decision that the Montana coal severance tax did not violate the commerce clause or the supremacy clause of the United States Constitution.

The case was of special note for both political and legal reasons. Politically it represented a clash between the taxing state and a combination of coal mining interests and their customers; federalism was the legal vehicle that carried them to the court. The root controversy was between Montana and mining and utility interests and not between Montana and the federal government. Such a litigation strategy is not unusual, as "it would be the grossest fiction to ignore the fact that the issues of federalism are contests between persons in private character and others in official character, whatever the legal framework of the controversy."[57] In arguments to the Court, accordingly, it was asserted that the utilities and coal companies used "nebulous allegations of constitutional infirmity [to] support their own economic interest."[58]

These private concerns were carried to the Court in the context of constitutional interpretations that represented departures from established doctrine. Four legal arguments were sufficiently novel that they enhanced the political interest of the case. The constitutional frontiers the Court explored were the questions of (1) whether the case fell within the scope of commerce clause scrutiny, (2) whether state exploitation of a favorable market position constituted discrimination against interstate commerce, (3) whether a state general revenue measure should be tested by standards traditionally applied to user fees, and (4) whether a vague national policy should be held to preempt the "hostile forces of localism."[59] The Supreme Court ruled in favor of the appellants on the threshold issue of commerce clause jurisdiction, but the Court rejected the appellants' formulations regarding a coal cartel, a strict benefit–for–payment rationale, and an activist judicial role in preemption cases.

The first issue the Court took up in *Commonwealth Edison* was whether the Court's jurisdiction extended to a state tax on mining. Montana had argued that mining was a purely local activity and did not fall within the scope of the interstate commerce clause. Both appellants and the United States, as *amicus curiae*, took the position that the Court could subject the tax to commerce clause scrutiny because mining activities affected interstate commerce. Their argument drew an analogy between the direct and negative implications of the commerce clause: Since Congress was authorized by the clause to regulate local activities that were closely related to interstate commerce, then the clause also could be used to negate a state tax or regulation of an interstate operation that had a burdensome effect on interstate commerce.

The Supreme Court had not directly passed on this jurisdictional question since 1922. In that year it upheld a state tax on coal mined in the state, saying that the tax was not on goods in interstate commerce even though there were fixed plans for shipping the coal out of state.[60] The appellants in *Commonwealth Edison* argued that the Supreme Court had ceased to follow the 1922 decision and in support cited two recent cases. *Lewis* v. *B.T. Investment Managers, Inc.* (1980) concerned a Florida statute that prohibited the out–of–state ownership of investment advisory services.[61] The *obiter dicta* of that case included the following comment: "This Court has observed that the same interstate attributes that establish Congress' power to regulate commerce also support Constitutional limitations on the powers of the states."[62] In *Philadelphia* v. *New Jersey* (1978), the Court declared unconstitutional a New Jersey statute that prohibited the importation of solid or liquid waste from outside the state.[63] The pertinent part of the Court's reasoning was as follows:

> Although the Constitution gives Congress the power to regulate commerce among the States, many subjects of potential federal regulation under that power inevitably escape congressional attention "because of their local character and their number and diversity". . . . In the absence of federal legislation, these subjects are open to control by the States so long as they act within the restraints imposed by the Commerce Clause itself. . . .The bounds of these restraints appear nowhere in the words of the Commerce Clause, but have emerged gradually in the decisions of this Court giving effect to its basic purpose. . . .
>
> The opinions of the Court through the years have reflected an alertness to the evils of "economic isolation" and protectionism while at the same time recognizing that incidental burdens on interstate commerce may be unavoidable when a State legislates to safeguard the health and safety of its people. Thus, where simple economic protectionism is effected by state legislation, a virtually per se rule of invalidity has been erected. . . . The clearest example of such legislation is a law that overtly blocks the flow of interstate commerce at a State's borders.[64]

The prior cases relied upon by the appellants were both representative of the "clear examples" mentioned in the *Philadelphia* case. Florida and New Jersey had enacted "overt blocks" which left no doubt as to their effect on the flow of commerce. The Montana case did not present such a clear situation. There was no prohibition on out–of–state shipments of coal, and there did exist an arguably legitimate state purpose. *Commonwealth Edison*, therefore, may have been anticipated by the *dicta* of the cited precedents, but it did not fall within the rules of the two cases. When the Court in 1981 held that "a state severance tax is not immunized from Commerce Clause scrutiny by a claim that the tax is imposed on goods prior to their entry into the stream of interstate commerce,"[65] new constitutional ground was broken. For the first time the Supreme Court ruled that the dormant commerce power, i.e., the restrictions the commerce clause imposes on the states, can apply to a local tax or an intrastate activity that allegedly has some effect on interstate commerce.

Having settled the jurisdictional question of the commerce clause's applicability, the Court turned to the issue of whether that clause prohibited the Montana tax. The argument of the appellants was that Montana discriminated against interstate commerce and out–of–state coal consumers by using its monopoly position to accumulate an excessive amount of revenue. The Court said that this claim amounted to an injection of antitrust law into commerce clause doctrine and represented a "departure from the rationale of our prior discrimination cases."[66] The majority found that this novel interpretation was unwarranted and ruled that the tax policy did not amount to unconstitutional discrimination.

The Supreme Court's rejection of the exploitation argument was based more on the judicial role it necessitated than on the merits of the formulation. Prior cases concerning the limits of the commerce clause had required the Court to balance local regulation with the burden on interstate activities and to strike down undue preference for local matters and prohibitions on state exportation. The rationale of these discrimination cases was that the founding fathers had intended state boundaries to be of no consequence to commerce and the Court to be the guardian of the open borders. The Court's work was done when it found that the Montana tax was assessed on each ton of coal mined in the state, whatever its destination, and that the coal was free to pass to distant consumers. However, for the Court to determine whether Montana was also shipping its tax burden to other states involved calculations of a different type. The Court, consequently, wished to avoid "complex factual inquiries about such issues as elasticity of demand for the product and alternate sources of supply."[67]

The appellees were successful in their advocacy with the theme that Congress, and not the Supreme Court, is the proper forum for resolving sticky tax policy conflicts. It was stressed both that the required analysis was beyond the competence of the judicial process and that the Court would be overwhelmed with such difficult litigation if one case were to be decided against the states. On the latter point it was argued that challenges would be inevitable regarding

> various taxes on Washington's timber, Alaska's salmon, California's grapes, New York's stock exchange transactions, Florida's tourist industry, and many other variants of wealth, which due to the climate, geography, or accidents of fortune originate in that particular state and are exported to others.[68]

On the former point, a noted tax economist recently commented on the elementary state of the science of tax exporting and warned that the legislative process, let alone the path of adjudication, would have a most difficult time arriving at a reliable conclusion:

> The analysis of tax exporting is sufficiently complicated that attempting to base constitutionality on estimates of tax exporting is fraught with danger. . . . It seems extremely difficult to gain sufficient agreement on the likelihood of exporting of various taxes to provide the basis for adjudication. Basing legislation on estimates of tax exporting is, of course, somewhat more practical, since Congressional hearings provide a forum in which to consider such estimates—though not one particuarly conducive to determination of truth.[69]

Presented with this situation the Court chose to exercise restraint. Justice White wrote, "the better part of both wisdom and valor is to respect the judgment of the other branches of the Government."[70]

In the third part of its *Commonwealth Edison* opinion, the Supreme Court again considered an argument that the Montana tax treated out–of–state consumers unfairly. The previous allegations of discrimination were based upon arguments of monopolistic exploitation and exportation of taxes. The appellants' remaining charge under the commerce clause was that the Montana tax was a "tailored tax," that is, that it was designed in such a way that interstate commerce paid Montana far in excess of what the commerce cost the state.[71] The remedy that the appellants sought was an order for a trial in order to prove their claim. The Montana district court had granted the defendants' motion to dismiss prior to the compiling of a factual record, and this judgment was sustained by the Montana Supreme Court.

The United States Supreme Court, in turning down the request for a trial, followed several lines of argument to the common conclusion that judicial restraint was in order. The underlying position of the Court was that the "appropriate level or rate of taxation is essentially a matter for legislative, and not judicial, resolution."[72] For the Supreme Court to have voided a state tax because of its rate would have been an unprecedented and poorly conceived action. The Court recognized that it had institutional competence to determine whether or not a state has authority to tax a specific operation, but it concluded that the political process is better suited for the complex fact–finding and weighing of estimates and values that determine a state's policy of taxation.

The principal departure in constitutional law that was urged upon the Court concerning the rate of taxation was that the Montana coal severance tax should be viewed as a user fee instead of a general revenue measure. The nature of a user fee requires that there be a direct relationship between the amount of the fee collected and the value of the service or opportunity provided by the government. The test used to determine the constitutionality of a user fee is whether or not the assessment is "manifestly disproportionate" to the cost of the service.[73] Montana, it was argued, in placing 50 percent of the severance tax proceeds in a trust fund, was turning a *de jure* general revenue tax into a *de facto* user fee. The surplus collections over and above what was currently needed to meet general revenue requirements should be given the careful scrutiny that a trial would provide.[74]

The Supreme Court upheld the lower court's conclusion that the coal tax was a general revenue measure and "put to one side those cases in which the Court reviewed challenges to " user ' fees."[75] The Montana tax, accordingly, was judged from the perspective of its furtherance of the general welfare and not from the narrower rationale of a *quid pro quo*. Seemingly of consequence in the Court's reasoning was the argument of an *amicus curiae*:

> Except in the case of user fees, "a tax is not an assessment of benefits. . . . The only benefit to which the taxpayer is constitutionally entitled is that derived from his enjoyment of the privileges of living in an organized society, established and safeguarded by the devotion of taxes to public purposes.". . . In this case there is no allegation that the taxes are not devoted to public purposes.[76]

Another *amicus curiae* observed that "[n]o taxpayer can expect an accounting from his government as to the equivalence between the taxes he pays and the benefits he receives" because the purpose of government is to provide "public goods" and not to market discretely priced commodities.[77] These themes were evident in the majority's opinion. Montana, the Court said, was free to set its own tax policy as long as the tax is related to the provision of "an orderly society."[78]

The specific test the Supreme Court applied to the Montana tax concerning the relationship of benefit to burden assumed that the tax as enacted was valid but analyzed whether or not its implementation unduly affected interstate commerce. Both parties acknowledged the appropriateness of the test but disagreed as to its meaning. The disputed language was the fourth consideration in a four–part rationale used by the Court in *Complete Auto Transit, Inc.* v. *Brady* (1977).[79] There the court said that prior decisions

> have considered not the formal language of the tax statute, but rather its practical effect, and have sustained a tax against Commerce Clause challenge when the tax is applied to an activity with a substantial nexus with the taxing state, is fairly apportioned, does not discriminate against interstate commerce, and is fairly related to the services provided by the State.[80]

The "fair relationship" portion of the "practical effect" test, in the judgment of the appellants, required treating the Montana tax as if it were a user fee. This position, as discussed above, called for specialized and complicated factual analysis in a trial to see if Montana was requiring interstate commerce to bear more than its share of general governmental costs. The majority's interpretation of the "fair relationship" standard rejected the formulation that measured the taxes paid against the benefits received. The Court's approach focused on the "measure" of the tax. The proper inquiry, the majority said, was determination of the relationship between the presence of the taxpayer in the state and the amount of taxes paid. The tax would be valid if its "measure" was "reasonably related to the extent of the contact."[81] This analysis was applied to the challenged tax:

> Because it is measured as a percentage of the value of the coal taken, the Montana tax is in "proper proportion" to appellants' activities within the State and, therefore, to their "consequent enjoyment of the opportunities and protections which the State has afforded" in connection to those activities.[82]

Justice Blackmun's dissenting opinion said that this interpretation "emasculates the fourth prong" of the *Complete Auto Transit* test and would justify the unacceptable result of a tax of 1,000 percent of value and Montana's elimination of all other taxes on its citizens.[83] The majority's position was that the Court was the wrong body to check an exorbitant state tax, and not that there was no constitutional means of imposing a limitation:

> Questions about the appropriate level of state taxes must be resolved through the political process. Under our federal system, the determination is to be made by state legislatures in the first instance, and, if necessary, by Congress, when particular state taxes are thought to be contrary to federal interests.[84]

The Court, therefore, shunned an activist interpretation of the "fair relationship" test—an approach that would have had the Court treating a general revenue measure as a user fee and setting state tax rates rather than passing upon tax authority. No prior case supported use of the "fair relationship" test for scrutinizing state tax rates, and this Court refused to produce such a precedent.[85]

After the Supreme Court disposed of the commerce clause argument that the coal tax was discriminatory and unrelated to benefits received, it turned to the final constitutional challenge—that the revenue measure was in violation of the supremacy clause. Again the appellants would have had the Court adopt a new path of constitutional analysis, this time regarding the role of the judiciary in finding federal preemption of state action. The position of the appellants was that the Court must become the assertive defender of the federal scheme when state programs frustrate any Congressional policy. As was argued in the brief of one *amicus curiae*,

> appellants would have this Court rule that a broad national policy only vaguely defined by Congress should serve to preempt specific state legislation to which Congress has never addressed itself. Such an interpretation of the Supremacy Clause would be a dramatic departure from the Court's prior rulings in this area.[86]

Both the majority and minority opinions were in agreement that the appellants' theory of preemption should be rejected. The Court's analysis relied on the preemption rationale that state action stands unless it can be shown that it has been explicitly excluded by federal action or that its subject matter unmistakably lies in a field where federal interest is dominant. The appellants' approach stressed a third and independent ground for preemption:

> The issue here is whether, in the absence of any specific conflict apparent on the face of a federal statute, a state statute which "substantially frustrates" the objectives of federal legislation is preempted by the Supremacy Clause.[87]

Appellants insisted that the Court should infer preemption from the obstructive operation of a state program, even in the absence of clear Congressional purpose. The Court said that preemption can be based only on express and specific preemptive language in an act of Congress.

In this portion of the majority's opinion, as in other segments, positions of the appellee and of other interests arguing as *amici curiae* appeared to anticipate the

Court's mood. The logic of the appellees was that taxation is an essential component of governmental sovereignty, that limitation on routine use of this power must proceed from Congress and not the Court, and such congressional restriction must be clearly expressed. These assertions played upon a disposition of the Court to defer normally to the judgments of the federal and state elected assemblies. Especially critical in the advocacy of the respondents were the following points: (1) *state sovereignty*—the appellees argued that the Supreme Court has always recognized that the "states' "power of taxation is indispensable to their existences . . .   and thus to the federal scheme that the framers created";[88] (2) *judicial restraint*—Senator Max Baucus as *amicus curiae* argued that the factors which go into the determination of which intrastate activities are subject to taxation "are political factors which are appropriately considered by the state legislature, not the federal courts";[89] (3) *fear of centralization*—the Western Conference of the Council of State Governments stressed that the appellants' position would result in "further erosion of state and local decision making and a concomitant solidification of power at the national level";[90] (4) *absence of precedent*—North Dakota and West Virginia argued that none of the cases relied upon by the appellants involved preemption of a state tax statute and that in the seminal preemption case "this Court emphasized the special status of state tax laws";[91] (5) *clear federal intent required*—North Dakota and West Virginia also emphasized that the cases relied upon by the appellants involved preemption of a state statute "because of conflict between it and very specific federal enactments occupying the same field";[92] and (6) *vague national policy*—the appellees stressed that "no state law, much less a state tax law, has been struck down on the basis of general national policies of the sort that underlie appellants' claim here."[93]

The majority opinion of the Supreme Court dismissed the supremacy clause contention in relatively brief fashion. The Court found that Congress, as early as 1920, anticipated and accommodated state severance taxes. There was, consequently, no basis to conclude that the subject matter of mineral taxation implied a dominant national interest. The Court then failed to infer from general congressional language regarding use of coal a federal intent to preempt. Arguing *a fortiori*, the Court said:

> Since PIFUA [Powerplant and Industrial Fuel Use Act of 1978] is the only federal statute that even comes close to providing a specific basis for appellants' claims that the Montana statute "substantially frustrates" federal energy policies, this aspect of appellants Supremacy Clause argument must also fail.[94]

In his dissenting opinion, Justice Blackmun resorted to a footnote to resolve the preemption question: "I agree with the Court that the appellants' Supremacy Clause claims are without merit."[95]

The special interest of *Commonwealth Edison* is that the case fits into the recent pattern of Court decisions grounded in deference to Congress and the states. The Court refused to develop new law that would have positioned it, rather than Congress, as the referee of economic interests contending under the banner of federalism. The policy issues so presented, dressed up in commerce clause and supremacy clause finery, were neither sufficiently tempting nor deceiving. The question of equitable state tax policy was so replete with complexities and uncertainties that it was better left to local legislatures. And the matter of state thwarting of national policies was exactly the kind of controversy that the representative scheme of Congress was designed to address. The Court said[96] and Congress knew that a congressional resolution was possible.[97] An activist posture in realigning state and federal energy, environmental, and taxation policies simply had no appeal to the Supreme Court.

V. Conclusion

This study of the Montana coal severance tax has described its political contours and has analyzed the constitutional issue of which level of government in our federal system should set natural resources taxation policy. In the case of Montana coal, both the United States Supreme Court and Congress have scrutinized the work of the Montana legislature and judiciary to see whether the state has exceeded the boundaries of its authority. The resulting answer to the question of who should ultimately define resource federalism—i.e., which level of government should control natural resource policy—appears to be the United States Congress.

The theory and practice of federalism is one of the central concerns of the American governmental system. Federalism has both legal and political perspectives that refuse to stay separate and sharp. The tale of Montana's coal severance tax is a good example of these blurred lines. Private business concerns presented their political differences with the Montana tax to the United States Supreme Court in the context of legal doctrines. As the arguments for federal control and state autonomy unfolded before the Court, the realization spread that Montana was not the only potential loser. The taxing discretion and practice of most states was at stake. The Court's response was that the coal tax controversy was complicated, loaded with implications and ramifications for the nation and states, and the type of problem that Congress should handle. The Court, therefore, deferred to Congress.

The Supreme Court's decision in the Montana coal tax case attests to the perceptiveness of America's early statesmen. The instruction is that the representative scheme of Congress—a political process that gives direct voice to the states—should make the calls in state-federal contests. The alternative is the federal judiciary's traditional disposition to resolve issues pitting local concerns against national concerns by a balancing test, which can be a disguise for the play of personal values. Congress' accountability ordinarily makes it the better branch for dealing with estimates and making subjective judgments. If natural resource development becomes a matter of crisis, and not an issue of a firm's market advantage or a state's comparative fiscal capacity, then the nation must look to Congress for a valid national consensus.

Notes

1. Bureau of Land Management, *Final Environmental Statement, Federal Coal Management Program*, United States Department of Interior, at 3 (April, 1979).
2. See Hearings on HR 6625, HR 6654, and HR 7163 Before the Subcommittee on Energy and Power of the House Committee on Interstate and Foreign Commerce, 96th Cong., 2d Sess., at 22 (1980).
3. Office of the Montana Attorney General, "Summary of Major Issues Concerning the Montana Coal Severance Tax," at 14–15 (1981).

4. State of Montana, "State Tax Fairness: Montana's Coal Tax in the Context of State Resource Taxation in Our Federal System," at 14 (1981).

5. Id., at 7.

6. Brief for the Western Conference of the Council of State Governments as *Amicus Curiae* at 17, *Commonwealth Edison Co.* v. *Montana,* 69 L. Ed. 2d 884 (1981).

7. *Denver Post,* December 11, 1981.

8. Hearings *op. cit.*

9. Sec. 1, Ch. 525, *Laws of Montana 1975,* codified as *Montana Code Ann.* § 15–35–101 (1979).

10. Id., § 15–23–701.

11. Id., § 15–38–101.

12. Id., § 15–6–138 (b).

13. Id., § 15–31–101.

14. Id., § 82–4–338.

15. *The Missoulian,* January 11, 1975.

16. Id.

17. Statement to Accompany the Report of the Free Joint Conference Committees on Coal Taxation, 44th Montana Legislature, at 1 (1975).

18. Rand Corporation, *Coal Development and Government Regulation in the Northern Great Plains: A Preliminary Report* 148 (August, 1976), cited in Jurisdictional Statement at 2, *Commonwealth Edison Co.* v. *Montana.*

19. Montana Coal Board, "Local Impact Assistance Grant Program, 1975–1980," at 1–2 (1981).

20. *Montana Const.,* art. IX, sec. 5.

21. *Montana Code Ann.,* § 15–35–108 (1979).

22. Brief for Appellants at 11–12, *Commonwealth Edison Co.* v. *Montana.*

23. T. Cohea, "Mineral Tax Trust Funds and Special Funds in the States," A Report of the Montana Governor's Office of Budget and Program Planning (Helena, Montana, 1981), at 1–10. The following discussion of state trust funds is derived from the Cohea study.

24. Bureau of Census, *State Government Tax Collections in 1979,* United States Department of Commerce (1980), Table 3.

25. R. Watson, "The Expenditure of Royalty and Severance Tax Revenues," National Conference of State Legislatures, Legislative Finance Paper 10, at 2 (November, 1981).

26. B. Weinstein, "Texas Tax Intake Threatened by Yankees," *Texas Business,* at 29 (1981).

27. Watson, *op. cit.,* at 1–8.

28. Weinstein, *op. cit.*

29. *New York Times, June 5, 1981.*

30. *Watson, op. cit.,* at 8.

31. Montana Attorney General, *op. cit.,* at 10.

32. Brief for North Dakota and West Virginia as *Amici Curiae* at 3, *Commonwealth Edison Co.* v. *Montana.*

33. Id.

34. Brief for Wyoming, Colorado, Nevada, Idaho, Washington, and Oregon as *Amici Curiae* at 11, *Commonwealth Edison Co.* v. *Montana.*

35. R. Stern, "The Commerce Clause and the National Economy," 59 *Harv. L. Rev.* 645 (1946), reprinted in *Selected Essays on Constitutional Law* (St. Paul, Minn., 1963), 218–279.

36. Author cannot locate the source.

37. J. Briggs, "State Rights," 10 *Iowa L. Bull.* 297 (1925), reprinted in *Selected Essays on Constitutional Law* (Chicago, 1938), 12–25.

38. 53 U.S. 299, 318 (1851).

39. N. Dowling, "Interstate Commerce and State Power—Revised Version," 27 *Va. L. Rev.* 1 (1940), in *Selected Essays on Constitutional Law* 281 (1963).

40. See, for example, *Philadelphia* v. *New Jersey,* 98 S. Ct. 2531 (1978), where the Supreme Court struck down as a "protectionist measure" a New Jersey statute prohibiting the importation into the state of solid or liquid waste destined for landfills in the state.

41. H. Wechsler, "Political Safeguards of Federalism," *Principles, Politics, and Fundamental Law* (1961), in *Selected Essays on Constitutional Law* 200 (1963).

42. Dowling, *op. cit.*, at 282.

43. 27 L. Ed. 2d 669, 675 (1971).

44. 312 U.S. 100 (1941).

45. Stern, *op. cit.*, at 279.

46. 49 L. Ed. 2d 245 (1976).

47. *Id.*, at 254, quoting *Coyle* v. *Smith*, 221 U.S. 559, 580 (1911).

48. *Id.*, at 257.

49. *Id.*, at 258.

50. *Id.*, at 259.

51. *Hammer* v. *Dagenhart*, 247 U.S. 251 (1918).

52. *National League of Cities* v. *Usery*, 49 L. Ed. at 252, quoting *Fry* v. *United States*, 421 U.S. 542, 547 (1975).

53. See J. Lopach, "The New Federalism of the Supreme Court: Diminished Expectations of *National League of Cities*," 43 *Montana Law Review* 181–196 (1982), for a discussion of the pattern of decisions among the *National League of Cities* progeny.

54. *Reeves, Inc.* v. *Stake*, 447 U.S. 429 (1980).

55. 69 L. Ed. 2d 884 (1981).

56. *Commonwealth Edison Co.* v. *Montana*, 615 P.2d 847 (1980).

57. P. Freund, "Umpiring the Federal System," 54 *Colum. L. Rev.* 561 (1954), in *Selected Essays on Constitutional Law* 213 (1963).

58. Brief for Western Governors' Policy Office as *Amicus Curiae* at 2, *Commonwealth Edison Co.* v. *Montana*.

59. Brief for Apellants at 32, *Commonwealth Edison Co.* v. *Montana*, quoting L. Tribe, *American Constitutional Law* 319 (1978).

60. *Heisler* v. *Thomas Colliery Co.*, 260 U.S. 245 (1922).

61. 64 L. Ed. 2d 702 (1980).

62. *Id.*, at 713.

62. 57 L. Ed. 2d 475 (1978).

64. *Id.*, at 481.

65. *Commonwealth Edison Co.* v. *Montana*, 69 L. Ed. 2d at 894.

66. *Id.*, at 895.

67. *Id.*, at 896, n. 8.

68. Brief for The Western Conference of the Council of State Governments as *Amicus Curiae* at 7, *Commonwealth Edison Co.* v. *Montana*.

69. C. McLure, "Tax Exporting and the Commerce Clause: Reflections on *Commonwealth Edison*," Hoover Institution Working Papers in Economics No. E–81–13, at 2–3, 27 (1981).

70. *Commonwealth Edison Co.* v. *Montana*, 69 L. Ed. 2d at 907 (White, J., concurring.)

71. Brief for Appellants at 27, *Commonwealth Edison Co.* v. *Montana*.

72. *Commonwealth Edison Co.* v. *Montana*, 69 L. Ed. 2d at 901.

73. *Id.*, at 897 n. 12, quoting *Clark* v. *Paul Gray, Inc.*, 306 U.S. 583, 599 (1939).

74. Brief for Appellants at 30, *Commonwealth Edison Co.* v. *Montana*.

75. *Commonwealth Edison Co.* v. *Montana*, 69 L. Ed. 2d. at 897.

76. Brief for Honorable Max Baucus *et al.* as *Amici Curiae* at 18, *Commonwealth Edison Co.* v. *Montana*, quoting *Carmichael* v. *Southern Coal and Coke Co.*, 301 U.S. 495, 522 (1937).

77. Brief for the Western Conference of the Council of State Government as *Amicus Curiae* at 12, *Commonwealth Edison Co.* v. *Montana*.

78. *Commonwealth Edison Co.* v. *Montana*, 69 L. Ed. 2d at 899, quoting *Wisconsin* v. *J.C. Penney Co.*, 311 U.S. 435, 444 (1940).

78. 51 L. Ed. 2d 326 (1977).

80. *Id.*, at 331.

81. *Commonwealth Edison Co.* v. *Montana*, 69 L. Ed. 2d.

82. *Id.*, at 900, quoting *General Motors Corp.* v. *Washington*, 377 U.S. 436, 441 (1964).

83. *Id.*, at 912 (Blackmun, J., dissenting).

84. *Id.* at 901.

85. Brief for Honorable Max Baucus *et al.* as *Amici Curiae* at 14, *Commonwealth Edison Co.* v. *Montana.*

86. Brief for North Dakota and West Virginia as *Amici Curiae* at 8, *Commonwealth Edison Co.* v. *Montana.*

87. Brief for Appellants at 42, *Commonwealth Edison Co.* v. *Montana.*

88. Brief for Appellees at 44, *Commonwealth Edison Co.* v. *Montana*, quoting *Gibbons* v. *Ogden*, 22 U.S. 1, 199 (1824).

89. Brief for Honorable Max Baucus *et al.* as *Amici Curiae* at 2, *Commonwealth Edison Co.* v. *Montana.*

90. Brief for Western Conference of the Council of State Governments as *Amicus Curiae* at 13, *Commonwealth Edison Co.* v. *Montana.*

91. Brief for North Dakota and West Virginia as *Amici Curiae* at 5, *Commonwealth Edison Co.* v. *Montana.*

92. *Id.,* at 6.

93. Brief for Appellees at 41, *Commonwealth Edison Co.* v. *Montana.*

94. *Commonwealth Edison Co.* v. *Montana.* 69 L. Ed. 2d. at 906.

95. *Id.,* at 917 n. 21 (Blackmun, J., dissenting).

96. *Id.,* at 901 n. 18.

97. The congressional attack on Montana's coal tax occurred at the same time as the judicial challenge. In 1979 bills were introduced in both houses (HR 6625, HR 6654, HR 7163, S 1778, S 2695), and two years later the 97th Congress took up similar measures (HR 1313, S 178).

Conclusion: Public Lands, Natural Resources, and the Shaping of American Federalism

JOHN G. FRANCIS

RICHARD GANZEL

In his essay introducing a report on a 1980 Conference on the Future of Federalism (ACIR 1981), David R. Beam sensibly warns against a tendency to project the past into the future. He provides several illustrations of major unanticipated problems that had major impacts on federalism in the 1960s and 1970s. The contributors to this collaborative volume have not consciously sought to predict future trends. Nevertheless, their studies of selected facets of public land and natural resource policy underscore Beam's message. Many policies formulated by the federal government or by states were deflected in unintended directions by the force of new events. Other initiatives turn out to have had implications quite different from those foreseen or intended by their proponents, or from those initially perceived by academic interpreters or those affected by those actions.

If one looks at current commentaries on trends in federalism, their preoccupation with fiscal and social policy dimensions and stresses is apparent (Walker 1981). The Reagan Administration's version of "New Federalism" shares those preoccupations, and associates a reinvigoration of public support for government with reducing the functions performed by government in general and devolving some responsibility for what remains to the states. Modifications of the intergovernmental relationships associated with public land and natural resource policy have thus far remained outside the central debate over the respective future roles of national, state, and local governments. These relationships need to be brought into the larger debate.

Public land and natural resource policies and relationships are of national importance (Price 1982; Portney 1982; Welch and Miewald 1983), as well as of central concern to the West (Ganzel 1983). They involve key underlying factors that shape fiscal capacities, individual opportunities for employment and amenities, demographic shifts, interregional competition, and even the reputations of presidents. Public land and natural resource initiatives led to a depiction of Jimmy Carter as anti–West; they may earn Ronald Reagan the sobriquet of anti–environmentalist. Though neither characterization may be completely fair, each is rooted in an inadequately appreciated interrelatedness of policies and institutions. American public land and resource policies are implemented through a complex pattern of federal, regional, state, and local institutions. Policy initiatives conceived in isolation will be frustrated by that interwoven reality. Goals that are not formulated and articulated in full recognition of their institutional impacts will be distorted in practice and yield unexpected opposition from apparent beneficiaries.

The question of whether the 1980s and 1990s promise continued intense federal conflict or greater harmony will be answered by the sense of vision that is brought to the task of crafting new policies. Clearly, new policies are needed to deal more adequately with goals sought in the public lands and natural resources area. How should energy, environmental, recreational, timber, water, wildlife, wilderness, and other closely related policies be approached as policy reformulation is undertaken? What sense of vision on these issues is contained in this volume?

Obviously there are important differences in perspective among the contributors, yielding ample room for productive debate. But there is also much that is common. That common perspective stresses renewed attention to the task of ascertaining proper proportions in the respective roles played by different levels of government. There is much appreciation of the flexibility and creativity of multiple institutions, as well as the contemporary or inherent limitations of those institutions. Consequently, there is a shared, even matter–of–fact, commitment to responsible stewardship of the institutions of federalism as well as of the American public land and natural resource heritage.

The other face of this vision of the need to craft carefully even major policy reforms is a shared impatience with tidy prognoses of contemporary ills and simple recipes for their cure. That vision may be surprising in a volume that self–consciously addresses itself to the values that are to be pursued in public land and resource policy, especially from contributors who frequently offer sharp criticisms of the policies and institutions that now exist. The explanation lies in the style of analysis undertaken, which emphasizes specific problems and portrays them within a context of competing goals and evolving practices. No retreat to turn–of–the–century progressivism, embrace of an unshackled capitalism, or ritualistic assertion of the virtues of rational central direction or "close–to–the–folks" state direction commends itself to problems explicated in that thorough fashion.

As one moves from broad generalizations to specific analyses, the vision takes firmer shape. The authors urge a refined and constrained conception of the public land interests of the federal government that preserves distinctive resources and traditional responsibilities, such as trustee roles toward Indian tribes. The federal role in commodity production is seen as open to debate that balances conservation and production goals. There is a broad willingness to entertain selective deference to state and local domination of the substantive content of some policy decisions, along with an appreciation of the role of larger responsibilities for environmental management that transcend state boundaries. Finally, there is openness toward the devolution of administrative responsibility for particular planning functions that involve public lands and even toward responsibility for devolved administration of resource use and land sales.

When the focus is upon specific resource policies, there is an emphasis upon the diversity of existing reality and upon fundamental turning points that have already been reached or are on the immediate horizon. States in the West have moved from relatively undeveloped, even frontier, status to capable, assertive, proponents of their own interests. In part, their capacity rests upon successfully defended taxation of the severance of natural resources, as they follow the example of their counterparts who have achieved a degree of independence with oil and natural gas revenues. At the same time, the era of federal responsibility for water project development is drawing to a close; and there is little evidence to indicate that a new federal role in water law administration will be forthcoming or would be welcome. The federal role will not disappear, for an increasing burden of facility maintenance will be thrust upon it. That is unexciting stuff for politicians seeking issues to enhance their reputations. No doubt, however, politicians will find considerable new opportunity in coping with the growing problems of emissions and wastes and in assisting with the interstate implications of energy production and distribution.

The future, therefore, would seem to entail considerable rearrangement of responsibilities among governments, without a sharp shift toward any level within the federal system. State roles, and some local roles, seem destined to grow substantially, continuing current trends. But the federal role will shift rather than shrink. Whether attention is upon toxic or nuclear waste, air or water quality, or the stimulation of commodity production, recreation enjoyment, or urban expansion in the West, the federal role in policy formulation will remain crucial. With increased state capacities and a growing appreciation of the respective attributes of federal, regional, state, and local governments, as well as of the private sector, there is room for a different federal role. Instead of a thrust toward central direction and control or even preemption, the need is to formulate effective facilitative policies that will stimulate local and private efforts while protecting distinctive national values. This is indeed a provocative task in an age of centrifugal federalism.

References

Advisory Commission on Intergovernmental Relations. 1981. *The Future of Federalism in the 1980s*. Washington: ACIR.

Ganzel, R., ed. 1983. *Resource Conflicts in the West*. Reno: University of Nevada Public Affairs Institute.

Portney, P.R., ed. 1982. *Current Issues in Natural Resource Policy*. Baltimore: Johns Hopkins University Press for Resources for the Future.

Price, K.A., ed. 1982. *Regional Conflict & National Policy*. Baltimore: Johns Hopkins University Press for Resources for the Future.

Walker, D.A. 1981. *Toward a Functioning Federalism*. Cambridge, Mass.: Winthrop Publishers.

Welch, S., and Miewald, R., eds. 1983. *Scarce Natural Resources: The Challenge to Public Policymaking*. Beverly Hills: Sage Publications.

Appendix A   Data Sources

Variables	Sources
State population; Population density; Percent employed in manufacturing; Personal income per capita; E.P.A.; Waste treatment facilities construction	*Statistical Abstract of the United States*, various years.
Median family income; Percentage of poor families	*Statistical Abstract of the United States*, 1975.
Governors' appointive powers for environmental protection	*Book of the States*, various years.
Bureaucratic consolidation for environmental protection	*Book of the States*, 1976-77 (updated through telephone interviews with officials of state legislative councils and responsible environmental agencies).
Jones index of pollution potential	Herbert Jacob and Kenneth Vines, *Politics in the American States: A Comparative Analysis* (Boston: Little, Brown, 1976).
Democratic Party strength	*Book of the States*, various years.
Legislative professionalism	Douglas Dobson, "Social, Economic and Political Systems: The State Context of Aging Policies," Northern Illinois University: Program for Applied Policy Research (Working Paper No. 3, 1976).
Total state budgetary expenditures	*Statistical Abstract of the United States*, various years.
Environmental quality control expenditures	State and Local Government Special Studies No. 83, 1975, U.S. Department of Commerce, Bureau of the Census, *Environmental Quality Control*, (Washington, D.C.: Government Printing Office), and various years thereafter.

Index

Administrative Procedure Act, 123
Agriculture: governmental role in, 11; private land ownership in, 116–17
Air quality, 276–77
Alaska: federal land in, 29; land settlement decision regarding, 88, 102, 135; Sagebrush Rebellion in, 32; school lands in, 98
Alaska National Interest Lands Conservation Act of 1980, 88, 102, 135
Allen-Warner Valley Energy System, 268–79; general aspects of, 268–71; regulatory action in, 271–75
Allotment management plans, 56–58, 83
American Mining Congress, 9, 121
Andrus, Cecil, 88, 114, 274
Animal unit months, 48, 53–54, 60, 61, 65, 163
Arizona: Sagebrush Rebellion in, 32, 37, 38, 39, 40, 44; school lands in, 98; water resource development in, 242, 244, 245
Army Corps of Engineers, 216, 217–18, 220, 222, 225
Aspinall, Wayne, 82, 222
Asset Management Program, 114, 119, 138
Atomic Energy Act of 1954, 177, 180
Atomic Energy Commission, 177, 178, 270
Audubon Society, 9

Babbitt, Bruce, 20, 44, 109n., 120
Bonneville Power Administration, 248, 249, 250, 251, 252, 253, 260, 261–63
Boulder Canyon Project Act of 1928, 218
Brown, Jerry, 32
Bureau of Land Management: Alaska land settlement and, 88; changing orientation of, 68–70; Classification and Multiple Use Act and, 82–83; environmental impact statements by, 58, 64–66, 67–68, 69–70, 83, 154; federal authority of, 80, 87–90; governmental intervention by, 11, 13; grazing land values and, 135; legislative proposals by, 82; livestock industry, relations with, 80–81; public land classification by, 82–83; public land management by, 149–50; rangeland investments analysis by, 55–63; range management objectives of, 49–50; as *rentier*, 113; responsibility of, 18, 155; Sagebrush Rebel-

lion and, 35, 79, 80–90; statutory authority for, 81, 83, 87, 88–89, 135, 154, 156–57; total land acreage managed by, 48, 119, 155. *See also* Rangeland management
Bureau of Reclamation, 222, 225

Calhoun, John C., 159, 216
California: energy system for, 268–79; nuclear power plant siting in, 86; Sagebrush Rebellion in, 32, 33, 37, 38, 39; water resource development in, 243, 244
California Gold Rush, 11, 16
California Public Utilities Commission, 271
Canal Era, 216
Carey Act of 1894, 216
Carter Administration: energy policy of, 1, 2, 7, 264–66; natural resource policy of, 19; nuclear waste policy of, 183–84; public land policy of, 25; water resource policy of, 1, 223, 224; water rights policy of, 86; western state resource development and, 31
Central Arizona Project, 220
Chafee, John, 114
Church, Frank, 33
Classification and Multiple Use Act, 82–83, 135
Clean Air Act, 12, 13, 278
Clean Water Act, 12, 19, 233
Cleveland Administration, 16
Coal: severance tax on, 282, 284–88, 291–99; western deposits of, 283
Coastal Zone Management Act, 101, 108n.–9n.
Colorado: Sagebrush Rebellion in, 32, 33, 37, 38, 39, 40; water resource development in, 220, 222, 228–29, 242–43
Colorado Basin Project Act of 1968, 220, 222
Colorado River Compact of 1922, 218
Commonwealth Edison v. *Montana*, 96, 292–302
Congress: land resource value legislation by, 133–36
Conservation: in energy planning, 256–57, 259, 263–64; land management and, 47–48